T0353585

HIGHER MECHANICS

Cambridge University Press
Fetter Lane, London

New York
Bombay, Calcutta, Madras
Toronto
Macmillan

Tokyo
Maruzen-Kabushiki-Kaisha

Gyroscope (Art. 33)

HIGHER MECHANICS

BY

HORACE LAMB, Sc.D., LL.D., F.R.S.
HONORARY FELLOW OF TRINITY COLLEGE, CAMBRIDGE;
LATELY PROFESSOR OF MATHEMATICS IN THE
VICTORIA UNIVERSITY OF MANCHESTER

SECOND EDITION

CAMBRIDGE
AT THE UNIVERSITY PRESS
1929

CAMBRIDGE UNIVERSITY PRESS
Cambridge, New York, Melbourne, Madrid, Cape Town, Singapore, São Paulo, Delhi

Cambridge University Press
The Edinburgh Building, Cambridge CB2 8RU, UK

Published in the United States of America by Cambridge University Press, New York

www.cambridge.org
Information on this title: www.cambridge.org/9780521744492

First edition 1920
Second edition 1929
This digitally printed version (with corrections) 2009

A catalogue record for this publication is available from the British Library

ISBN 978-0-521-74449-2 paperback

PREFACE TO THE SECOND EDITION

THIS book treats of three-dimensional Kinematics, Statics, and Dynamics in what is I think a natural, as I have found it to be a convenient, order. It may be regarded as a sequel to two former treatises* to which occasional reference is made; but it is not dependent on these, and can, I trust, be readily followed by students who are conversant with ordinary two-dimensional Mechanics.

The subject is, of course, a very wide one, and some principle of selection was necessary. I have tried to confine myself to matters of genuine kinematical or dynamical importance, avoiding developments whose interest, often considerable, is purely mathematical or now mainly historical.

I have taken advantage of a second edition to revise the work throughout, and to introduce a number of changes in the later chapters. Many sections have been re-written and, I hope, improved; the arrangement has in many cases been modified; and some additions of importance have been made. The book, however, still claims only to be an elementary one, regard being had to the nature of the subject. Further developments must be sought in books like Thomson and Tait, the works of Routh, which abound in interesting historical references, Rayleigh's *Theory of Sound*, Whittaker's *Analytical Dynamics*, or the various continental treatises, the latest of which is the very elegant and systematic *Lezioni di meccanica razionale* by Professors Levi-Civita and Amaldi. To this I am indebted for one or two interesting examples.

H. L.

Cambridge, 1929.

* *Statics*, 3rd ed., Cambridge, 1928, and *Dynamics*, 2nd ed., Cambridge, 1923.

CONTENTS

CHAPTER I

KINEMATICS OF A RIGID BODY. FINITE DISPLACEMENTS

CHAPTER II

INFINITESIMAL DISPLACEMENTS

CHAPTER III

STATICS

CHAPTER IV

MOMENTS OF INERTIA

CHAPTER V

INSTANTANEOUS MOTION OF A BODY (KINEMATICS)

CHAPTER VI

DYNAMICAL EQUATIONS

CHAPTER VII

FREE ROTATION OF A RIGID BODY

CHAPTER VIII

GYROSTATIC PROBLEMS

CHAPTER IX

MOVING AXES

CHAPTER X

GENERALIZED EQUATIONS OF MOTION

CHAPTER XI

THEORY OF VIBRATIONS

CHAPTER XII

VARIATIONAL METHODS

CHAPTER 1

KINEMATICS OF A RIGID BODY. FINITE DISPLACEMENTS

1. Degrees of Freedom.

In forming the equations of motion of any dynamical system we proceed in the first instance by infinitesimal steps. The configuration and state of motion at a given instant t being regarded as known, we calculate the changes which take place in a time δt in consequence of the forces and constraints to which the system is subject. This leads, in the manner which will in principle be already familiar to the reader, to the differential equations of motion. Accordingly, an analysis of the possible *infinitesimal* displacements of the system is all that is really required in the way of kinematical preliminary.

In the case of a rigid body, however, the study of *finite* (as distinguished from infinitely small) displacements leads to some theorems so simple and elegant that it is usual to devote a little space to them in the first instance. A similar procedure, it may be recalled, is usually adopted in the particular case of two-dimensional displacements [S. 30]*.

The position of a rigid body moveable in three dimensions is fixed when we know the positions of any three points A, B, C of it which are not in the same straight line. For if P be any fourth point of the body, the tetrahedron $PABC$ is of invariable dimensions. The coordinates (Cartesian or other) of the three points A, B, C relative to any fixed frame of reference are nine in number, but they are not independent, being connected by the three relations which express that the distances BC, CA, AB have given values. The number of independent variables, or 'coordinates' (in a generalized sense), which are necessary and sufficient to specify the position of the body is therefore *six*. A rigid body whose

* References in this form are to the author's *Statics*. The number indicates the Article.

position is not restricted in any way is accordingly said to have six 'degrees of freedom.'

The generalized coordinates referred to may be chosen in various ways, and speaking generally we may say that *any* six independent kinematical conditions will restrict the body to one or other of a series of definite positions, none of which can be departed from without violating the conditions in question.

This principle, simple as it is, has useful practical applications. For instance, the position of a theodolite relative to its stand is (in most forms) determined by the fact that its three rounded feet rest in three V-shaped grooves, there being thus six contacts. In another method, one foot rests in a trihedral hollow, a second in a V-shaped groove, whilst the third rests on a flat surface.

Again, a rigid structure may be fixed relatively to the earth by six links, provided certain 'critical' configurations of these are avoided, analogous to those met with in the two-dimensional theory of structures [S. 13, 15].

A rigid body which is subject to *five* conditions only has one degree of freedom. An instance is that of a body carried by three rounded feet, two of which are in contact with the sides of a V-shaped groove cut in a plane surface, whilst the third rests on the plane. The body has thus freedom to move parallel to the groove, but in no other way so long as all five contacts are maintained. This arrangement, called a 'geometric slide,' is adopted in some modern physical instruments.

Again, *four* conditions leave two degrees of freedom, as when a rifle-barrel rests on two forks with rounded arms, one in front of the other. The rifle can still be moved lengthways; it can rotate about its length; and it can of course be made to execute any combination of these movements. Another instance is the telescope of an altazimuth instrument. The geometrical conditions here are that a certain point in the body, viz. the intersection of the axis of the telescope with the axis of the pivots, is fixed, and (further) that the axis of the pivots is restricted to a horizontal plane.

A simple example of a body subject to *three* conditions, and therefore having three degrees of freedom, is that of a body with three rounded feet resting on a fixed surface. If this surface be plane we have the type of two-dimensional freedom discussed under Plane Kinematics [S. 13]*.

2. Displacements about a Fixed Point. Euler's Theorem.

The displacement of a rigid body from one given position to another may be effected in various ways. In particular we may

* The practical importance of the geometrical theory of freedom and constraint was first duly emphasized by Lord Kelvin (Thomson and Tait, *Natural Philosophy*, 2nd ed., Art. 198).

imagine that by a pure translation, in which all points of the body describe equal and parallel straight paths, some arbitrarily chosen point of the body is brought to its final position O (say), and that the body is then turned about O as a fixed point until two other points A, B of it, not in a straight line with O, are brought into their final positions. By a theorem due to Euler*, this second operation is equivalent to a *rotation* about some fixed axis through O.

To see this, we imagine two coincident spherical surfaces to be described about O as centre. One of these surfaces is conceived as fixed in space; the other as fixed in the body and therefore moveable with it. The displacement about O may be regarded as due to a sliding of the latter surface over the former, as in an idealized 'ball and socket' joint; and the theory is accordingly very similar to that of a plane figure moving in its own plane [S. 14]. Suppose that as a result of the displacement a point of the moveable sphere is brought from the position A in space to the position B, whilst the point which was at B is brought to C. The plane ABC cuts the fixed sphere in a circle, usually (but not necessarily) a 'small' circle. If J be either pole of this circle, the isosceles spherical triangles AJB, BJC are congruent, since $AB = BC$, these being two positions of the same great-circle arc of the moving sphere. Hence AB can be brought into coincidence with BC by a rotation about OJ as axis, through an angle AJB†.

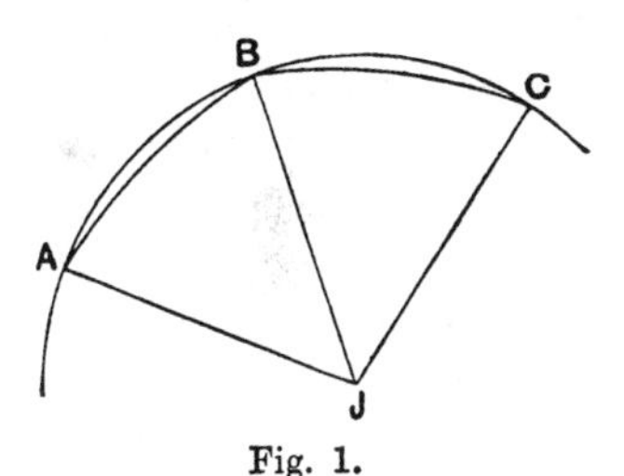

Fig. 1.

It appears, as was independently evident, that a rigid body one point of which is fixed has three degrees of freedom, and that three independent coordinates are therefore necessary to specify a particular position. These might be for instance, the two angular coordinates of the axis by rotation about which the body can be brought from

* Leonhard Euler, b. Bâle 1707, d. St Petersburg 1783. The date of the theorem is 1776.

† The case where C coincides with A may be treated as a limiting case, but it is plain that the axis of rotation then bisects the arc AB, the angle of rotation being π. If A and B be antipodal points, the construction becomes indeterminate, and some other point must be chosen in place of A.

some standard position to the position in question, together with the angle of the rotation. Another system of coordinates, more convenient for purposes of calculation, will be explained later (Art. 33).

3. Composition of Finite Rotations.

It is now necessary to distinguish between the two senses in which rotation can take place about a given axis. If O, A be any two points on the axis, a rotation will be reckoned as a positive rotation about OA when it is related to the direction *from* O *to* A in the same way as the sense of the rotation is related to that of advance in a right-handed screw. A rotation in the opposite sense will be accounted as negative. Thus if N and S denote the north and south poles of the earth, the diurnal rotation may be reckoned either as a positive rotation about SN, or as a negative rotation about NS.

The following construction for finding the result of a rotation α about an axis OA, followed by a rotation β about an axis OB, appears to have been first given by O. Rodrigues (1840)*. We denote by A, B the points where the respective axes meet the fixed spherical surface having O as centre. On AB as base we construct the spherical triangles ACB, $AC'B$ having the angles at A each equal to $\frac{1}{2}\alpha$, and the angles at B each equal to $\frac{1}{2}\beta$. It is plain from the figure that the rotation α about OA will bring a point of the body from C to C', and that the subsequent rotation β about OB will bring it back to C. Hence OC is the axis of the rotation equivalent to the two former ones. We notice that if the order of the two rotations about OA, OB had been inverted, the result would have been equivalent to a rotation about OC'. We have here an instance of operations which do not follow the 'commutative law.'

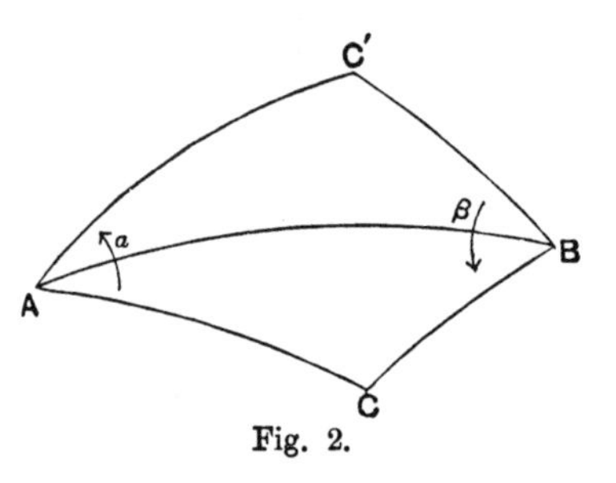

Fig. 2.

* Subsequently by Sylvester, *Phil. Mag.* (3), t. xxxvii (1850), and Hamilton, *Lectures on Quaternions* (1853), p. 332.

To find the angle of the resultant rotation about OC, we construct on the opposite side of BC a triangle $A'BC$ symmetrically equal to ABC. Since the rotation about OA does not affect a point initially at A, whilst the rotation about OB would bring it to A', the required angle is ACA', or $2(\pi - C)$. Since a rotation through four right angles does not affect the position of a body, this is equivalent to a rotation $-2C$.

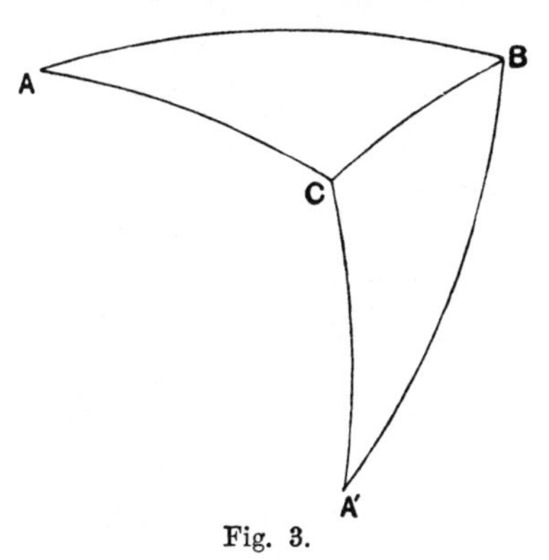

Fig. 3.

We derive the following theorem, apparently first stated explicitly (with some extensions) by Hamilton in 1844*: If ABC be any spherical triangle, and O the centre of the sphere, the effect of three successive rotations of amounts $2A$, $2B$, $2C$ about OA, OB, OC, respectively, will be to restore the body to its initial position, provided the sense of the rotations be the reverse of that indicated by the order of the letters A, B, C.

4. Donkin's Theorem.

A displacement of a rigid body about a fixed point has been regarded in the preceding investigation as determined by the axis and the angle of the equivalent rotation. A rotation about a diameter of a fixed unit sphere may however also be specified by an arc of the great circle whose plane is perpendicular to this diameter. Thus if P, Q be two points on this circle, a rotation which would bring a point from P to Q may be specified by the arc PQ. The order of the letters is of course important, but the precise position of the arc PQ on its great circle is immaterial†.

On this understanding we have the following theorem, dis-

* Sir W. R. Hamilton (1805–1865), Royal Astronomer of Ireland 1827–1865. See his *Lectures on Quaternions*, p. 334.

† Arcs used in this sense, and here distinguished by Roman type, have a certain analogy to localized vectors *in plano*, although the law of combination is different, as the theorem which follows shews.

covered independently by Donkin*, and Hamilton†: If ABC be any spherical triangle, three successive rotations represented by 2.BC, 2.CA, 2.AB will restore a body to its original position. The equivalence of this theorem to that given at the end of Art. 3 may be established by means of the polar triangle, but the following proof, given by Donkin‡, is direct.

The sides of the triangle ABC are produced as in the figure, so as to make

$$C_2B = BC = CB_3,\ A_3C = CA = AC_1,\ B_1A = AB = BA_2.$$

The triangles AB_1C_1, A_2BC_2, A_3B_3C are therefore directly equal to one another, and 'symmetrically' equal to ABC. The rotation

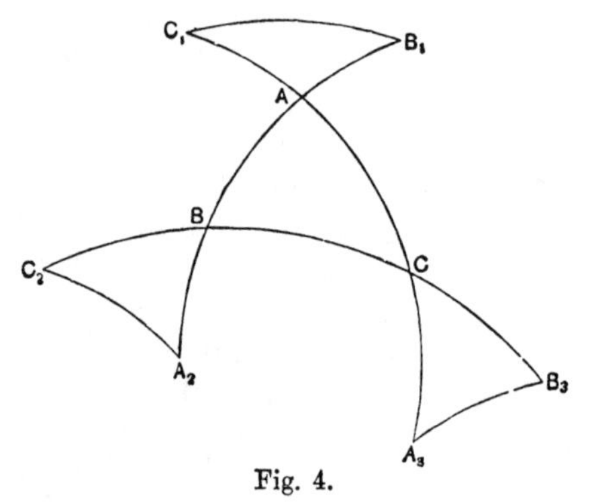

Fig. 4.

2.BC would bring the triangle A_2BC_2 into the position A_3B_3C, the rotation 2.CA would then bring it into the position AB_1C_1, and the rotation 2.AB would finally bring it back to its original position A_2BC_2.

It follows that two successive rotations represented by two great-circle arcs AB, BC are equivalent to a rotation 2.XY, where X, Y are the middle points of AB, BC, respectively.

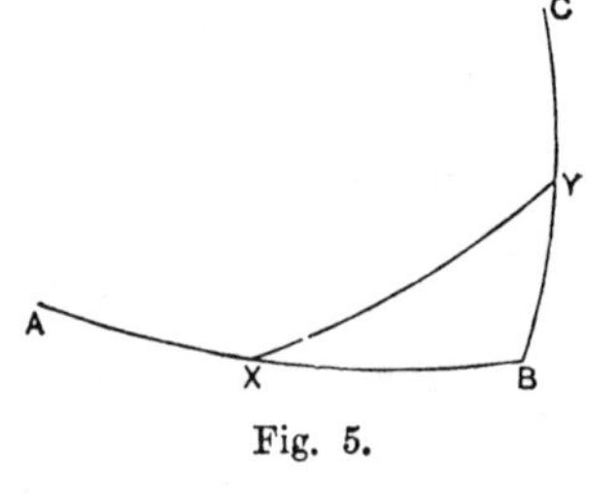

Fig. 5.

On the other hand it is to be noted that three successive rotations represented by the arcs AB, BC, CA

* W. F. Donkin (1814–1869), Savilian professor of Astronomy at Oxford, 1842–1869. The theorem was published in 1850.

† *Lectures on Quaternions*, p. 330.

‡ *Phil. Mag.* (4), t. i (1851).

themselves (not their doubles) are equivalent to a rotation about OA through an angle equal to the *spherical excess* of the triangle ABC*. That the point which starts from A is finally restored to its initial position is obvious. To find the angle of the resultant rotation we need only consider the successive positions assumed by the arc originally coincident with AB. The first rotation (AB) places it along BX in the figure; the second shifts it to the position CY, such that the angle BCY is equal to the angle B of the spherical triangle; whilst the third rotation brings it to AZ, so that $ZAC = \pi - ACY = \pi - B - C$. The angle between the first and last positions is therefore

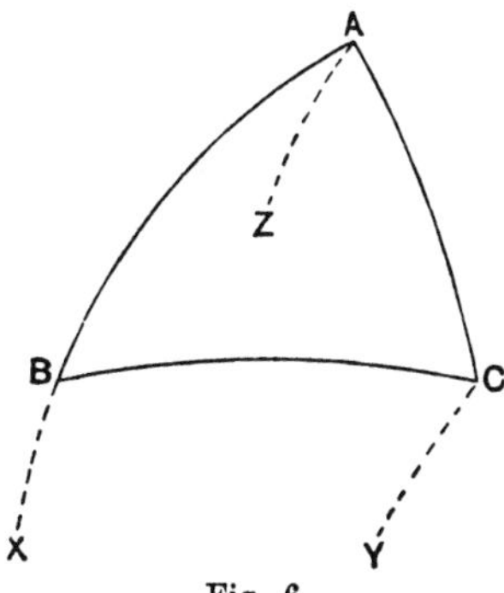

Fig. 6.

$$BAZ = A + B + C - \pi.$$

The theorem of Hamilton and Donkin has an interesting application to the kinematics of the eye. The movements considered are of course relative to the head, which for the present purpose may be regarded as fixed. The organ rotates very approximately about a fixed point O, and so far as the muscular equipment is concerned, it has the usual three degrees of freedom. In its normal operation, however, the movements are so coordinated that there is virtually freedom only of the *second* order, the position of the eye as a whole being completely determined by the direction of the visual axis, i.e. of the line joining O to that point of the field which is the object of direct vision. This limitation is essential in order that the same object seen in the same position relative to the head should affect always the same elements of the retina whenever the gaze is directed to the same point of it†.

The law which defines the position of the eyeball in terms of the direction of the visual axis was formulated by Listing (1857). The various directions of this axis may be indicated in the usual manner by their intersections with a fixed spherical surface (of arbitrary radius) described about O. There is a certain 'primary' position A on this surface, to which all others are referred‡. The law in question is that when the visual axis is transferred from A to any 'secondary' position P, the displacement of the eyeball is equivalent to a

* Hamilton, *Lectures on Quaternions*, p. 335. The simple proof in the text is by H. Hankel (1867).

† This is known as Donders' law (1847).

‡ This is the position assumed when, standing upright with head erect, we look towards a distant point of the horizon straight in front.

rotation represented by the great-circle arc AP. It is not necessary, of course, that the transition should actually be effected in this particular manner, but the result must be the same, whatever the intervening positions, in virtue of Donders' law.

It follows that the transition from one secondary position P to another Q is equivalent to a rotation XY, where X, Y are the middle points of the arcs AP, AQ, respectively. For the transition may be supposed made, first from P to A, and then from A to Q. Hence the transition from P to any other

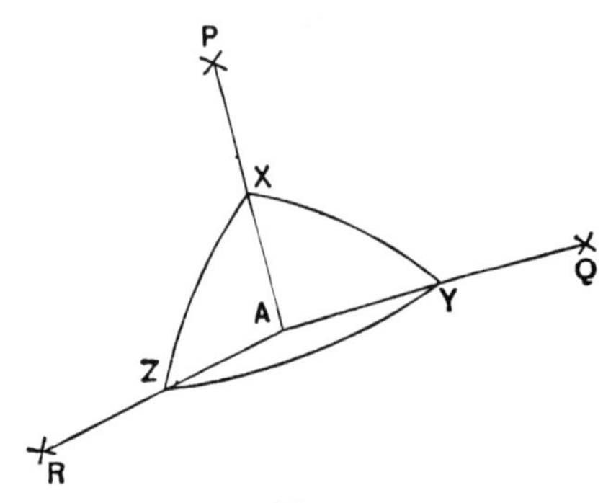

Fig. 7.

position whatever (such as Q or R in the figure) is represented by some great-circle arc through X, i.e. it is equivalent to a rotation about some axis at right angles to OX. This line OX is therefore called (by Helmholtz) the 'atropic line' for the position P.

5. The most general Displacement of a Rigid Body.

We have seen that any displacement of a rigid body may be resolved into (i) a pure translation by which an arbitrarily chosen point is brought from its initial to its final position O, and (ii) a rotation about some axis through O; and it is evident that the order in which these operations are performed is indifferent. The direction and amount of the translation will vary with the particular point chosen; but the direction of the axis of rotation, and the angle of rotation, will be independent of this choice.

There is one set of planes, fixed relatively to the body, which remain throughout the above operations parallel to their original positions, viz. the planes perpendicular to the axis of rotation. Let σ, σ' denote any figure in one of these planes, in its initial and final positions respectively. By a pure translation parallel to the axis of rotation σ may be brought into the plane of σ'; let σ''

denote its position after this operation. It is known that σ'' may now be brought into coincidence with σ' by rotation about a certain point I in its plane [S. 14]. Hence the most general displacement of a rigid body may be resolved into (i) a translation parallel to a certain axis, and (ii) a rotation about this axis*. This is the same as if the body had been rigidly attached to a nut revolving on a screw of suitable pitch. As special cases we may have a pure rotation, or a pure translation. In the latter case the point I in the above argument is at infinity.

There are various other types of elementary operation into which a displacement of a rigid body may be resolved. One of the most interesting of these is the 'half-turn,' i.e. a rotation through two right angles about a given axis†. This is specially simple in that the axis only need be indicated, the sense of the rotation being indifferent.

The method rests on the following lemmas:

1°. A half-turn about an axis a, followed by a half-turn about a *parallel* axis b, is equivalent to a translation parallel to the shortest distance between

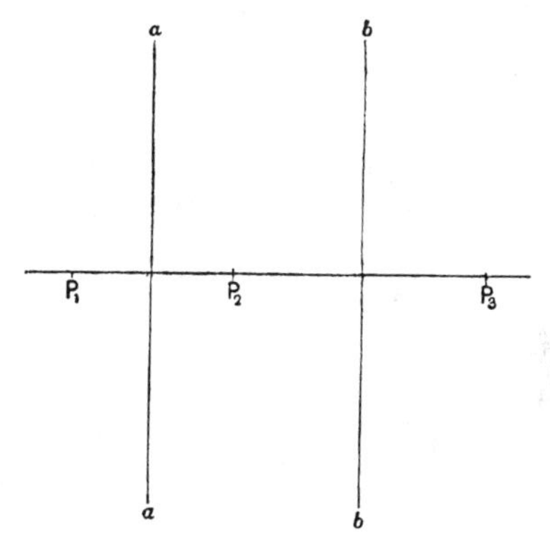

Fig. 8.

a and b, of amount equal to twice this distance, in the direction from a to b. This is proved at once by considering the displacement of any point P in the plane of a, b.

* The theorem was given in 1830 by M. Chasles (1793–1880), distinguished for his writings on Modern Geometry. The proof in the text is given by Thomson and Tait.

† This was employed by Hamilton as regards rotation about a fixed point. Further developments are due to W. Burnside (1889) and H. Wiener (1890)

2°. A half-turn about an axis a, followed by a half-turn about an *intersecting* axis b, is equivalent to a rotation about the common perpendicular to these

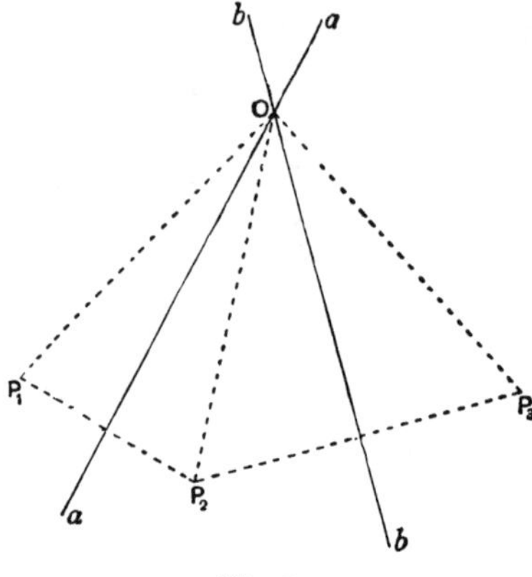

Fig. 9.

axes, of amount equal to twice the angle between them, in the direction from a to b. This is proved in a similar manner.

3°. To find the effect of successive half-turns about two *skew* axes a, b, let AB be the shortest distance between these axes, and let a' be the line through B parallel to a. Nothing is changed if we interpose two half-turns about a'. Now a half-turn about a, followed by a half-turn about a', gives a translation parallel to AB, of amount 2AB. And a half-turn about a', followed by a half-turn about b, gives a rotation about AB, of amount equal to twice the angle between a and b. We infer that any displacement of a rigid body is equivalent to successive half-turns about two axes.

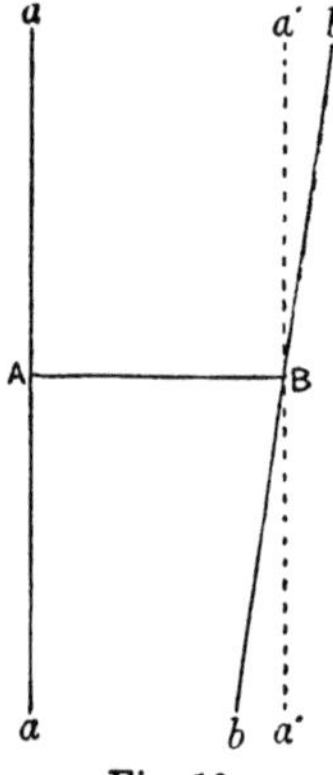

Fig. 10.

Since the specification of two skew axes involves eight elements, whereas the number of degrees of freedom of a rigid body is only six, the above resolution can be effected in a doubly infinite number of ways. All that is essential for the representation of a given displacement is that the axes a, b should meet at right angles the axis of the equivalent screw, at the proper interval, with the proper mutual inclination.

EXAMPLES. I.

1. Prove that in a three-dimensional frame of n joints the number of bars just sufficient for rigidity is $3n-6$, and that there must be at least one joint at which not more than five bars meet.

Prove that a frame of jointed bars represented by the edges of a polyhedron whose faces are all triangular is just rigid.

2. A body is rotated through 90° about each of two axes which intersect at an angle of 60°. Find the axis and the angle of the equivalent single rotation. [The angle is 104° 28′.]

3. Prove that if the telescope of an altazimuth instrument be turned in altitude through an angle α, and in azimuth through an angle β, the angle (ϕ) through which it has been rotated is given by

$$\cos\tfrac{1}{2}\phi=\cos\tfrac{1}{2}\alpha\cos\tfrac{1}{2}\beta.$$

Also that the altitude (θ) of the axis of the resultant rotation is given by

$$\cos\theta=\sin\tfrac{1}{2}\alpha/\sin\tfrac{1}{2}\phi.$$

4. If OA, OB be axes fixed in a rigid body, and moving with it, prove that a rotation $2A$ about OA, followed by a rotation $2B$ about OB, is equivalent to a rotation $-2C'$ about OC', where OC' is related to the initial positions of OA and OB as in Fig. 2, p. 4.

5. Shew that successive half-turns about three intersecting mutually perpendicular axes will restore a body to its original position.

6. Prove that successive half-turns about two skew axes at right angles are equivalent to a half-turn about the shortest distance, together with a translation represented by twice this shortest distance.

7. Deduce Donkin's theorem (Art. 4) from the theory of half-turns. (Burnside.)

8. If $ABC \ldots K$ be any fixed spherical polygon, prove that successive rotations of a rigid body about the centre O of the sphere, represented by AB, BC, ..., KA, will be equivalent to a rotation about OA through an angle proportional to the area of the polygon. (Hamilton.)

9. Also that successive rotations represented by the doubles of the arcs AB, BC, ..., KA, will restore the body to its original position. (Hamilton.)

10. Also that successive rotations about $OA, OB, OC, \ldots, OK$, through double the angles A, B, C, ..., K of the polygon, will restore a body to its original position. (Hamilton.)

11. Deduce the rule for composition of half-turns about intersecting axes from Rodrigues' construction (Art. 3).

12. Examine the proof of Rodrigues' theorem (Art. 3) in the case where the two rotations α, β have opposite signs.

13. Assuming Listing's Law (Art. 4) prove that if the eyeball rotates about a fixed axis, the visual axis traces on the spherical field an arc of a circle through the 'occipital point,' i.e. the point diametrically opposite to the primary position A.

14. If vectors be drawn from a fixed point O to represent the displacements of the various points of a rigid body, their other extremities will lie in a plane.

15. Prove directly that a straight line AB of given length can be brought into any other position $A'B'$ by a rotation about some axis, and hence that any given displacement of a rigid body can be effected by two finite rotations.

16. Prove from the theory of half-turns, or otherwise, that any displacement of a rigid body may be resolved into two rotations, and that the axis of one of these may be chosen arbitrarily.

17. Suppose that in any given displacement of a rigid body the point which was at A is brought to B, that the point which was at B is brought to C, and that the point which was at C is brought to D. Let HK be the shortest distance between the bisectors of the plane angles ABC, BCD. Prove that the displacement is equivalent to a translation HK together with a rotation about HK through an angle equal to the angle between BH and CK.

(Crofton.)

18. Having given that three points of a rigid body are brought from the positions A, B, C, to A', B', C', respectively, construct the equivalent screw-displacement.

19. Prove the following construction for finding the resultant of two finite screw-displacements about given axes a, b. Let AB be the shortest distance between these axes; let PQ be a line meeting a at right angles, such that *half* the given displacement about a would bring PQ to AB; and let RS be a line meeting b at right angles, such that *half* the given displacement about b would bring AB to RS. Further, let QS be the shortest distance between PQ and RS. Then QS is the axis of the resultant screw-displacement; the translation is $2QS$; and the rotation is twice the inclination of RS to PQ. (Burnside.)

20. A rigid body is turned through an angle θ about an axis through the origin whose direction is (l, m, n). If P and P' be the initial and final positions of any point, and ξ, η, ζ, the coordinates of the middle point (Q) of PP', prove that the direction-cosines of PP' are

$$\frac{m\zeta - n\eta,\ n\xi - l\zeta,\ l\eta - m\xi}{QN},$$

where QN is the perpendicular from Q on the axis of rotation.

If x, y, z be the coordinates of P, and x', y', z' those of P', prove that

$$x' - x = 2(\mu\zeta - \nu\eta),\quad y' - y = 2(\nu\xi - \lambda\zeta),\quad z' - z = 2(\lambda\eta - \mu\xi),$$

where $$\lambda = l\tan\tfrac{1}{2}\theta,\quad \mu = m\tan\tfrac{1}{2}\theta,\quad \nu = n\tan\tfrac{1}{2}\theta.$$

21. Deduce the equations

$$x' + \nu y' - \mu z' = x - \nu y + \mu z,$$
$$y' + \lambda z' - \nu x' = y - \lambda z + \nu x,$$
$$z' + \mu x' - \lambda y' = z - \mu x + \lambda y,$$
$$\lambda x' + \mu y' + \nu z' = \lambda x + \mu y + \nu z,$$

and prove that x', y', z' are expressed in terms of x, y, z by the following scheme, where $\rho^2 = 1 + \lambda^2 + \mu^2 + \nu^2$:

	x	y	z
$\rho^2 x'$	$1+\lambda^2-\mu^2-\nu^2$	$2(\lambda\mu-\nu)$	$2(\lambda\nu+\mu)$
$\rho^2 y'$	$2(\lambda\mu+\nu)$	$1-\lambda^2+\mu^2-\nu^2$	$2(\mu\nu-\lambda)$
$\rho^2 z'$	$2(\lambda\nu-\mu)$	$2(\mu\nu+\lambda)$	$1-\lambda^2-\mu^2+\nu^2$

(Rodrigues. The method indicated is due to Darboux.)

CHAPTER II

INFINITESIMAL DISPLACEMENTS

6. Rotations as Localized Vectors.

From this point onwards we consider only infinitesimal displacements. The theory is included to a certain extent in that of finite displacements, but is simpler in that the effect of successive operations can be found by superposition, without reference to the order. This is in virtue of the general principle of superposition of small variations.

An infinitesimal rotation is determinate when we know the axis about which it takes place, and the angle, regard being had of course to the *sense*. It may therefore be indicated completely by a segment AB* measured along the axis, of a length proportional, on some magnified scale†, to the angle, the direction from A to B being chosen so that the rotation is right-handed with respect to it (Art. 3). The exact position of AB along the axis is immaterial.

The utility of this convention rests on the fact that a rotation, as thus represented, has, like a force in Statics, the properties of a 'localized vector.' In particular, rotations about intersecting axes may be compounded by addition of the corresponding vectors.

To prove this, we note in the first place that the displacement of a point O due to a rotation AB will be normal to the plane OAB, and of amount $AB\,.\,OM$, where OM is the perpendicular drawn from O to AB. In the annexed figure it will be towards or from the reader, and may therefore be reckoned as positive or negative, according as O lies to the left or right of AB. The

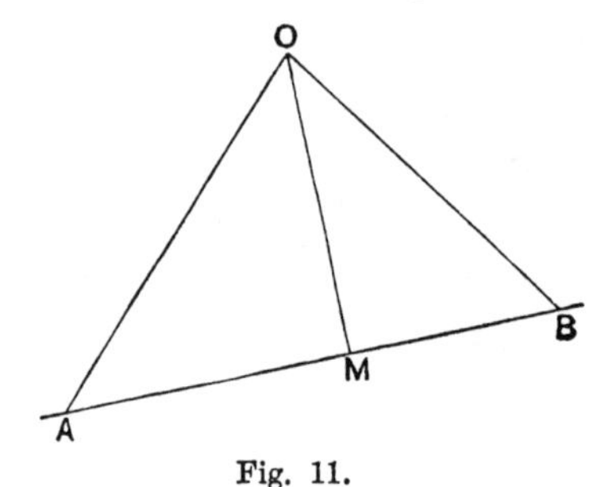

Fig. 11.

* Roman type is used to distinguish the *vector* AB from the mere *length* AB.

† Thus we may suppose that the angle is $\epsilon\,.\,AB$, where ϵ is some infinitesimal constant, of the nature of the reciprocal of a line. The factor ϵ is however omitted in what follows.

displacement of O is thus given by the same rule, both as to magnitude and sign, as the *moment* of a *force* AB about O in Plane Statics. It follows that the effect on a rigid body of two infinitesimal rotations represented by AB, AC will be the same as that of a rotation AD, the geometric sum of AB, AC. For we can infer from Varignon's theorem of moments that this is true as regards the displacements of all points in the plane ABC. The displacements of three such points (not in the same line) are however sufficient to determine the altered position of the body.

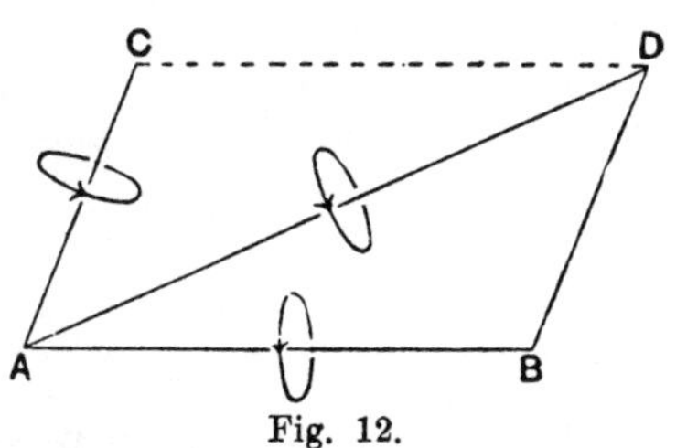

Fig. 12.

There is thus a complete analogy, as regards mathematical relations, between infinitesimal rotations on the one hand, and forces in Pure Statics on the other. A rotation, like a force, is associated with a definite straight line, but has no reference to any special point in the line; it admits of opposite senses; and equal and opposite rotations about the same line neutralize one another; moreover, rotations about intersecting axes are compounded by the law of vector addition.

Since the assumptions corresponding to these are all that is required as a basis for Pure Statics, we may infer that every proposition in the latter subject has an exact analogue in the kinematical theory of the infinitesimal displacements of a rigid body, and *vice versâ*. A theorem established in either connection is at once known to be true when translated into the language of the other application.

For example, from the known rules for the composition of parallel forces we may deduce the result of rotations about parallel axes. The ordinary statical proof may in fact be adapted step by step.

An independent proof is easily given. Thus, suppose we have two infinitesimal rotations p, q (reckoned positive when right-handed) about the parallel axes AA', BB'.

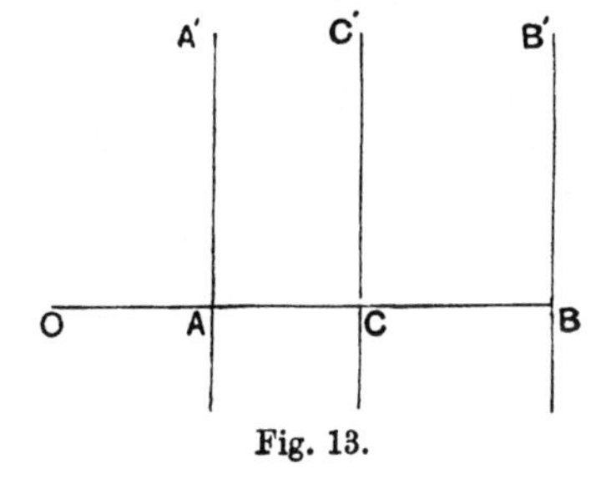

Fig. 13.

Let O be any point in the plane of these axes, and draw the perpendicular OAB. The resulting displacement of a point at O will be normal to the plane of the figure, of amount $p\,.\mathrm{OA} + q\,.\,\mathrm{OB}$, if suitable signs be attributed to OA, OB according to their direction. If in AB we take a point C such that, in the vector interpretation,

$$p\,.\,\mathrm{CA} + q\,.\,\mathrm{CB} = 0, \quad \text{....................(1)}$$

we have

$$p\,.\mathrm{OA} + q\,.\,\mathrm{OB} = p\,(\mathrm{OC} + \mathrm{CA}) + q\,(\mathrm{OC} + \mathrm{CB}) = (p + q)\,\mathrm{OC}. \text{...(2)}$$

The displacement of O is therefore such as would have been produced by a rotation $p + q$ about a parallel axis through C*

A special case arises, as in Statics, when $p + q = 0$. We then have

$$p\,.\,\mathrm{OA} + q\,.\,\mathrm{OB} = p\,.\,(\mathrm{OA} - \mathrm{OB}) = p\,.\,\mathrm{BA}, \text{........(3)}$$

the same for all positions of O in the plane. Hence equal and opposite rotations about parallel axes are equivalent to a translation normal to the plane of the axes, of amount equal to the product of either rotation into the distance between the axes. The *sense* of the translation is ascertained by taking O on one of the axes. We thus learn, as a detail of the analogy above referred to, that a translation corresponds to a *couple* in Statics, and the amount of the translation to the moment of the couple.

7. Composition of Rotations about Skew Axes.

It was shewn in Art. 5 that the most general displacement of a rigid body is equivalent to a rotation about a definite axis, together with a translation parallel to that axis. It is therefore described as a 'twist' about a certain screw. By a 'screw' is here meant merely the *geometrical form* which indicates the *type*, but not the *amount*, of the particular displacement. It is completely specified when we know the axis, and the ratio of the translation to the rotation. This latter ratio, which is of the nature of a line, is called the 'pitch,' and is reckoned positive or negative, according as the relation between the rotation and the translation is right or left-handed (Art. 3). When the axis and the pitch are known, the amount of the twist is specified by the angle of rotation, which

* The figure is drawn for the case where p, q have the same sign; but the proof is general.

may be positive or negative according to the particular convention adopted.

The composition of rotations about axes in the same plane has already been dealt with. We proceed to investigate the twist which is equivalent to rotations p, q about two *skew* axes. Let ω be the rotation which would be equivalent to rotations p, q about two intersecting axes respectively parallel to the given ones; and let α, β be the angles which the axes of p and q make with that of ω on opposite sides. These quantities ω, α, β may be supposed found by an auxiliary construction. Now let AB be the shortest

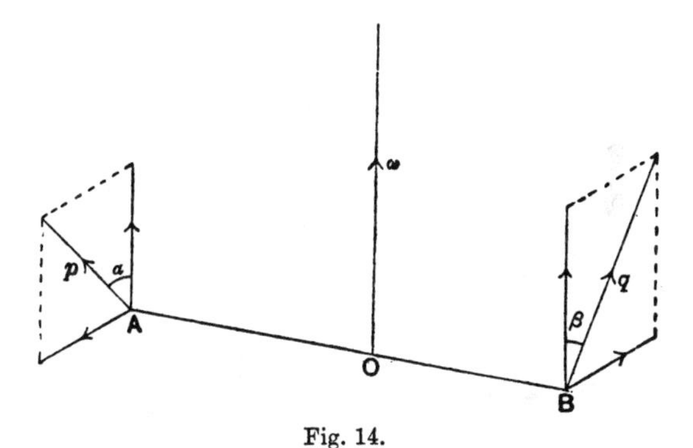

Fig. 14.

distance between the given axes, and let a plane be drawn through AB parallel to the axis of ω. The rotation p may be resolved at A into two components $p\cos\alpha$, $p\sin\alpha$, about axes in and perpendicular to this plane; and q will have corresponding components $q\cos\beta$, $q\sin\beta$. The components $p\sin\alpha$, $q\sin\beta$ are equal in magnitude, but of opposite sense, and are therefore equivalent to a translation

$$\tau = p\sin\alpha\,.\,AB = q\sin\beta\,.\,AB, \quad \text{..............(1)}$$

parallel to the axis of ω. The components $p\cos\alpha$, $q\cos\beta$ are equivalent to a rotation

$$\omega = p\cos\alpha + q\cos\beta \quad \text{....................(2)}$$

about a parallel axis through a point O in AB* such that

$$p\cos\alpha\,.\,AO = q\cos\beta\,.\,OB. \quad \text{.................(3)}$$

* The figure is drawn for the case where the angles α, β are both acute. If one of them be obtuse the point O will fall in the prolongation of AB in one direction or the other.

We have thus determined the axis of the resultant twist, and the amounts τ, ω of the translation and rotation composing it.

From (2) and (3) we have

$$\frac{p\cos\alpha}{OB}=\frac{q\cos\beta}{AO}=\frac{\omega}{AB}. \quad \text{.................(4)}$$

The pitch of the screw about which the twist takes place is therefore

$$\varpi=\tau/\omega=OB\tan\alpha=AO\tan\beta, \quad \text{..............(5)}$$

whence
$$\varpi=\frac{\sin\alpha\sin\beta}{\sin(\alpha+\beta)}.AB. \quad \text{....................(6)}$$

It is apparent, conversely, from the preceding investigation that any infinitesimal displacement of a rigid body can be resolved into two rotations in an endless variety of ways. We can shew, further, that the axis (but not the amount) of one of these may be chosen quite arbitrarily.

Suppose, for example, that the axis of p in the above figure is assigned. We construct the shortest distance AO between this axis and that of the given twist. The quantities ω, τ, α, and the magnitude of AO, are known; and we have to find the position of B, the angle β, and the rotations p, q. The values of OB and β are given by (5); and those of p, q then follow from (4).

That the resolution of a twist into two rotations is possible, with one of the axes arbitrary, might have been expected. For when the axis of p (say) has been assigned, we have still six adjustable elements at our disposal, viz. the four quantities necessary to fix the axis of q, and the magnitudes of p and q.

Two lines which are related in the above manner with respect to a given screw are said to be 'conjugate.' It will be noted that the orthogonal projections of two conjugate lines on a plane perpendicular to the axis of the screw are parallel (Fig. 14).

Finally, we have

$$\begin{aligned} pq\sin(\alpha+\beta).AB &= p\sin\alpha.AB.q\cos\beta+q\sin\beta.AB.p\cos\alpha \\ &= \omega\tau, \quad \text{.......................................(7)} \end{aligned}$$

by (1) and (2). The expression in the first member is six times the volume of a tetrahedron having two opposite edges in the axes of p and q, respectively, of lengths proportional to these quantities.

If the two rotations be represented, in the manner already explained, by localized vectors HK, LM, it appears from (7) that in whatever way the above resolution be effected, the volume of the tetrahedron of which HK, LM are opposite edges is constant, being proportional to the product $\omega\tau$.

8. Null-Lines, Points, and Planes.

A 'null-line' with respect to a given screw may be defined from the present point of view as a straight line such that the infinitesimal displacements of all points of the body which lie in it are at right angles to the length. It is known that if this condition is fulfilled for any one point of the line it will hold with respect to the rest [S. 50, 51]. It is evident at once that any null-line which meets one of a pair of conjugate lines must also meet the other.

The conception of null-lines will be met with again in the analogous theory of systems of forces in Statics, in which connection it was indeed first introduced; see Art. 16. It has however an importance from the present kinematical standpoint. For instance, if one or more points of the body be constrained to lie on given surfaces, the normals to the surfaces at these points are null-lines. In particular, if a structure be connected to the earth by one or more links, freely jointed at their extremities, the axes of these links are null-lines.

The whole assemblage of null-lines, since it consists of lines in space which are subject to one condition, form a three-fold infinity. In the language of Line Geometry they constitute a 'complex.' In the present case, those lines of the complex which pass through any assigned point O will lie in a plane, for being null-lines they must be perpendicular to the direction of displacement of that point of the body which is at O. The complex is therefore of the type called 'linear,' or 'of the first degree*.' The plane which is the locus of the null-lines through O is called the 'null-plane,' or 'polar plane,' of O.

Conversely, the null-lines which lie in a given plane will in general meet in a point†. For if two null-lines in the plane meet

* In general, those lines of a complex which pass through a given point generate a cone. If this surface is algebraic, its degree fixes the 'degree' of the complex.

† In the general case of a complex of the nth degree, the lines which lie in a given plane envelope a curve of the nth 'class.'

in O the displacement at O is normal to the plane, and all lines in the plane which pass through O will be null-lines. And there are in general no others. For if three null-lines in the plane could form a triangle, the displacement at each intersection would be normal to the plane, and the displacement of the solid would then evidently consist merely of a rotation about some finite or infinitely distant line in the plane. This special case is to be excepted from the preceding statement. The point O in which the null-lines of a plane meet is called the 'null-point,' or 'pole' of the plane.

The matter may also be looked at from another interesting point of view. Consider the points of the body which lie originally in a plane ϖ, and let ϖ' be their plane after an infinitesimal displacement. Let σ be any figure in the plane ϖ, and let σ' denote its displaced position. If σ' be projected orthogonally on the plane ϖ, we obtain a figure σ'' which may be regarded as congruent in all respects with σ, since we neglect small quantities of the second order. The figures σ, σ'' will not in general coincide, but may be brought into coincidence by a rotation about a point O in their plane [S. 14, 15]. Let m denote the normal to ϖ at O, and n the ultimate intersection of the planes ϖ and ϖ'. It is evident that the displacement of the body may be regarded as made up of rotations of suitable amounts about these axes m, n. It follows that all the null lines in the plane must intersect m as well as n, and must therefore pass through O. It will be noticed that we have in m, n a pair of conjugate lines which are at right angles. The line n is called the 'characteristic' of the plane*.

If the pole A of a plane α lies in a plane β, the pole B of β will lie in α. This follows at once from the fact that AB is a null-line. Also, the poles of all planes through AB will lie in the intersection of the planes α, β. This intersection is conjugate to AB.

Some further properties of null systems will be noticed in the Chapter on Statics.

9. Analytical Formulæ.

Whenever in the course of this book we employ rectangular Cartesian coordinates, we shall assume that the axes form a 'right-handed' system. That is, we assume the positive directions Ox, Oy, Oz to be so related that a positive rotation of $\frac{1}{2}\pi$ about Ox would bring Oy into the former position of Oz, and so on in cyclical order of the letters. Thus, if Ox, Oy are in their usual relative position in the plane of the paper, Oz will be *above* this plane. If the

* This theory is due to Chasles.

direction of *one* of the axes be reversed we obtain a 'left-handed' system. One set of axes Ox, Oy, Oz can be brought into coincidence with another set $O'x'$, $O'y'$, $O'z'$ by a mere displacement, so that Ox' shall coincide with Ox, Oy' with Oy, and Oz' with Oz, only if both systems are right-handed, or both left-handed*.

In proceeding to an analytical discussion of the small displacements of a rigid body, we begin with the case where one point O of the body is supposed fixed. We have seen (Art. 2) that any small displacement is then equivalent to a rotation ω about some axis OJ. Taking rectangular axes through O, this rotation can be resolved into three components p, q, r about Ox, Oy, Oz, respectively; thus if λ, μ, ν be the direction-cosines of OJ we have

$$p = \lambda\omega, \quad q = \mu\omega, \quad r = \nu\omega, \qquad (1)$$

and

$$p^2 + q^2 + r^2 = \omega^2. \qquad (2)$$

To find the component displacements, parallel to the axes, of a point P whose coordinates are (say) x, y, z, we make use of the principle that the displacements of any two points A, B, of the body when resolved in the direction AB are equal, since the length AB is unaltered [S. 50, 51]. We draw PL normal to the plane yz,

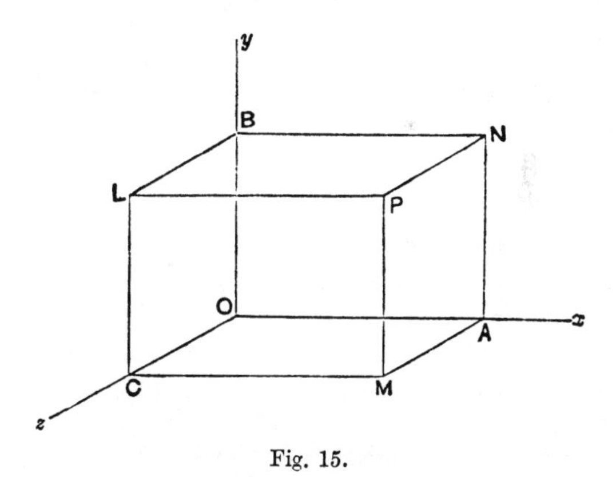

Fig. 15.

and LB, LC perpendicular to Oy, Oz, respectively. The displacement of P parallel to Ox will be equal to that of L, which is made

* The relation between two systems of opposite character is that of an object and its image in a plane mirror. The geometrical process of passing from the one to the other is called 'perversion.'

up of qz due to the rotation about Oy, and $-ry$ due to the rotation about Oz. Hence, and by similar considerations, we find that the component displacements of P are

$$\delta x = qz - ry, \quad \delta y = rx - pz, \quad \delta z = py - qx. \quad \text{.........(3)}$$

We note that

$$x\delta x + y\delta y + z\delta z = 0, \quad \text{.............................(4)}$$

$$p\delta x + q\delta y + r\delta z = 0, \quad \text{.............................(5)}$$

which verify that the displacement of P is perpendicular to both OP and OJ. We have also

$$\begin{aligned}(\delta x)^2 + (\delta y)^2 + (\delta z)^2 &= (p^2+q^2+r^2)(x^2+y^2+z^2) - (px+qy+rz)^2 \\ &= \omega^2 . OP^2 . \sin^2 JOP. \quad \text{............................(6)}\end{aligned}$$

This is otherwise evident*, for if PQ be drawn perpendicular to OJ, the displacement of P is

$$\omega . PQ = \omega . OP . \sin JOP.$$

To express the most general displacement of a rigid body relative to fixed coordinate axes, we have only to superpose a uniform translation whose components l, m, n (say)† are the component displacements of that point of the body which was at O. Thus

$$\left.\begin{aligned}\delta x &= l + qz - ry, \\ \delta y &= m + rx - pz, \\ \delta z &= n + py - qx.\end{aligned}\right\} \quad \text{.......................(7)}$$

The locus of the points which are displaced parallel to the axis of rotation is determined by

$$\frac{\delta x}{p} = \frac{\delta y}{q} = \frac{\delta z}{r}, \quad \text{........................(8)}$$

or

$$\frac{l + qz - ry}{p} = \frac{m + rx - pz}{q} = \frac{n + py - qx}{r}. \quad \text{.........(9)}$$

These are the equations of a straight line; and since they are unaltered when we write $x + kp$ for x, $y + kq$ for y, and $z + kr$ for z, whatever the value of k, we see that the line is parallel to the axis of the rotation. The displacement of the body is therefore equivalent to a rotation about this line, together with a translation

* If $OP = \rho$, we have

$$\cos JOP = \frac{p}{\omega} . \frac{x}{\rho} + \frac{q}{\omega} . \frac{y}{\rho} + \frac{r}{\omega} . \frac{z}{\rho}.$$

† This notation has been chosen for the sake of conformity with the analogous formulæ in Statics (Art. 19).

parallel to it*. In our previous terminology we have a twist about a certain screw, of which the line (9) is the axis.

The six elements l, m, n, p, q, r on which the specification of a particular infinitesimal displacement of a solid depends, or any six quantities proportional to them, determine by the five *ratios* which they bear to one another the axis and the pitch of the screw about which the body is displaced, and may be called the 'coordinates' of the screw.

The displacement of any point, resolved parallel to the axis of the screw, is

$$\tau = \frac{p}{\omega}\,\delta x + \frac{q}{\omega}\,\delta y + \frac{r}{\omega}\,\delta z = \frac{lp + mq + nr}{\omega}; \quad \text{.........(10)}$$

and the pitch is therefore

$$\varpi = \frac{\tau}{\omega} = \frac{lp + mq + nr}{p^2 + q^2 + r^2}. \quad \text{...............(11)}$$

If we were to refer the same displacement to any other congruent system of rectangular axes, the value of $p^2 + q^2 + r^2$, or ω^2, would of course be unaltered. Also, since the pitch is fixed, the same statement must hold with regard to the expression $lp + mq + nr$. Hence in the transformation from one right-handed set of axes to another, the expressions

$$p^2 + q^2 + r^2 \text{ and } lp + mq + nr$$

are absolute invariants.

If the displacement reduces to a pure rotation, we have $\varpi = 0$, or

$$lp + mq + nr = 0. \quad \text{.....................(12)}$$

The six quantities l, m, n, p, q, r, *when connected by this relation*, determine a *line*, viz. the axis of rotation. This system of 'line-coordinates,' as they are termed, is employed in many geometrical investigations†. From this point of view we are concerned only with their *ratios*; they are accordingly equivalent to only four independent quantities, in virtue of (12).

If (x_1, y_1, z_1) be any point on the line, the values of l, m, n are obtained from (7) by momentarily transferring the origin to (x_1, y_1, z_1), and writing $-x_1, -y_1, -z_1$ for x, y, z, respectively. Thus

$$l = ry_1 - qz_1, \quad m = pz_1 - rx_1, \quad n = qx_1 - py_1, \quad \text{...............(13)}$$

* This theorem was given in 1827 by A. L. Cauchy (1789–1857). As already stated the extension to finite displacements is due to Chasles.

† It was introduced by Cayley in 1860, and subsequently employed by Plücker in his researches on Line Geometry.

which verify (12), as they ought. The condition that a line (l, m, n, p, q, r) should be a null-line with respect to the screw (l', m', n', p', q', r') can now be expressed. Since the direction-cosines of the line are proportional to p, q, r, the condition that the longitudinal displacement of a point at (x_1, y_1, z_1) should vanish is

$$p(l'+q'z_1-r'y_1)+q(m'+r'x_1-p'z_1)+r(n'+p'y_1-q'x_1)=0, \quad \text{...(14)}$$

or

$$lp'+mq'+nr'+l'p+m'q+n'r=0, \quad \text{....................(15)}$$

by (13). This is the equation of the linear complex formed by the null-lines of the screw (l', m', n', p', q', r').

We shall meet with a relation of this form as the more general condition that two screws should be 'reciprocal' (Art. 22). The quantities l, m, n, p, q, r are then no longer subject to the condition (12).

10. Constraints.

Whenever the small displacements of a rigid body are subject to some one geometrical condition, the analytical expression of this leads in general to a homogeneous linear relation between the six variable quantities l, m, n, p, q, r, that is, to a relation of the form

$$Al + Bm + Cn + Fp + Gq + Hr = 0, \quad \text{............(1)}$$

where the coefficients are constants. For example, if a point (x_1, y_1, z_1) be constrained to lie on a given surface, the displacements of this point are subject to the condition

$$\lambda\delta x + \mu\delta y + \nu\delta z = 0, \quad \text{....................(2)}$$

where λ, μ, ν are the direction-cosines of the normal to the surface at the point (x_1, y_1, z_1). Hence

$$\lambda(l + qz_1 - ry_1) + \mu(m + rx_1 - pz_1) + \nu(n + py_1 - qx_1) = 0, \text{...(3)}$$

which is of the above form.

If the body is subject to *six* relations of the above type, the only solution is in general $l = m = n = p = q = r = 0$. The body is then fixed; i.e. it cannot be moved without violating one or other of the given conditions.

Thus, as already pointed out, a rigid structure which is anchored down by six links is in general fixed.

It may, however, happen that owing to the special configuration of the links the body is still capable of an infinitesimal displacement. The analogous but much simpler question in Plane Kinematics [S. 15] will be remembered.

Let the line-coordinates of the six links be denoted by

$(l_1, m_1, n_1, p_1, q_1, r_1)$, $(l_2, m_2, n_2, p_2, q_2, r_2)$, ..., $(l_6, m_6, n_6, p_6, q_6, r_6)$, where the symbols in each group are supposed subject to a relation of the form (12) of Art. 9. By hypothesis, the lines in question are null-lines with respect to some displacement (l', m', n', p', q', r'), and must therefore satisfy the equation

$$lp' + mq' + nr' + l'p + m'q + n'r = 0. \quad \text{.....................(4)}$$

In words, they must all belong to the same linear complex*, of which this is the equation. Analytically, the condition is

$$\begin{vmatrix} l_1, & m_1, & n_1, & p_1, & q_1, & r_1 \\ l_2, & m_2, & n_2, & p_2, & q_2, & r_2 \\ \cdots & \cdots & \cdots & \cdots & \cdots & \cdots \\ l_6, & m_6, & n_6, & p_6, & q_6, & r_6 \end{vmatrix} = 0. \quad \text{........................(5)}$$

Cf. Art. 19 (26).

If we have *five* independent relations of the type (1), these will give unique values to the five ratios $l:m:n:p:q:r$. The body has now one degree of freedom, viz. it can twist (to an arbitrary extent) about the screw thus determined. The displacement of any point of the body can now take place only in a definite direction.

If we have *four* independent constraints, the ratios $l:m:n:p:q:r$ are indeterminate; but if (l', m', n', p', q', r') and $(l'', m'', n'', p'', q'', r'')$ be any two independent solutions of the four equations, any other solution can be expressed in the form

$$\left.\begin{aligned} l &= \lambda l' + \mu l'', \quad m = \lambda m' + \mu m'', \quad n = \lambda n' + \mu n'', \\ p &= \lambda p' + \mu p'', \quad q = \lambda q' + \mu q'', \quad r = \lambda r' + \mu r''. \end{aligned}\right\} \quad \text{...(6)}$$

For the given conditions are obviously satisfied by the six quantities

$$l - \lambda l' - \mu l'', \quad m - \lambda m' - \mu m'', \quad n - \lambda n' - \mu n'',$$
$$p - \lambda p' - \mu p'', \quad q - \lambda q' - \mu q'', \quad r - \lambda r' - \mu r''.$$

We may determine λ and μ so as to make two of these vanish, and the equations can then only be satisfied by zero values of the remaining four. The body has now two degrees of freedom, since it can twist to arbitrary and independent amounts about the two screws (l', m', n', p', q', r'), $(l'', m'', n'', p'', q'', r'')$. These two screws may themselves be chosen in an infinity of ways. Since the directions of motion of any point of the body, due to twists about the two screws, are independent, the point is capable of motion in two dimensions. It follows that the case of a rigid body, four of whose points are constrained to move on given surfaces, embraces all possible types of freedom of the second order.

* Six lines so related are said by Sylvester to be 'in involution.' The theory was developed from the statical standpoint by Cayley, Spottiswoode, and Sylvester.

Three constraints, again, leave the body free to twist about three independent screws. Each point of the body has now, in general, liberty to move (but not of course independently) in three dimensions. Hence, although the case of a body in contact at three points with fixed surfaces is an important instance, it is not sufficiently general to be typical.

The remaining cases, where a body is subject only to *two* conditions, or *one*, and has therefore four or five degrees of freedom, respectively, are of less interest, but will be referred to later (Art. 22)*.

11. Two Degrees of Freedom. The Cylindroid.

We proceed to consider more fully the case of a body having two degrees of freedom. It has been pointed out that this is typified, with complete generality, by a solid in contact at four points with fixed surfaces, and that the normals at these points are null-lines with respect to any possible displacement. These normals are met by two real or imaginary transversals†, which are evidently conjugate lines. When they are real, any possible displacement can be resolved into two *rotations* about them as axes; but since they may be imaginary, this statement does not hold in general. We have therefore to ascertain the result of twists of arbitrary amounts about any two screws whose axes and pitches are given, and to examine the configuration of the singly-infinite system of screws thus obtained.

We begin, somewhat indirectly, with the case where the axes of the two given screws are at right angles, and intersect. We take these as the axes of x and y, and denote the respective pitches by a, b. If p, q be the rotations about them, we have, in the notation of Art. 9,

$$l = ap, \quad m = bq, \quad n = 0, \quad r = 0. \qquad \text{(1)}$$

* The theory of infinitesimal displacements was developed from the above point of view by Sir R. S. Ball (1840–1913), Lowndean professor of Astronomy at Cambridge 1892–1913, in a series of papers dating from 1871. See his *Theory of Screws*, Dublin, 1876, and Cambridge, 1900.

† Any three skew lines determine a hyperboloid of one sheet, on which they are generators of the same system. The fourth line meets this surface in two real or imaginary points. The generators of the opposite system which pass through these points are the transversals in question.

The equations of the axis of the resultant twist (Art. 9 (9)) become

$$\frac{ap+qz}{p}=\frac{bq-pz}{q}=\frac{py-qx}{0}, \quad \text{...............(2)}$$

whence

$$x/p=y/q, \quad z(p^2+q^2)=(b-a)pq. \quad \text{...........(3)}$$

Hence, whatever the values of p, q, the axis of the resultant twist is a generator of the conoidal surface

$$z(x^2+y^2)=(b-a)xy. \quad \text{...................(4)}$$

The coordinates of any point on this surface may be expressed in the forms

$$x=\rho\cos\theta, \quad y=\rho\sin\theta, \quad z=c\sin 2\theta, \quad \text{.........(5)}$$

where

$$c=\tfrac{1}{2}(b-a). \quad \text{........................(6)}$$

The generators meet the axis of z at right angles, and intersect a circular cylinder described about this axis in a curve which, when

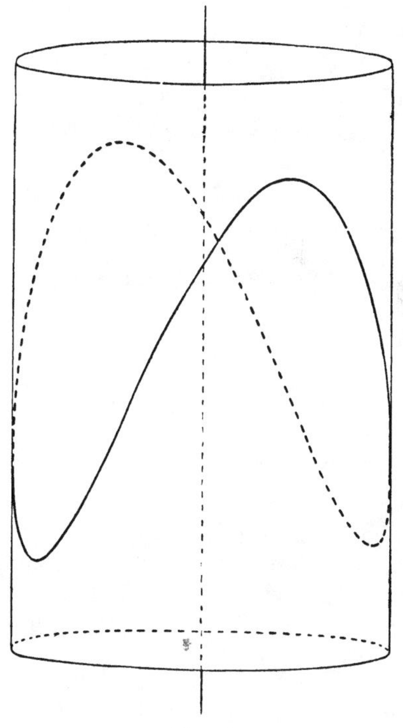

Fig. 16.

developed into a plane, is a curve of sines, the perimeter including two complete periods. The surface is called the 'cylindroid*'; its scale is determined by the parameter c.

The pitch of the screw whose azimuth is θ is, by Art. 9 (11),

$$\varpi = \frac{lp + mq}{p^2 + q^2} = a\cos^2\theta + b\sin^2\theta = \tfrac{1}{2}(a+b) + \tfrac{1}{2}(a-b)\cos 2\theta. \qquad \text{......(7)}$$

The distribution of pitch among the various screws of the system is indicated by the conic

$$ax^2 + by^2 = C, \qquad \text{........................(8)}$$

where C is any constant; for if ρ be the radius drawn in the direction θ, we have

$$\varpi = C/\rho^2. \qquad \text{..........................(9)}$$

The curve (8) is accordingly called the 'pitch conic.' We notice that not more than two screws of the system have the same pitch.

It is important to notice that the cylindroid itself is unchanged if the pitches a, b be increased by equal amounts, since the parameter c is unaltered. The only modification is that the pitches ϖ of the screws associated with the various generators of the surface are all increased by the same amount, in virtue of (7).

If we distinguish by suffixes the quantities relating to any two screws of the system, we have from (5)

$$z_1 - z_2 = 2c\sin(\theta_1 - \theta_2)\cos(\theta_1 + \theta_2), \qquad \text{............(10)}$$

and from (7)

$$\varpi_1 - \varpi_2 = 2c\sin(\theta_1 - \theta_2)\sin(\theta_1 + \theta_2), \qquad \text{.........(11)}$$

whence

$$\tan(\theta_1 + \theta_2) = \frac{\varpi_1 - \varpi_2}{(z_1 - z_2)}. \qquad \text{..................(12)}$$

It appears from these formulæ that a system of screws having the above configuration can be constructed so as to include any two screws whatever. For, the perpendicular distance $z_1 - z_2$ between the two axes, the mutual inclination $\theta_1 - \theta_2$, and the difference $\varpi_1 - \varpi_2$ of the pitches being regarded as given, the angle $\theta_1 + \theta_2$ is determined by (12), and either of the equations (10), (11) then serves to determine the parameter c.

* The name was given by Cayley. The surface presents itself in the geometrically cognate subject of the Linear Complex, as investigated by Plücker and others.

We may assert, then, that whenever a body has two degrees of freedom, the axes of the various screws about which it can twist will lie on a certain cylindroid, and that the distribution of pitch among these screws will be given by a formula of the type (7).

As a particular case, the cylindroid may of course degenerate into a plane, the parameter c being zero. The various screws have then all the same pitch, as is otherwise evident.

12. Three Degrees of Freedom.

When a body has three degrees of freedom, as defined by three independent screws having given axes and pitches, the axis and the pitch of any other screw about which it can twist are determined by the two ratios between the three angles of rotation. The number of such screws is accordingly doubly infinite, and their axes form what is called in Line Geometry a 'congruence.' To ascertain the configuration, and the distribution of pitch, we begin, as in the preceding Art., with a special assumption, and afterwards proceed to shew that it can be made to cover all possible cases.

We assume that the axes of the three fundamental screws intersect at right angles. We take these as coordinate axes, and denote the corresponding pitches by a, b, c. If, in any small displacement, p, q, r be the angles of rotation about these axes, we have

$$l = ap, \quad m = bq, \quad n = cr, \quad \text{..................(1)}$$

and the pitch of the resultant screw is

$$\varpi = \frac{ap^2 + bq^2 + cr^2}{p^2 + q^2 + r^2} \text{. ....(2)}$$

The equations of its axis are, by Art. 9 (9),

$$\frac{ap + qz - ry}{p} = \frac{bq + rx - pz}{q} = \frac{cr + py - qx}{r}, \quad \text{......(3)}$$

and it is to be noted that each of these fractions is equal to ϖ. Hence, to find the locus of the axes of all the screws having a given pitch, we have to eliminate the ratios $p : q : r$ between the equations

$$\left.\begin{array}{rrrl} (a - \varpi)\, p & + zq & - yr & = 0, \\ - zp & + (b - \varpi)\, q & + xr & = 0, \\ yp & - xq & + (c - \varpi)\, r & = 0. \end{array}\right\} \quad \text{.........(4)}$$

The result is

$$(a-\varpi)x^2+(b-\varpi)y^2+(c-\varpi)z^2+(a-\varpi)(b-\varpi)(c-\varpi)=0. \quad \text{......(5)}$$

The locus, when it is real, is accordingly a hyperboloid of one sheet. It will be imaginary if ϖ exceeds the greatest, or is less than the least, of the three quantities a, b, c.

In particular, the locus of the axes of the screws of *zero* pitch is the quadric

$$ax^2+by^2+cz^2+abc=0, \quad \text{................(6)}$$

which is, however, imaginary if a, b, c have all the same sign.

By varying the value of ϖ in (5) we get a series of hyperboloids, containing the axes of all possible screws of the system.

If ρ be the radius of the quadric (6) drawn parallel to any one of these axes, it results from a comparison of (2) and (6) that the corresponding pitch is

$$\varpi=-\frac{abc}{\rho^2}. \quad \text{......................(7)}$$

The surface (6) is therefore called the 'pitch-quadric.'

We have still to shew that a system of screws of the above kind, based on three screws whose axes intersect at right angles, represents the most general case of freedom of the third order.

In the first place, if the possible screws include three screws of zero pitch, i.e. three axes of pure rotation, these determine a certain hyperboloid, and the generators of the opposite system will be null-lines. Any other axis of pure rotation must meet all the null-lines, and must therefore lie on the aforesaid hyperboloid. The equation of this surface, when referred to its principal axes, will be of the form

$$\frac{x^2}{\alpha^2}+\frac{y^2}{\beta^2}-\frac{z^2}{\gamma^2}=1, \quad \text{......................(8)}$$

and, to identify this with (6), we have only to make

$$bc=-\alpha^2, \quad ca=-\beta^2, \quad ab=\gamma^2. \quad \text{..............(9)}$$

Taking α, β, γ as positive, we find

$$a=-\beta\gamma/\alpha, \quad b=-\gamma\alpha/\beta, \quad c=\alpha\beta/\gamma. \quad \text{........(10)}$$

The present case arises whenever the body is in contact at three points with fixed surfaces, but, as already pointed out, this example is not sufficiently general.

We have seen, however, that the configuration of the axes of the various screws which result from arbitrary twists about two primitive screws is unchanged if the pitches of these be altered by the same amount. The only effect is that the pitches throughout the resulting system are all altered in the same way. This conclusion must hold when we add a third primitive screw, provided its pitch be similarly altered.

Take now the most general case of three degrees of freedom, and suppose that ϖ is the pitch of some screws in the system. If we add $-\varpi$ to all the pitches, the configuration of the axes is unchanged. Hence the axes of all the screws whose pitch was ϖ will lie on a hyperboloid whose equation, referred to the principal diameters, is of the form

$$a'x^2 + b'y^2 + c'z^2 + a'b'c' = 0. \quad \text{..............(11)}$$

The system of (altered) screws is that due to arbitrary twists about three screws whose axes intersect at right angles, and whose pitches are a', b', c', respectively. The original system can therefore be based on three screws about intersecting and mutually perpendicular axes, whose pitches are

$$a = a' + \varpi, \quad b = b' + \varpi, \quad c = c' + \varpi. \quad \text{.........(12)}$$

In terms of a, b, c the equation (11) takes the form (5), as before. The locus of the axes of the screws of zero pitch is the quadric (6), but this is now not necessarily real. If it be imaginary, the distribution of pitch may be found by reference to the conjugate hyperboloid obtained by reversing the sign of the last term in the equation*.

EXAMPLES. II.

1. Deduce the rule for compounding infinitesimal rotations about parallel axes from Rodrigues' construction for the case of finite rotations (Art. 3).

2. Prove from first principles that three infinitesimal rotations represented by BC, CA, AB are equivalent to a translation normal to the plane ABC, of amount proportional to the area of the triangle ABC.

* The substance of Arts. 11, 12 is derived mainly from Ball, *l.c. ante* p. 26. References are given by him to parallel investigations of other writers.

3. Verify by the usual formulæ for transformation of rectangular coordinates that the expressions

$$p^2+q^2+r^2 \quad \text{and} \quad lp+mq+nr$$

of Art. 9 are absolute invariants.

4. Prove that the equations (Art. 9 (9)) of the axis of the twist which is equivalent to any given infinitesimal displacement may be written

$$\frac{x-(qn-rm)/\omega^2}{p}=\frac{y-(rl-pn)/\omega^2}{q}=\frac{z-(pm-ql)/\omega^2}{r}.$$

5. In any infinitesimal displacement of a rigid body, the straight lines drawn through the various points of a given straight line, in the directions of the displacements of these points, are generators of a paraboloid.

6. If a line be parallel to the axis of a screw, its conjugate is at infinity.

If it be perpendicular to the axis, its conjugate intersects the axis.

If a line coincides with its conjugate, it is a null-line.

7. The null-planes of the various points of a straight line intersect in the conjugate line.

8. In any infinitesimal displacement of a rigid body, the straight lines drawn through the various points of it in the directions of the respective displacements are perpendicular to their conjugates.

9. Any two pairs of lines which are conjugate with respect to a given twist lie on a hyperboloid.

10. If the axis of z be the axis of a screw of pitch ϖ, the equation of the polar plane of the point (x_1, y_1, z_1) is

$$xy_1-yx_1=\varpi\,(z-z_1).$$

Prove that the line conjugate to

$$x=\lambda z+a, \quad y=\mu z+\beta,$$

is

$$x=\frac{\varpi}{\beta\lambda-a\mu}(\lambda z+a), \quad y=\frac{\varpi}{\beta\lambda-a\mu}(\mu z+\beta).$$

11. Prove that the lines which are perpendicular to their conjugates form a complex of the second order, whose equation in line-coordinates is

$$lp+mq+\varpi\,(p^2+q^2)=0,$$

if the axis of z be that of the screw.

12. Prove that the locus of points whose directions of displacement all pass through a given point O is a twisted cubic lying on a circular cylinder which has the axis of the twist and a parallel line through O as opposite generators.

13. If a body is subject to one constraint only, a certain line of particles in it is incapable of motion in the direction of the length, unless this is accompanied by a rotation about the line in fixed proportion to the longitudinal motion. (Thomson and Tait.)

14. If a solid ellipsoid is in contact with fixed surfaces at three points A, B, C which are extremities of conjugate diameters, then in any infinitesimal displacement the centre moves parallel to the plane ABC.

EXAMPLES. III.

(Cylindroid, etc.)

1. Prove that a plane through a generator of the cylindroid cuts the surface again in an ellipse whose projection on the plane xy is a circle.

2. Prove that if a body twist about a screw belonging to the cylindroid

$$z(x^2+y^2)=2cxy$$

the displacement at the origin is parallel to that diameter of the pitch conic which is conjugate to the direction of the axis of the screw.

3. A circle of radius c revolves with uniform angular velocity about an axis AB in its plane. A point P describes the circle with twice this angular velocity. Prove that the perpendicular PN on AB generates a cylindroid of parameter c, and that the length NP gives the pitch of the corresponding screw, in a possible distribution. (Lewis.)

4. Prove that the sum of the pitches of the two screws which pass through any point on the axis of a cylindroid is constant.

5. If a body receives twists about three screws of a cylindroid, and if the amount of each twist be proportional to the sine of the angle between the other two screws, the body will occupy the same position as at first.

6. A body receives independent infinitesimal *rotations* about two axes which are at right angles, but do not intersect. Prove that the equation of the cylindroid thus determined is

$$z(x^2+y^2)=c(x^2-y^2),$$

the axes of x, y being parallel to the given lines, and the origin half-way between them.

7. The axes of two screws are

$$y=0,\ z=c, \quad \text{and} \quad x=0,\ z=-c,$$

and their pitches are $\varpi, -\varpi$, respectively; prove that the corresponding cylindroid is

$$z(x^2+y^2)=c(x^2-y^2)-2\varpi xy.$$

8. A body receives arbitrary rotations about the three lines

$$y=b,\ z=-c;\quad z=c,\ x=-a;\quad x=a,\ y=-b.$$

Prove that the locus of the axes of all screws of pitch ϖ is

$$2ayz+2bzx+2cxy+2abc+\varpi(x^2+y^2+z^2-a^2-b^2-c^2)+\varpi^3=0.$$

CHAPTER III

STATICS

13. Statics of a Particle.

The extension to three dimensions of the familiar propositions in this subject presents no difficulty. The resultant of any number of forces acting on a particle may be found by a polygon of forces. The fact that the sides of the polygon are not assumed to lie in the same plane makes no difference to the proof [S. 7]. For equilibrium, the polygon must be closed.

We may also recall the theorem that if G be the mass-centre of particles $m_1, m_2, \ldots, m_n$ situate at $P_1, P_2, \ldots, P_n$, and O any other point, we have, in vectors,

$$(m_1 + m_2 + \ldots + m_n)\,\mathrm{OG} = m_1 \,.\, \mathrm{OP}_1 + m_2 \,.\, \mathrm{OP}_2 + \ldots + m_n \,.\, \mathrm{OP}_n. \quad \ldots\ldots(1)$$

Hence forces represented by $m_1 \,.\, \mathrm{OP}_1$, $m_2 \,.\, \mathrm{OP}_2, \ldots, m_n \,.\, \mathrm{OP}_n$ have a resultant $\Sigma(m) \,.\, \mathrm{OG}$. The scalar quantities $m_1, m_2, \ldots, m_n$ need not all have the same sign, but it is assumed that $\Sigma(m) \neq 0$*.

Analytically, the process of reduction of a system of forces acting on a particle is as follows. Let OP be a vector representing a force F acting at the origin; and draw through P three planes parallel to the coordinate planes. These will enclose with the latter a parallelepiped. Then, referring to Fig. 15, p. 21, we have the vector equation

$$\mathrm{OP} = \mathrm{OA} + \mathrm{AN} + \mathrm{NP} = \mathrm{OA} + \mathrm{OB} + \mathrm{OC}. \quad \ldots\ldots\ldots(2)$$

The force F is therefore equivalent to three forces X, Y, Z along the coordinate axes, viz.

$$X = F\lambda, \quad Y = F\mu, \quad Z = F\nu, \quad \ldots\ldots\ldots\ldots\ldots(3)$$

where λ, μ, ν are the direction-cosines of OP.

A system of forces acting at O is reduced in this way to three forces

$$P = \Sigma(X), \quad Q = \Sigma(Y), \quad R = \Sigma(Z), \quad \ldots\ldots\ldots\ldots(4)$$

* The theorem is attributed to Leibnitz (1646–1716); it is an immediate consequence of the modern vector definition of the mass-centre [S. 64].

along Ox, Oy, Oz, respectively. If l, m, n be the direction-cosines of the resultant S, we have

$$P = Sl, \quad Q = Sm, \quad R = Sn, \quad \text{..............(5)}$$

and therefore

$$S^2 = P^2 + Q^2 + R^2 \text{........................(6)}$$

The formulæ (5) and (6) determine S and its direction (l, m, n).

For equilibrium we must have $S = 0$, and therefore

$$P = 0, \quad Q = 0, \quad R = 0. \quad \text{.................(7)}$$

That is, the sum of the resolved parts of the given forces in each of three mutually perpendicular directions must vanish.

14. Statics of a Rigid Body. Definition of a Moment.

A force, acting on a rigid body is conceived, in Pure Statics, as resident in a certain line, and as having a definite magnitude and sense, but it has no necessary reference to any particular point on the line, the position of the point of application being indifferent. Moreover, it is assumed that two forces in intersecting lines are equivalent to a single force through the intersection, derived from them by the law of vector addition. Also that equal and opposite forces in the same line neutralize one another. All this is summed up in the statement that a force has the properties of a 'localized vector.'

In consequence of the analogy pointed out in Art. 6 there is an exact parallelism between the theory of systems of forces and the kinematical theory of the small displacements of a rigid body. The various statical theorems might, indeed, be inferred without further proof, but it is at all events instructive to investigate them from the new point of view. It may be added that in the historical order of development the statical theorems came first.

The definition of a 'moment' employed in two-dimensional Statics now requires to be extended. To find the moment of a force PQ about any axis AB, we project PQ orthogonally on a plane perpendicular to AB. The moment is then given, as to *magnitude*, by the product of the projection (P′Q′, say) into its distance from the nearest point A (say) of AB. It will be noticed that the corresponding magnitude in Kinematics is the longitudinal displacement of any point in AB, due to a rotation PQ. This analogy gives perhaps the most convenient rule as to the *sign* of the moment. The convention which we adopt is that the moment is positive or negative, according as a right-handed rotation about $P'Q'$ would displace a point at A in the direction from A to B, or the reverse. Thus moments with respect to AB and BA are distinguished by the opposition of sign.

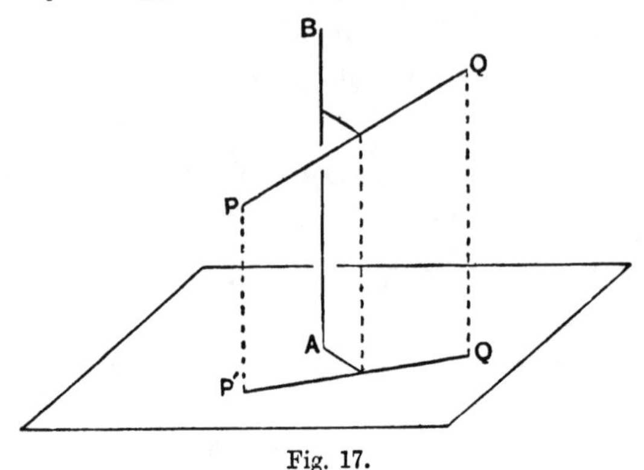

Fig. 17.

A geometrical expression for the magnitude of the moment is obtained as follows. If AB be a *unit* length measured along the axis, and if θ be the inclination of PQ to AB, and MN their shortest distance, the moment is equal in absolute magnitude to

$$PQ \sin \theta \,.\, MN \,.\, AB, \quad \ldots\ldots\ldots\ldots\ldots\ldots(1)$$

i.e. to six times the volume of the tetrahedron of which PQ and AB are opposite edges*.

It easily follows from the definition that the sum of the moments of two intersecting forces about any axis AB is equal to the moment of their resultant. For if we project orthogonally on a plane perpendicular to AB we obtain two intersecting forces and their resultant. We have then only to apply Varignon's theorem [S. 20] to these, taking moments about the point in which the plane meets AB.

The proof can be extended to the case of parallel forces.

15. Reduction of a three-dimensional System of Forces. Poinsot's Theorem.

By repeated applications of the principles stated at the beginning of Art. 14 a given system of forces may be replaced by

* M. Chasles (1847).

another system in a great variety of ways. But in no such transformation is any change made at any stage, either in the geometric sum of the forces, or in the sum of their moments about any given axis. It is convenient, in what follows, to suppose that by means of a polygon of forces [S. 31], or otherwise, a free vector **R** is constructed which represents, as far as magnitude and direction are concerned, the geometric sum of the given forces. We will suppose for the present that **R** is not zero.

There are certain modes of reduction which are of special interest.

In the first place, consider a fixed plane normal to the direction of **R**. We may resolve each force of the given system, at its intersection with this plane, into two components, viz. a force **P** parallel to **R**, and a force **Q** in the plane. The forces **P** form a parallel system, and have a resultant obviously equal to **R**, acting in a definite line. The geometric sum of the forces **Q**, on the other hand, must vanish, so that they are equivalent to a couple.

Hence a three-dimensional system of forces is in general reducible to a single force acting in a definite line, together with a couple in a plane perpendicular to that line*. The line in question is called the 'central axis' of the system; and the above combination of force and couple is called a 'wrench.' The *type* of a given wrench may be specified, apart from its intensity, by a screw whose axis is the central axis, and whose pitch is the linear magnitude which is the ratio of the couple to the force.

The above reduction is moreover unique except in so far that the plane of the couple may be any plane perpendicular to the axis. For the central axis is necessarily parallel to the vector **R**, and since the sum of the moments about any line which meets it at right angles must vanish, it can only have one position. The moment of the couple, again, is determinate, since it must be equal to the sum of the moments of the given forces about the central axis.

It appears that the statical effect of any system of forces depends on *six* independent elements. These may be, for instance, the four quantities necessary to determine the central axis,

* This theorem is due to L. Poinsot (1777–1859), *Élements de Statique*, Paris, 1804.

together with the magnitudes of the force and couple. We infer that in order that the system may be in equilibrium six independent conditions must be satisfied. Also that any system of forces which involves six arbitrary and independent elements can be adjusted so as to be equivalent to any given wrench. In particular a wrench may be resolved into six forces acting in six assigned lines. As a special case, it may be resolved into six forces acting in the edges of a given tetrahedron. And such resolutions are in general determinate.

It has been seen, for example, that a three-dimensional structure can in general be anchored down in a definite position relative to the earth by means of six links (Art. 1). We now learn that the tensions or thrusts evoked in these links in consequence of any given forces acting on the structure are in general determinate. There are possible cases of exception, when the links have certain 'critical' configurations. (Arts. 10, 19).

16. Null-Systems.

The determination of the wrench equivalent to two given forces, and the converse proof that a given wrench can be resolved into two forces, one of which may act in any arbitrary line, follow so closely the steps of the corresponding kinematical investigations that it is unnecessary to go through them again. See Art. 7. The lines of action of the two forces are said to be 'conjugate' with respect to the given wrench.

A 'null-line' is now defined as a line such that the moment of the wrench about it is zero*. If a null-line meets one of a pair of conjugate lines it must also meet the other.

The geometrical relations between the various lines are the same as in the kinematical analogue, but we may briefly indicate alternative proofs.

If through a point O we draw any line, the null-lines through O must meet the conjugate line, and must therefore lie in a plane. This is the 'null-plane,' or 'polar-plane,' of O. The conjugates of the various lines through O will all lie in the null-plane of O. If the line through O be normal to the null-plane, we have the

* It was in the present statical connection that the theory of null-lines and planes was developed by A. Möbius (1790–1868), in his *Lehrbuch der Statik*, 1837.

analogue of a case discussed in Art. 8. The conjugate line is then the 'characteristic' of the plane.

If in any plane we take a straight line, the null-lines which lie in the plane must all pass through the point where the conjugate line meets the plane. This is the 'null-point,' or 'pole,' of the plane.

If a point A lies in the polar plane of B, then B lies in the polar plane of A; and the intersection of these planes is the line conjugate to AB.

The theory of conjugate lines leads to a demonstration of an important theorem in Graphical Statics.

Two plane figures, each composed of a number of polygons, are said (in this connection) to be 'reciprocal' when to every line in one figure there corresponds a parallel line in the other, in such a way that to lines which meet in a point in either figure correspond lines forming a closed polygon in the other. The theorem in question is that a figure of the above kind will admit of a reciprocal if it is the orthogonal projection of the edges of a closed polyhedron with plane faces [S. 34].

If we consider any such polyhedron, the poles of its various faces with respect to a given wrench will be the vertices of a second polyhedron, whose edges are the lines joining the poles of adjacent faces of the first figure. Thus, to a vertex A of the first figure where m faces meet there will correspond in the second figure a polygonal face of m sides, whose plane is the polar plane of A, in virtue of the property above proved. Moreover, the sides of the polygon will be respectively conjugate to the m edges of the original figure which meet in A. It appears that the relation between the two figures is mutual, each being derived from the other by the same process*.

It has been remarked (Art. 7) that the orthogonal projections of conjugate lines on a plane perpendicular to the central axis are parallel. Hence if the two polyhedra are thus projected we obtain two plane figures which are 'reciprocal' in the sense above defined†.

17. Theory of Couples.

It was pointed out in Art. 6 that there is a complete mathematical analogy between a couple and a pair of equal and opposite infinitesimal rotations about parallel axes. Since such a pair of rotations is equivalent to a translation normal to the plane of the

* Since the polar plane of A contains A, the vertices of each polyhedron lie in the planes of the corresponding faces of the other. Hence either polyhedron may, in a sense, be said to be both inscribed and circumscribed to the other. The possibility of such a relation between two solid figures was pointed out by Möbius.

† The theorem is due to Maxwell. The above proof was given by Cremona, *Le figure reciproche nella statica grafica*, 2nd ed., Milan, 1872.

two axes, and since translations can be represented by free vectors, being compounded by the usual process of vector addition, we may infer that couples can be similarly represented. More precisely, if we draw a vector (anywhere) normal to the plane of the couple, of length proportional to the moment of the couple, and in the sense such that the moment of the couple about it is positive, according to the convention of Art. 14, the inference is that couples in different planes may be compounded by geometric addition of the representative vectors*.

We proceed to give an independent, statical, proof of this proposition.

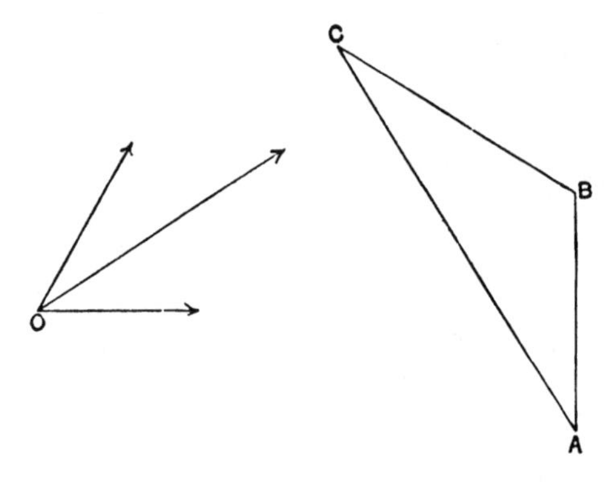

Fig. 18.

We imagine the plane of the paper to be perpendicular to the planes of the two couples to be compounded, and therefore normal to the intersection of these planes, which is represented by the point B in the diagram. In virtue of known theorems we may suppose each couple to be replaced by a pair of forces $\pm P$, normal to the plane of the paper, and so arranged that one force of each couple is in the common section (B). The arms AB, BC will then be proportional to the respective moments. The two forces in the line B will cancel, and there remains a couple of moment $P \, . \, AC$ in the plane AC. If we draw from any origin O three vectors to represent these couples, in the manner explained, they will be respectively perpendicular and proportional to the sides of the

* The theorem is due to Poinsot, *l.c.*

triangle ABC. Hence the third vector is the sum of the other two.

Ex. Four couples in the planes of the faces of a tetrahedron, of moments proportional to the areas of these faces, will be in equilibrium provided they are all right-handed (or all left-handed) with respect to normals drawn outwards from the respective faces.

It is known that three forces represented by the sides BC, CA, AB of any triangle ABC are equivalent to a couple whose moment is represented by twice the area of the triangle. Hence, in the present case, the couples acting in the faces of a tetrahedron $OABC$ can be replaced by forces in the edges which cancel in pairs. If we reverse the sense of the couple in ABC, it is equivalent to the couples in the other three planes; see Fig. 19.

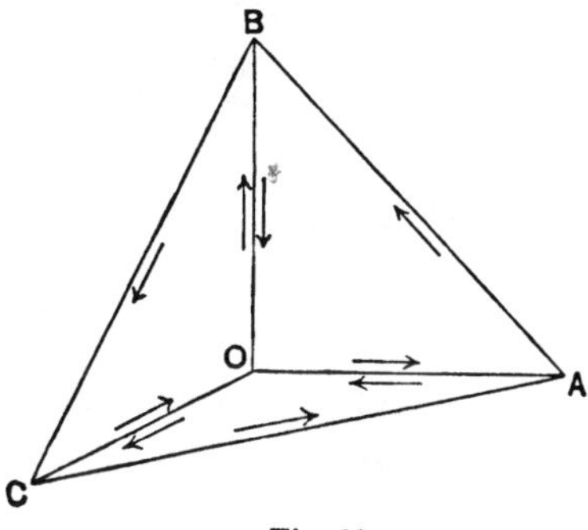

Fig. 19.

The particular case where the three edges OA, OB, OC are mutually perpendicular, as in the figure, is worth notice. If Δ be the area of the triangle ABC, and l, m, n the cosines of the angles which the normal to it from O makes with OA, OB, OC as coordinate axes, respectively, the areas of the projections of Δ on the coordinate planes will be $l\Delta$, $m\Delta$, $n\Delta$ respectively. Hence a couple G about an axis whose direction-cosines referred to rectangular axes are l, m, n is equivalent to couples lG, mG, nG about these axes, as is otherwise evident from Poinsot's theorem.

Another inference from the present theorem is that four *concurrent* forces whose directions are parallel to the normals drawn all outwards (or all inwards) from the faces of a given tetrahedron will be in equilibrium if their magnitudes are proportional to the areas of the corresponding faces.

18. Reduction of a System of Forces to a Force through an assigned point, and a Couple.

We can now shew that a three-dimensional system of forces may in general be replaced by a single force acting through *any* assigned point O, and a couple. For let P be any force of the system, and at O introduce two equal and opposite forces P_1, P_2, of which P_1 is parallel as well as equal to P. The original force P is thus equivalent to P_1 at O together with the couple formed by P and P_2. When all the forces of the given system have been dealt with in this way, we have a concurrent system of forces, such as P_1, at O, which may be compounded into a single resultant S, together with a series of couples which may be compounded into a single couple. The single resultant will be given as regards magnitude and direction by the geometric sum of the given forces. The plane of the couple will be parallel to the null-plane of O.

If the geometric sum vanishes, the system reduces to a couple, unless indeed it is in equilibrium.

If the system be previously reduced to a wrench consisting of a force S and a couple G_0, the force S can be transferred to any point O by the introduction of a couple of moment pS, where p is the distance of O from the central axis. This couple is in the plane of S and p; combining it with G_0 we have a couple of moment

$$G = \sqrt{(G_0^2 + p^2 S^2)}, \quad \text{.......................(1)}$$

whose axis makes an angle θ with the central axis, such that

$$\tan \theta = pS/G. \quad \text{..........................(2)}$$

When the system has been reduced to a force S at O and a couple G, these may be resolved according to three mutually perpendicular axes through O, viz. the force S into three components P, Q, R along these axes, and the couple into three components L, M, N about them. It is evident that P, Q, R are respectively the sums of the components of the original forces resolved parallel to the respective axes, and that L, M, N are the sums of the moments of the original forces about the axes. In the following Art. 19 this reduction is carried out analytically. For equilibrium it is essential that each of the six quantities P, Q, R, L, M, N should vanish.

Ex. 1. Four forces acting in the perpendiculars drawn from the vertices of a tetrahedron to the opposite faces, and proportional to the areas of these faces, are in equilibrium.

Let $OABC$ be the tetrahedron, and let the perpendiculars drawn from A, B, C to the opposite faces be projected orthogonally on the plane ABC, The projections will be the three perpendiculars of the triangle ABC. Hence a line drawn through the orthocentre of this triangle normal to its plane meets three of the given forces and is parallel to the fourth. It is therefore a null-line. Since four such null-lines can be drawn which are not parallel to one plane, the forces cannot reduce to a couple. And we have seen (Art. 17) that the geometric sum vanishes.

Ex. 2. Four forces perpendicular to the faces of a tetrahedron, at the mean centres of these faces, and proportional to the respective areas, will be in equilibrium provided they act all inwards, or all outwards*.

For if O', A', B', C' be the mean centres of the faces opposite O, A, B, C, respectively, the four forces will be related to the tetrahedron $O'A'B'C'$ as in the preceding Example.

19. Analytical Formulæ.

We adopt as usual a right-handed system of rectangular axes.

Let (x_1, y_1, z_1), (x_2, y_2, z_2), ... be the coordinates of any points P_1, P_2, ... on the lines of action of the respective forces. The force

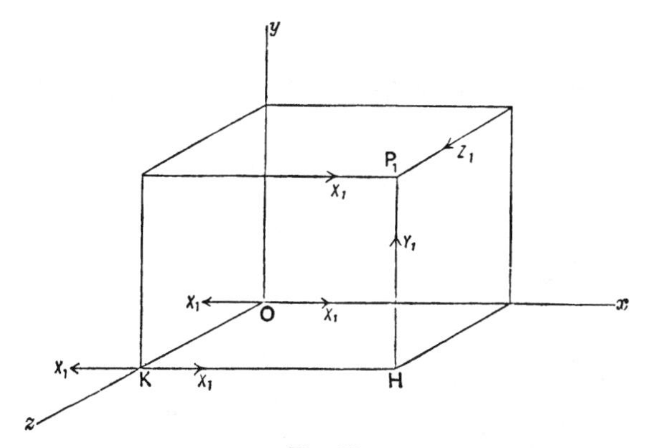

Fig. 20.

at P_1 may be supposed specified by its components X_1, Y_1, Z_1 parallel to the coordinate axes, that at P_2 by its components X_2, Y_2, Z_2, and so on. We draw P_1H normal to the plane zx, and HK perpendicular to Oz. If we introduce two equal and opposite forces $\pm X_1$ in KH, we see that the force X_1 at P_1 is equivalent to an equal and parallel force in KH together with a couple about Oz whose moment is $-y_1X_1$. Again, introducing equal and opposite forces $\pm X_1$ in Ox, we see that the force X_1 in KH is equivalent to an equal and parallel force X_1 in Ox, together with a couple about Oy whose moment is z_1X_1. Thus the force X_1 is transferred from P_1 to O, provided we introduce couples z_1X_1 and

* This follows from hydrostatic principles, but the statical proof is interesting.

$-y_1X_1$ about Oy and Oz, respectively. The forces Y_1, Z_1 can be transferred from P_1 to O in a similar manner.

We find that the force (X_1, Y_1, Z_1) may be transferred to O, provided we introduce couples about the axes Ox, Oy, Oz whose moments are

$$L_1 = y_1Z_1 - z_1Y_1,\ M_1 = z_1X_1 - x_1Z_1,\ N_1 = x_1Y_1 - y_1X_1, \ldots(1)$$

respectively. It is to be noticed that these are equal to the moments about the coordinate axes of the original force at P_1.

Dealing in this way with the several forces of the given system we obtain a single force S at O, whose components are

$$P = \Sigma(X),\ Q = \Sigma(Y),\ R = \Sigma(Z), \ldots\ldots\ldots\ldots(2)$$

and a couple G whose components are

$$L = \Sigma(yZ - zY),\ M = \Sigma(zX - xZ),\ N = \Sigma(xY - yX). \ldots(3)$$

Since P, Q, R are the rectangular projections of S, we have

$$S^2 = P^2 + Q^2 + R^2. \ldots\ldots\ldots\ldots(4)$$

Similarly,

$$G^2 = L^2 + M^2 + N^2, \ldots\ldots\ldots\ldots(5)$$

by Art. 17.

For equilibrium we must have $S = 0$, $G = 0$, which involve the six conditions

$$P = 0,\ Q = 0,\ R = 0,\ L = 0,\ M = 0,\ N = 0. \ldots\ldots(6)$$

In words:

The sum of the components of the forces in each of three mutually perpendicular directions must vanish; and

The sum of the moments of the forces about each of three mutually perpendicular axes must vanish.

If in the above process the origin be transferred to a point O' whose coordinates relative to O are (x', y', z'), the results will only differ in the substitution of the relative coordinates

$$(x_1 - x',\ y_1 - y',\ z_1 - z')$$

of P_1 in place of (x_1, y_1, z_1), and so on for the other points of application. The components P, Q, R of the force S will be unaltered, but the new components of couple will be

$$\begin{aligned} L' &= \Sigma\{(y - y')Z - (z - z')Y\} \\ &= \Sigma(yZ - zY) - y'\Sigma(Z) + z'\Sigma(Y), \text{ etc., etc.} \end{aligned}$$

Thus

$$\left.\begin{aligned} L' &= L - y'R + z'Q, \\ M' &= M - z'P + x'R, \\ N' &= N - x'Q + y'P. \end{aligned}\right\} \quad \text{.....................(7)}$$

By a suitable choice of the position of O' we can make the plane of the couple perpendicular to the direction of S. The conditions for this are

$$L'/P = M'/Q = N'/R, \quad \text{.......................(8)}$$

or

$$\frac{L - y'R + z'Q}{P} = \frac{M - z'P + x'R}{Q} = \frac{N - x'Q + y'P}{R}. \quad \text{......(9)}$$

These are the equations of the central axis of the system.

Since the direction-cosines of S are P/S, Q/S, R/S, the moment of the couple about the central axis is $(L'P + M'Q + N'R)/S$, and the pitch (Art. 15) of the equivalent wrench is therefore

$$\varpi = \frac{LP + MQ + NR}{P^2 + Q^2 + R^2}. \quad \text{....................(10)}$$

It appears, exactly as in Art. 9, that

$$P^2 + Q^2 + R^2 \text{ and } LP + MQ + NR$$

are absolute invariants.

The formal resemblance between the above results and those of Art. 9 are an illustration of the remarks made at the beginning of Art. 6.

Ex. 1. The application to the case of parallel forces may be noticed, although the result is a mere verification of a known theorem [S. 22, 65].

If the forces be F_1, F_2, ... acting in the direction (l, m, n), we have

$$X_1 = lF_1, \quad Y_1 = mF_1, \quad Z_1 = nF_1, \quad \text{....................(11)}$$

and so on. We find

$$P = l\Sigma(F), \quad Q = m\Sigma(F), \quad R = n\Sigma(F), \quad \text{.................(12)}$$

$$L = (n\bar{y} - m\bar{z})\,\Sigma(F), \quad M = (l\bar{z} - n\bar{x})\,\Sigma(F), \quad N = (m\bar{x} - l\bar{y})\,\Sigma(F), \quad \text{...(13)}$$

where

$$\bar{x} = \frac{\Sigma(Fx)}{\Sigma(F)}, \quad \bar{y} = \frac{\Sigma(Fy)}{\Sigma(F)}, \quad \bar{z} = \frac{\Sigma(Fz)}{\Sigma(F)}. \quad \text{....................(14)}$$

These results are the same as if we had a single force $\Sigma(F)$ acting at the point $(\bar{x}, \bar{y}, \bar{z})$, whose position is independent of the direction (l, m, n) of the parallel forces.

An exceptional case occurs when $\Sigma(F) = 0$. The system then either reduces to a couple, or is in equilibrium.

Ex. 2. To find the couple about the mass-centre of a body, due to the gravitational attraction of a distant particle M_0.

Taking the mass-centre as origin, let (ξ, η, ζ) be the coordinates of any particle m of the body. We have, then [S. 73],

$$\Sigma(m\xi)=0, \quad \Sigma(m\eta)=0, \quad \Sigma(m\zeta)=0. \qquad (15)$$

If the coordinates of M_0 be (x, y, z), the attraction of M_0 on m is a force $\gamma M_0 m/\rho^2$, where γ is the constant of gravitation [D. 74]*, and

$$\rho^2=(x-\xi)^2+(y-\eta)^2+(z-\zeta)^2. \qquad (16)$$

The direction-cosines of the line of action are $(x-\xi)/\rho$, $(y-\eta)/\rho$, $(z-\zeta)/\rho$, and the moment of the force about Ox is therefore

$$\frac{\gamma M_0 m}{\rho^3}\{y(z-\zeta)-z(y-\eta)\}=\frac{\gamma M_0 m}{\rho^3}(z\eta-y\zeta). \qquad (17)$$

If r denotes the distance of M_0 from the origin, so that

$$r^2=x^2+y^2+z^2, \qquad (18)$$

we have from (16)

$$\frac{1}{\rho^3}=\frac{1}{r^3}\left\{1-\frac{2(x\xi+y\eta+z\zeta)}{r^2}+\dots\right\}^{-\frac{3}{2}}$$

$$=\frac{1}{r^3}+\frac{3(x\xi+y\eta+z\zeta)}{r^5}+\dots \qquad (19)$$

We assume that ξ, η, ζ are so small compared with r that the remaining terms in the development may be neglected.

It will be shewn in Chap. IV that there is one set of rectangular axes through the origin which is such that

$$\Sigma(m\eta\zeta)=0, \quad \Sigma(m\zeta\xi)=0, \quad \Sigma(m\xi\eta)=0. \qquad (20)$$

These are called the 'principal axes of inertia' at O. Adopting them as axes of reference, substituting from (19) in (17), and prefixing the sign of summation with respect to m, we find, for the total couple about Ox,

$$L=\frac{3\gamma M_0}{r^5}\{\Sigma(m\eta^2)-\Sigma(m\zeta^2)\}yz. \qquad (21)$$

In the notation of Chap. IV we write

$$A=\Sigma m(\eta^2+\zeta^2), \quad B=\Sigma m(\zeta^2+\xi^2), \quad C=\Sigma m(\xi^2+\eta^2), \qquad (22)$$

these being the 'moments of inertia' of the body with respect to the principal axes at O. The required components of couple are therefore

$$L=\frac{3\gamma M_0}{r^5}(C-B)yz, \quad M=\frac{3\gamma M_0}{r^5}(A-C)zx, \quad N=\frac{3\gamma M_0}{r^5}(B-A)xy. \qquad (23)$$

These cannot all vanish unless the attracting particle lies in one of the principal axes at O.

If, as in the case of the earth, the body is symmetrical about an axis, then, taking this as axis of z we have $A=B$, and therefore $N=0$. The plane of the couple is accordingly that containing M_0 and the axis of symmetry, as is

* The reference is to the author's *Dynamics*.

otherwise obvious. Denoting by θ the angle which the axis of symmetry makes with OM_0, and putting $y=0$, $z=r\cos\theta$, $x=r\sin\theta$, we find, for the moment of the couple tending to increase θ, the expression

$$\frac{3\gamma M_0}{r^3}(C-A)\sin\theta\cos\theta. \qquad\text{.........................(24)}$$

This result is required in the theory of Precession (Art. 61).

Ex. 3. If (l, m, n, p, q, r) be the coordinates of a straight line (Art. 10), the components of the wrench due to a force F in this line will be

$$lF,\ mF,\ nF,\ pF,\ qF,\ rF,$$

respectively. Hence in order that forces F_1, F_2, ..., F_6 acting in six given straight lines should be in equilibrium we must have six equations of the forms

$$\left.\begin{array}{l} l_1F_1+l_2F_2+\ldots+l_6F_6=0, \\ \ldots\ldots\ldots\ldots\ldots\ldots\ldots\ldots \\ p_1F_1+p_2F_2+\ldots+p_6F_6=0. \\ \ldots\ldots\ldots\ldots\ldots\ldots\ldots\ldots \end{array}\right\} \qquad\text{......................(25)}$$

These cannot be satisfied by other than zero values of F_1, F_2, ..., F_6 unless

$$\begin{vmatrix} l_1, & l_2, & l_3, & l_4, & l_5, & l_6 \\ m_1, & m_2, & m_3, & m_4, & m_5, & m_6 \\ \ldots & \ldots & \ldots & \ldots & \ldots & \ldots \\ r_1, & r_2, & r_3, & r_4, & r_5, & r_6 \end{vmatrix} = 0. \qquad\text{................(26)}$$

By comparison with Art. 10 (5) we see that the six lines of action must belong to the same linear complex. This is, for instance, the condition that it should be possible for six links connecting a rigid structure to the earth to be self-stressed. If it is fulfilled, the stresses evoked in the links by any given forces acting on the structure are indeterminate.

The theory of the composition of wrenches about different screws is exactly analogous to the geometrical theory of composition of twists. For example, the result of two wrenches of given types, but arbitrary amounts, is a wrench about some screw belonging to a certain cylindroid. It is unnecessary to repeat the investigations.

20. Forces in Equilibrium.

It has been seen (Art. 19) that it is in general impossible for a system of forces acting in six arbitrarily assigned lines to be in equilibrium; and the conclusion holds *à fortiori* for any lesser number. There are however certain configurations of the lines which are exceptional.

In the case of *three* lines it is necessary that they should be in the same plane and concurrent (or, as an extreme case, parallel) [S. 48].

If we have *four* lines, any three of them determine a hyperboloid, of which they are generators. Every generator of the opposite system must meet the fourth line, for otherwise the sum of the moments of the four forces about it would not vanish. The four given lines must therefore be generators (of the same system) of a hyperboloid.

Conversely, we see that four forces acting in four given skew lines which are generators (of the same system) of a hyperboloid will be in equilibrium provided their geometric sum is zero. For the system cannot reduce to a couple, since the generators of the opposite system are null-lines and are not parallel to one plane.

Again, *five* forces in skew lines cannot be in equilibrium unless these lines all meet two transversals, viz. the two (real or imaginary) straight lines which can be drawn to meet any four of them. See Art. 11 (footnote)*.

21. The work done by a Wrench in an Infinitesimal Displacement.

The work done by a force in any infinitesimal displacement of a rigid body is defined as the product of the force into the orthogonal projection of the displacement of any point on its line of action on the direction of the force. It is immaterial, to the first order of small quantities, what point on the line of action is chosen [S. 51]. An alternative definition is that the work is the product of the displacement of any point on the line of action into the orthogonal projection of the force on the direction of the displacement. On either definition, the work is equal to $F \,.\, \delta s \cos \theta$, where F and δs are the magnitudes of the force and the displacement respectively, and θ is the angle between their directions.

The total work of two or more concurrent forces, in a small displacement of a rigid body, is equal to the work of the single force which is their geometric sum, acting at the point of concurrence. For if P, Q denote the two forces, R their resultant, and δs the displacement, the sum of the orthogonal projections of P and Q on the direction of δs is equal to the projection of R. The proof is the same as in two dimensions [S. 47], but it is understood now that δs is not necessarily in the plane of P and Q.

It follows that the total work of any system of forces, in any infinitesimal displacement of a rigid body, is equal to that of any other system which is statically equivalent to it. In particular it will be equal to the work done by the force and couple of the equivalent wrench.

* The above propositions are due to Möbius.

Suppose, for example, that the forces acting on the body have been reduced as in Art. 19 to a wrench (P, Q, R, L, M, N), and that the body receives a twist (l, m, n, p, q, r), the notation being that of Art. 9. The work done during this displacement is

$$Lp + Mq + Nr + Pl + Qm + Rn. \qquad \text{(1)}$$

For the work done by the force P at the origin in virtue of the translation (l, m, n) is Pl; that done by the couple L in consequence of the rotation (p, q, r) is Lp; and so on.

The formula (1) may be verified as follows. The work done by the force (X_r, Y_r, Z_r) acting at (x_r, y_r, z_r) is

$$\begin{aligned} X_r\delta x_r + Y_r\delta y_r + Z_r\delta z_r & \\ = X_r(l + qz_r - ry_r) &+ Y_r(m + rx_r - pz_r) \\ &+ Z_r(n + py_r - qx_r) \\ = X_r l + Y_r m + Z_r n &+ (y_r Z_r - z_r Y_r)p \\ &+ (z_r X_r - x_r Z_r)q + (x_r Y_r - y_r X_r)r \\ = X_r l + Y_r m + Z_r n &+ L_r p + M_r q + N_r r. \qquad \text{(2)} \end{aligned}$$

Summing up for all the forces we obtain the expression (1).

Hence if the forces be in equilibrium the total work will be zero in any infinitesimal displacement. For then we have

$$P, Q, R, L, M, N = 0.$$

Conversely, if the work vanishes for *all* infinitesimal displacements, these quantities must severally vanish, i.e. the system must be in equilibrium. This is the principle of 'Virtual Velocities,' as extended to the case of forces in three dimensions [cf. S. 52].

It is indeed sufficient for equilibrium if the total work is zero for each of *six* independent displacements, say

$$(l_1, m_1, n_1, p_1, q_1, r_1),\ (l_2, m_2, n_2, p_2, q_2, r_2), \ldots, (l_6, m_6, n_6, p_6, q_6, r_6).$$

For we have then six equations of the type

$$Lp_s + Mq_s + Nr_s + Pl_s + Qm_s + Rn_s = 0, \qquad \text{(3)}$$

which are in general incompatible unless L, M, N, P, Q, R all vanish.

A less analytical expression for the work done by a given wrench when the body receives a given twist can be found as follows. Let AB be the shortest distance between the axes of the wrench and screw, and let AA', BB' be the positive directions of the respective axes. Let the wrench consist of a force S along AA', and a couple G about AA', whilst the twist consists of a rotation ω about BB', and a translation τ along BB'. Let $AB = h$, and suppose that a right-handed rotation about AB through an angle θ would bring AA' into parallelism with BB'. These two quantities h, θ define the relative position of the two axes.

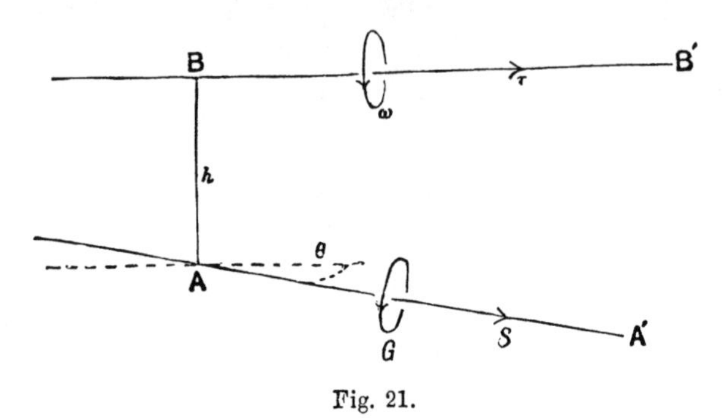

Fig. 21.

The couple G has the component $G \cos \theta$ about an axis parallel to BB'. The displacement of a point at A, resolved along AA', is

$$\tau \cos \theta - \omega h \sin \theta.$$

The total work is therefore

$$G\omega \cos \theta + S(\tau \cos \theta - \omega h \sin \theta).$$

If ϖ be the pitch of the wrench, ϖ' that of the screw, we have $G = \varpi S$, $\tau = \varpi' \omega$, and the expression becomes

$$S\omega \{(\varpi + \varpi') \cos \theta - h \sin \theta\}. \quad \text{...............(4)}$$

The factor of $S\omega$ in this expression is called by Ball the 'virtual coefficient' of the two screws.

22. Reciprocal Screws.

It follows from the expression (1) of Art. 21 that if the co-ordinates of two screws (l, m, n, p, q, r), (l', m', n', p', q', r') are connected by the relation

$$lp' + mq' + nr' + l'p + m'q + n'r = 0, \quad \text{............(1)}$$

a wrench of the type of either screw does no work on a body which is twisted about the other. Two screws so related are said to be 'reciprocal.' In terms of the relative configuration, the condition is

$$(\varpi + \varpi') \cos \theta - h \sin \theta = 0. \quad\quad (2)$$

The null-lines of Art. 16 will be recognized as a particular case. They are screws of zero pitch reciprocal to the screw which defines the type of the wrench. Putting $\varpi' = 0$ in (2), we find that the null-lines of a wrench of pitch ϖ which are at a distance h from its axis are inclined to this axis at an angle $\tan^{-1}(\varpi/h)$.

The null-lines which are at a given perpendicular distance h from a point O on the central axis will therefore form one system of generators of a hyperboloid of revolution; and by varying h we get a series of such hyperboloids with a common centre and axis. By displacing O along the central axis we obtain the whole complex of null-lines.

If two screws intersect, $h = 0$, and the formula (2) shews that they cannot be reciprocal unless they are at right angles, or unless the pitches are equal and opposite. This may be taken to include the statement that parallel screws cannot be reciprocal unless the sum of the pitches is zero. Two screws which are at right angles, but do not intersect, will be reciprocal if the sum of the pitches is infinite.

The notion of reciprocal screws is of interest for two reasons. In the first place a body which has n degrees (only) of freedom, where $n < 6$, can twist about n independent screws. If the constraints are 'frictionless,' i.e. the constraining forces do no work in any displacement, the body will remain at rest when subjected to a wrench of the type of any screw which is reciprocal to each of the n given screws. This is an immediate consequence of the principle of Virtual Velocities.

Again, we can give greater completeness to the analysis (Art. 10) of the various displacements which the constraints of a body admit of. The reciprocal screws form, in the above case, a system of order $6-n$, for we have n equations of the type (1), to be satisfied by the coordinates (l', m', n', p', q', r'), say, of any reciprocal screw.

Thus if $n=1$, the system of reciprocal screws will be of the fifth order. Any straight line in space is then a possible axis, the corresponding pitch being determined by (2).

If $n=2$, the axes of the reciprocal screws which pass through a given point P will generate a cone. For any line through P will meet the cylindroid which represents the freedom of the body (Art. 11) in three points, since the surface is of the third degree. If this line is the axis of a reciprocal screw of pitch ϖ, the pitch of any screw of the original system which it meets must be $-\varpi$, or else the intersection must be at right angles. Since not more than two screws on the cylindroid can have the pitch $-\varpi$, one intersection must be at right angles. The cone in question is therefore the locus of the perpendiculars drawn from P to the various generators of the cylindroid. We can now shew that the feet N of these perpendiculars lie on an ellipse, and thence that the cone is of the second degree. Let Q be the foot of that perpendicular which is parallel to the axis of the cylindroid. We denote the cylindrical coordinates of Q by (ρ_1, a, z_1), and those of any other position of N by (ρ, θ, z), the axes of reference being the same as in Art. 11. The Cartesian coordinates of N are accordingly

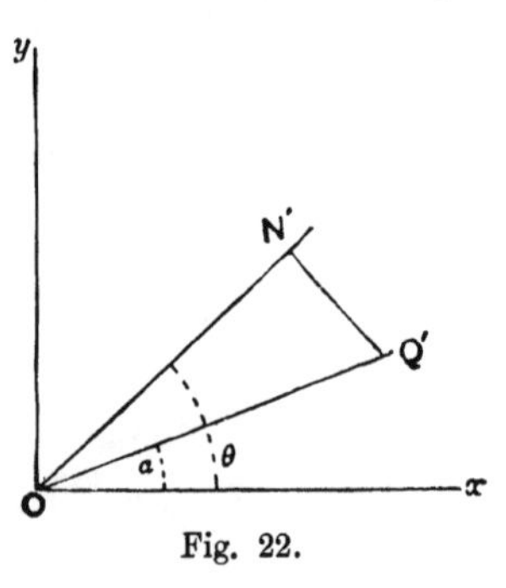

Fig. 22.

$$x=\rho\cos\theta, \quad y=\rho\sin\theta, \quad z=c\sin 2\theta, \qquad (3)$$

and those of Q

$$x_1=\rho_1\cos a, \quad y_1=\rho_1\sin a, \quad z_1=c\sin 2a. \qquad (4)$$

If Q', N' be the orthogonal projections of Q and N on the plane xy, it is easily seen that ON' is normal to the plane PQN, and therefore to $Q'N'$. Hence

$$\rho=\rho_1\cos(\theta-a), \qquad (5)$$

which is the equation of a circular cylinder whose axis is parallel to Oz and bisects OQ'. Again, eliminating ρ and θ, we find

$$x\sin a+y\cos a=\tfrac{1}{2}\frac{\rho_1}{c}(z+z_1), \qquad (6)$$

which is the equation of a plane. The locus of N is therefore the ellipse in which this plane meets the cylinder (5).

When $n=3$, the system of reciprocal screws is also of the third order, and the geometrical relation between the two systems is remarkable. We have seen (Art. 12) that the axes of the screws of any given pitch ϖ in the original system form one set of generators of a hyperboloid. The generators of the opposite set are the axes of reciprocal screws of pitch $-\varpi$, since the relation (2) is fulfilled. It is not difficult to see that the hyperboloid contains *all* the reciprocal screws of pitch $-\varpi$.

When $n=4$, the reciprocal screws form a system of the second order, and the mutual relations of the two systems are now the same as when $n=2$. The body can twist about any screw which is reciprocal to a certain cylindroid.

When $n=5$, there is only one reciprocal screw. The body can now twist about a screw whose axis may be any line whatever, provided the pitch be properly adjusted.

23. Vector Formulæ.

Several of the relations with which we have been concerned in this Chapter and the preceding have a concise expression in the language of Vector Analysis.

The 'scalar product' of two vectors **P**, **Q** is defined [S. 47] as the product of the absolute value of either into the orthogonal projection of the other on its direction. Thus if P, Q be the absolute values, and θ the angle between the directions,

$$\mathbf{PQ} = PQ\cos\theta = \mathbf{QP}. \quad \text{.....................(1)}$$

In particular, the scalar square of a vector is the square of its absolute value, whilst the scalar product of two perpendicular vectors is zero.

Hence if **i**, **j**, **k** denote unit vectors parallel to three mutually perpendicular lines we have

$$\left.\begin{array}{lll} \mathbf{i}^2 = 1, & \mathbf{j}^2 = 1, & \mathbf{k}^2 = 1, \\ \mathbf{jk} = \mathbf{kj} = 0, & \mathbf{ki} = \mathbf{ik} = 0, & \mathbf{ij} = \mathbf{ji} = 0. \end{array}\right\} \text{............(2)}$$

The formula (1) shews that the commutative law of multiplication is satisfied. The distributive law

$$\mathbf{P}(\mathbf{Q} + \mathbf{R}) = \mathbf{PQ} + \mathbf{PR} \quad \text{.....................(3)}$$

also holds, since the sum of the orthogonal projections of **Q** and **R** on the direction of **P** is equal to the projection of the sum **Q** + **R**.

If **r** denote the position vector, relative to the origin O, of a point A whose coordinates are x, y, z, we have

$$\mathbf{r} = x\mathbf{i} + y\mathbf{j} + z\mathbf{k}, \quad \text{................................(4)}$$

and, for an infinitesimal displacement,

$$\delta\mathbf{r} = \delta x\mathbf{i} + \delta y\mathbf{j} + \delta z\mathbf{k}. \quad \text{.............................(5)}$$

The work done by a force **P** acting at A is by definition $\mathbf{P}\delta\mathbf{r}$. If X, Y, Z be the components of **P** parallel to the coordinate axes, we have

$$\mathbf{P} = X\mathbf{i} + Y\mathbf{j} + Z\mathbf{k}, \text{................................(6)}$$

and therefore

$$\mathbf{P}\delta\mathbf{r} = (X\mathbf{i} + Y\mathbf{j} + Z\mathbf{k})(\delta x\mathbf{i} + \delta y\mathbf{j} + \delta z\mathbf{k}).$$

Developing the product in accordance with the commutative and distributive laws, and having regard to (2), we verify that

$$\mathbf{P}\delta\mathbf{r} = X\delta x + Y\delta y + Z\delta z. \quad \text{...........................(7)}$$

There is another type of product which is recognized in Vector Analysis. If as in Fig. 23 we draw OA, AC in succession to represent any two given vectors **P**, **Q**, respectively, the 'vector product' of **P** into **Q** is defined as a vector normal to the plane OAC, whose magnitude and sense are determined by the parallelogram $OACB$ constructed on OA, AC as adjacent sides. The absolute magnitude of the vector is given by the area of the parallelogram, and its sense is such that the circuit $OACBO$ is right-handed with respect

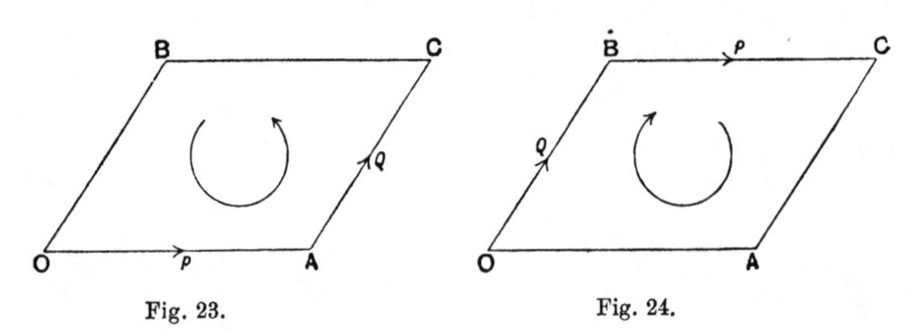

Fig. 23. Fig. 24.

to it. The vector thus obtained is denoted by $[\mathbf{PQ}]$. If $\mathbf{n}$ be a unit vector drawn normal to the plane OAC in the sense above explained we have

$$[\mathbf{PQ}] = PQ \sin\theta \,.\, \mathbf{n}, \quad \text{.....................(8)}$$

if θ be the angle ($< \pi$) which the direction of $\mathbf{Q}$ makes with that of $\mathbf{P}$.

In particular

$$[\mathbf{P}^2] = [\mathbf{PP}] = 0, \quad \text{.....................(9)}$$

since θ now $= 0$.

Hence if $\mathbf{i}$, $\mathbf{j}$, $\mathbf{k}$ denote, as before, unit vectors in the directions of the coordinate axes Ox, Oy, Oz, respectively, we have

$$\left.\begin{array}{lll} [\mathbf{i}^2] = 0, & [\mathbf{j}^2] = 0, & [\mathbf{k}^2] = 0, \\ [\mathbf{jk}] = -[\mathbf{kj}] = \mathbf{i}, & [\mathbf{ki}] = -[\mathbf{ik}] = \mathbf{j}, & [\mathbf{ij}] = -[\mathbf{ji}] = \mathbf{k}, \end{array}\right\} \text{...(10)}$$

provided the axes form a right-handed system.

To construct the vector $[\mathbf{QP}]$, where the order is inverted, we should begin by drawing OB, BC to represent $\mathbf{Q}$ and $\mathbf{P}$, respectively, as indicated in Fig. 24. The direction of the circuit, viz. $OBCAO$, is now reversed, and the sense of the resulting vector is therefore the opposite to what it was before. Hence

$$[\mathbf{QP}] = -[\mathbf{PQ}]. \quad \text{.....................(11)}$$

The commutative law therefore does not apply to vector products.

The distributive law, viz.

$$[\mathbf{P}(\mathbf{Q} + \mathbf{R})] = [\mathbf{PQ}] + [\mathbf{PR}], \quad \text{...............(12)}$$

is however valid. To prove this draw OA to represent $\mathbf{P}$, and AB, AC, AD to represent $\mathbf{Q}$, $\mathbf{R}$, and $\mathbf{Q}+\mathbf{R}$, respectively. Let us first suppose that AB, AC, and therefore also AD, are perpendicular to OA. The lines drawn from A to represent the vectors $[\mathbf{PQ}]$, $[\mathbf{PR}]$, and $[\mathbf{P}(\mathbf{Q}+\mathbf{R})]$ will then lie in the plane ABC, normal to OA, and will be perpendicular and proportional to AB, AC, AD, respectively. Hence in this particular case the relation (12) evidently holds. To prove it for the general case, let B', C', D' be the orthogonal projections of B, C, D on the plane through A perpendicular to OA. If we put

$$\mathrm{AB}'=\mathbf{Q}',\ \mathrm{AC}'=\mathbf{R}',\ \text{and therefore } \mathrm{AD}'=\mathbf{Q}'+\mathbf{R}',\ \ldots(13)$$

we have

$$[\mathbf{P}(\mathbf{Q}'+\mathbf{R}')]=[\mathbf{PQ}']+[\mathbf{PR}'],\ \ldots\ldots\ldots\ldots\ldots(14)$$

by the former case. But from the definition of a vector product it appears that

$$[\mathbf{PQ}']=[\mathbf{PQ}],\ [\mathbf{PR}']=[\mathbf{PR}],\ [\mathbf{P}(\mathbf{Q}'+\mathbf{R}')]=\mathbf{P}(\mathbf{Q}+\mathbf{R}).$$

The relation (12) follows.

We easily infer, having regard to (11), that

$$[(\mathbf{Q}+\mathbf{R})\,\mathbf{P}]=[\mathbf{QR}]+[\mathbf{QP}].\ \ldots\ldots\ldots\ldots\ldots(15)$$

The notion of a vector product may be illustrated from the theory of small rotations. If a rigid body receives a small rotation, represented by a vector $\boldsymbol{\omega}$, about an axis through the origin O, the displacement of a point A whose position vector is $\mathbf{r}$ is evidently given both as to magnitude and direction by the vector product $[\boldsymbol{\omega}\mathbf{r}]$. And if $\boldsymbol{\omega}'$ denote a small rotation about any other axis through O, the combined effect is

$$[\boldsymbol{\omega}\mathbf{r}]+[\boldsymbol{\omega}'\mathbf{r}],\ \text{or}\ [(\boldsymbol{\omega}+\boldsymbol{\omega}')\,\mathbf{r}],\ \ldots\ldots\ldots\ldots\ldots\ldots\ldots\ldots(16)$$

by (12). This verifies the law of composition of small rotations about intersecting axes (Art. 6).

If we write

$$\mathbf{r}=x\mathbf{i}+y\mathbf{j}+z\mathbf{k},\quad \boldsymbol{\omega}=p\mathbf{i}+q\mathbf{j}+r\mathbf{k},\ \ldots\ldots\ldots\ldots\ldots\ldots\ldots(17)$$

we find

$$[\boldsymbol{\omega}\mathbf{r}]=[(p\mathbf{i}+q\mathbf{j}+r\mathbf{k})\,(x\mathbf{i}+y\mathbf{j}+z\mathbf{k})]$$
$$=(qz-ry)\,\mathbf{i}+(rx-pz)\,\mathbf{j}+(py-qx)\,\mathbf{k},\ \ldots\ldots\ldots\ldots(18)$$

in virtue of the relations (9), (12), and (15); cf. Art. 9. The result may also be written

$$[\boldsymbol{\omega}\mathbf{r}]=\begin{vmatrix}\mathbf{i}, & \mathbf{j}, & \mathbf{k}\\ p, & q, & r\\ x, & y, & z\end{vmatrix}.\ \ldots\ldots\ldots\ldots\ldots\ldots\ldots\ldots\ldots\ldots(19)$$

A statical illustration is furnished by the moment of a couple. If $\mathbf{r}$ be the position vector of a point A with respect to the origin O, the vector $[\mathbf{rP}]$ evidently represents, on the convention of Art. 17, the couple formed by forces $\mathbf{P}$ at A and $-\mathbf{P}$ at O. If we write

$$\mathbf{P}=X\mathbf{i}+Y\mathbf{j}+Z\mathbf{k}, \quad \text{..........................(20)}$$

the rest of the notation being as before, we have

$$[\mathbf{rP}]=(x\mathbf{i}+y\mathbf{j}+z\mathbf{k})(X\mathbf{i}+Y\mathbf{j}+Z\mathbf{k})$$
$$=(yZ-zY)\mathbf{i}+(zX-xZ)\mathbf{j}+(xY-yX)\mathbf{k}. \quad \text{...(21)}$$

The coefficients of $\mathbf{i}$, $\mathbf{j}$, $\mathbf{k}$ are the moments of the couple about the coordinate axes; cf. Art. 19.

The scalar product of a vector $\mathbf{R}$ and a vector product $[\mathbf{PQ}]$ has a simple geometrical meaning. If we draw OA, OB, OC to represent $\mathbf{P}$, $\mathbf{Q}$, $\mathbf{R}$, respectively, and denote by ϵ the angle (acute or obtuse) which OC makes with the normal to the plane AOB drawn in the sense of $[\mathbf{PQ}]$, we have

$$[\mathbf{PQ}]\,\mathbf{R}=PQR\sin\theta\cos\epsilon, \quad \text{..............(22)}$$

where P, Q, R denote the absolute values of the respective vectors. Apart from the sign, this is equal to the volume of the parallelepiped having OA, OB, OC as edges. The sign is positive or negative according as the cyclic order of the letters A, B, C is right- or left-handed as regards the normal drawn from O to the plane ABC.

For instance, if $\mathbf{P}$ be a force acting at a point whose position vector relative to the origin of coordinates is $\mathbf{r}$, the moment of this force about an axis through the origin is $[\mathbf{rP}]\,\mathbf{a}$, where $\mathbf{a}$ is a unit vector in the direction of this axis. Cf. Art. 14 (1). Thus if we take the scalar product of $\mathbf{i}$ into the last member of (21), we obtain $yZ-zY$, the moment about the axis of x.

24. Flexible Chains.

A few notes on the three-dimensional Statics of Flexible Chains may find a place here. The geometrical relations involved are similar to those met with later in the Dynamics of a Particle (Art. 35).

Consider an element PP' $(=\delta s)$. The tensions T and $T+\delta T$ acting in the tangent lines at P and P' may be reduced to a single force, and a couple whose moment is the product of $T+\delta T$ into the shortest distance between these lines. Since this distance is of the third order of small quantities the couple may ultimately be neglected. To find the force, we may draw vectors KO, OK′ to represent the tensions at P and P' respectively. The resultant is represented by KK′. If we draw $K'N$ perpendicular to OK, this resultant may be resolved into KN and NK′, whose magnitudes are ultimately δT and $T\delta\epsilon$, where $\delta\epsilon$ is the angle KOK'. The plane KOK', being parallel to two consecutive tangent lines, is ultimately parallel to the osculating plane of the chain at P. The tensions on the element δs are therefore equivalent ultimately to a force δT in the tangent line at P, and a force $T\delta\epsilon$ along the principal normal. If ρ be the radius of curvature at P, we have $\delta\epsilon = \delta s/\rho$, so that the normal component may also be denoted by $T\delta s/\rho$.

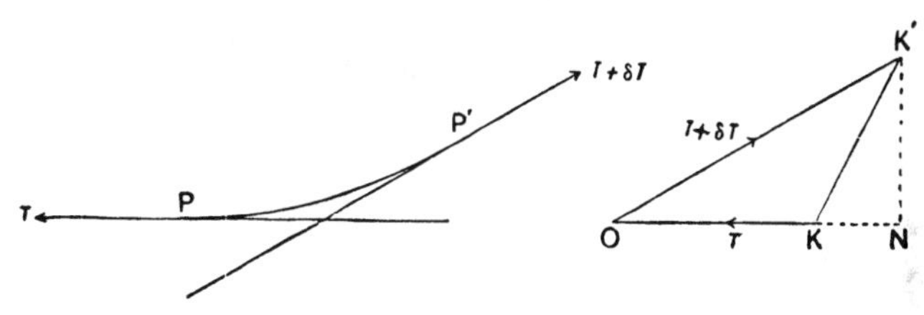

Fig. 25.

Hence if F, G, H are the components of the external force per unit length in the directions of the tangent line, the principal normal, and the binormal, respectively, we have

$$\delta T + F\delta s = 0, \quad \frac{T\delta s}{\rho} + G\delta s = 0, \quad H\delta s = 0,$$

or

$$\frac{dT}{ds} = -F, \quad \frac{T}{\rho} = -G, \quad H = 0. \quad \text{.............(1)}$$

The third equation expresses that the osculating plane contains always the direction of the resultant external force.

Hence in the case of a string stretched over a smooth surface, and subject to no external force except the normal reaction, the osculating plane at each point contains the normal to the surface. This is the condition for a 'geodesic,' or 'straightest' line, i.e. a line which is the shortest path between any two points of it which are not too far apart. For instance, a string wrapped round a circular cylinder takes the form of a helix. Again, since $F = 0$, the tension T is the same at all points.

If the surface be of revolution, the directions of the reactions at the various points all meet the axis of symmetry. Hence, considering any finite length of the string, the moments about this axis of the tensions on the two ends must be equal and opposite. If r denote distance from the axis, and ϕ the angle which the direction of the string makes with the circle of latitude, this gives

$$T \cos \phi \,.\, r = \text{const.},$$

or, since T is constant,

$$r \cos \phi = a \cos \alpha, \qquad \text{.......................(2)}$$

where a, α are any two corresponding values of r and ϕ. There is therefore, for a given geodesic, a lower limit to the value of r. For instance, in the case of an ellipsoid of revolution, a geodesic which cuts the equatorial circle ($r = a$) at an angle α is confined between two circles which it alternately touches, viz. the circles of radius $a \cos \alpha$.

In the case of a chain subject to gravity, whether hanging freely or in contact with a smooth surface, we have

$$F = -w \sin \psi, \qquad \text{.......................(3)}$$

where w is the weight per unit length, and ψ the inclination to the horizontal. If the axis of z be drawn vertically upwards, we have $\sin \psi = dz/ds$, and therefore, from (1),

$$\frac{dT}{ds} = w \frac{dz}{ds}. \qquad \text{..........................(4)}$$

Hence in the case of a uniform chain

$$T = wz + C, \qquad \text{..........................(5)}$$

as in two dimensions.

More generally, if the chain be subject to a field of force such that the potential energy per unit mass at any point is V, we have, if μ be the line-density,

$$\frac{dT}{ds} = \mu \frac{dV}{ds}, \qquad \text{..........................(6)}$$

and therefore, if μ be constant,

$$T = \mu V + C. \qquad \text{.......................(7)}$$

Ex. A uniform chain hangs in contact with a smooth cylinder (of any form of section) whose generating lines are vertical.

Resolving vertically the forces on an arc s beginning at the lowest point we have

$$T \sin \psi = ws. \qquad \text{..................................(8)}$$

Also, from (5),

$$T = wz, \quad \ldots\ldots(9)$$

if the origin of z be suitably adjusted. These relations are exactly the same as for a chain hanging in a vertical plane. It follows that in the present case the form assumed, when developed into a plane, would be the ordinary catenary.

The usual formulæ are easily deduced. Thus from (8) and (9)

$$s = z\sin\psi = z\frac{dz}{ds}, \quad \ldots\ldots(10)$$

whence, integrating,

$$z^2 = s^2 + a^2. \quad \ldots\ldots(11)$$

Again

$$\cos^2\psi = 1 - \frac{s^2}{z^2} = \frac{a^2}{z^2},$$

or

$$z = a\sec\psi. \ldots\ldots(12)$$

Combined with (10), this gives

$$s = a\tan\psi. \ldots\ldots(13)$$

If $R\delta s$ be the pressure on an element δs, we have, resolving along the normal to the cylinder,

$$\frac{T\delta s}{\rho}\cos\chi = R\delta s, \ldots\ldots(14)$$

where χ is the angle between the principal normal of the curve and the normal to the surface. By Meunier's theorem, $\rho = \rho'\cos\chi$, where ρ' is the radius of curvature of the normal section of the surface through the tangent line of the curve. And by Euler's theorem, $1/\rho' = \cos^2\psi/\rho_1$, where ρ_1 is the radius of curvature of the cross-section of the surface. Hence

$$R = \frac{T\cos^2\psi}{\rho_1}. \quad \ldots\ldots(15)$$

It follows that if the cylinder be of revolution the pressure per unit length varies inversely as the tension.

EXAMPLES. IV.

(Problems.)

1. A frame of six bars forming the edges of a regular tetrahedron rests on a smooth horizontal plane. Find the stresses in the several bars due to a weight W suspended from the highest point.

2. A uniform triangular plate hangs from a fixed point by three strings attached to the corners. Prove that the tensions of the strings are proportional to their lengths.

3. A smooth sphere rests on three pegs in a horizontal plane. Find the pressure on each peg.

4. A smooth sphere rests on a frame of three rods forming a horizontal triangle ABC. Find the pressure on each rod.

5. Three smooth spheres of equal radius a rest inside a smooth spherical bowl of radius b. A fourth equal sphere is placed symmetrically upon them. Prove that the lower spheres will not remain in contact unless $b/a < 7{\cdot}63$.

6. A uniform triangular plate whose sides are a, b, c rests in a smooth spherical bowl of radius r. Prove that the inclination (θ) of its plane to the horizontal is given by

$$(r^2 - R^2)\tan^2\theta = R^2 - \frac{1}{9}(a^2 + b^2 + c^2),$$

where R is the radius of the circumcircle of the triangle.

7. A sphere rests on three rough horizontal rods forming a triangle. Having given the coefficient of friction between the sphere and the rods, find the couple required to twist the sphere about its vertical diameter.

8. Three equal spheres are placed in contact on a rough horizontal plane, and a fourth equal sphere is placed upon them. Find the least coefficient of friction between the upper and lower spheres which will allow of equilibrium.

[$\mu = {\cdot}32$.]

9. A uniform solid sphere resting on a table is divided into thin wedges by planes meeting in a vertical diameter, and is held together by a string lying in a shallow groove along the horizontal great circle. Prove that the tension of the string must not be less than $\frac{3}{32}$ of the total weight.

10. A horizontal bar of weight W hangs by two vertical strings, each of length l, attached to it at equal distances a from the centre of gravity. Prove that the couple required to turn it through an angle ϕ about the vertical is

$$\frac{Wa^2 \sin\phi}{\sqrt{(l^2 - 4a^2 \sin^2 \frac{1}{2}\phi)}}.$$

11. The line of hinges of a door makes an angle α with the vertical. Prove that the couple necessary to turn the door through an angle θ from its equilibrium position is $Wa \sin\alpha \sin\theta$, where W is the weight, and a the distance of the centre of gravity from the line of hinges.

12. A solid whose lower surface is a paraboloid of revolution rests in equilibrium on a horizontal plane, the point of contact being the vertex. If the centre of gravity coincides with the centre of curvature at this point, prove that the equilibrium is stable.

13. Prove that a homogeneous ellipsoid can rest in equilibrium on three smooth pegs in a horizontal plane if the pegs are at the extremities of conjugate diameters.

Prove that the pressures on the pegs are proportional to the areas of the conjugate diametral sections of the ellipsoid.

14. Obtain the following equations for the equilibrium of a chain, where X, Y, Z are the components of external force per unit length, parallel to the coordinate axes:

$$\frac{d}{ds}\left(T\frac{dx}{ds}\right)+X=0,\quad \frac{d}{ds}\left(T\frac{dy}{ds}\right)+Y=0,\quad \frac{d}{ds}\left(T\frac{dz}{ds}\right)+Z=0.$$

15. Deduce from these equations that the osculating plane contains the direction of the resultant external force.

Also that in the case of a chain resting on a smooth surface of revolution whose axis is vertical

$$Tr\cos\phi=\text{const.},$$

in the notation of Art. 24.

16. Prove that a chain subject to no external forces can rotate in the form of a helix, about the axis; and that the tension is $\mu\omega^2 a^2/\sin^2\alpha$, where μ is the line-density, ω the angular velocity, a the radius of the helix, and α its slope.

17. A uniform chain rests on a smooth circular cone whose axis is vertical. Prove that the tangential polar equation of the curve, when developed into a plane, has the form

$$p(r+a)=\text{const.}$$

18. A chain hangs in contact with a smooth sphere. If p is the arc drawn from the highest point of the sphere perpendicular to the great circle tangential to the chain at any point P, and z the altitude of P above a certain horizontal plane, prove that

$$z\sin p=\text{const.}$$

EXAMPLES. V.

(Mainly Geometrical.)

1. Forces $P_1, P_2, \ldots, P_n$ act at a point; prove that the square of their resultant is

$$\Sigma(P_r^2)+2\Sigma\{P_rP_s\cos(P_rP_s)\},$$

where (P_rP_s) denotes the angle between the directions of P_r and P_s.

2. A particle is subject to a field of force whose components X, Y, Z are given functions of x, y, z. If it be constrained to lie on a smooth surface $\phi(x,y,z)=0$, prove that its positions of equilibrium are to be found from the equations

$$X=\lambda\frac{\partial\phi}{\partial x},\quad Y=\lambda\frac{\partial\phi}{\partial y},\quad Z=\lambda\frac{\partial\phi}{\partial z},$$

where λ is an undetermined multiplier. What is the meaning of λ?

Under what conditions is it possible for the particle to be in equilibrium at all points of the surface?

3. If a particle be constrained to lie on a smooth curve defined by $\phi(x,y,z)=0$, $\chi(x,y,z)=0$, prove that the positions of equilibrium are given by

$$X=\lambda\frac{\partial\phi}{\partial x}+\mu\frac{\partial\chi}{\partial x},\quad Y=\lambda\frac{\partial\phi}{\partial y}+\mu\frac{\partial\chi}{\partial y},\quad Z=\lambda\frac{\partial\phi}{\partial z}+\mu\frac{\partial\chi}{\partial z},$$

where λ, μ are undetermined multipliers.

4. Prove that four forces acting at the circumcentres of the four faces of a tetrahedron, in the directions of the inward (or outward) normals to these faces, will be in equilibrium if their magnitudes are proportional to the areas of the respective faces.

5. Forces act normally to the faces of a pyramid at the mean centres of the faces, and are proportional to the areas of those faces. Prove that if they all act inwards they are in equilibrium.

6. Prove that a three-dimensional frame consisting of 5 joints A, B, C, D, E and the 10 bars joining them can be self-stressed.

Also that if A' be the centre of the sphere circumscribing the tetrahedron $BCDE$, and so on, the stress in AB will be proportional to the area of the triangle $C'D'E'$, and so on. (Rankine.)

7. Find the conditions that forces acting along four generators of a paraboloid, of the same system, may be in equilibrium.

8. If forces P, Q, R act along the edges OA, OB, OC of a tetrahedron $OABC$, and P', Q', R' along BC, CA, AB, respectively, the system will reduce to a single force if

$$\frac{PP'}{OA\,.\,BC}+\frac{QQ'}{OB\,.\,CA}+\frac{RR'}{OC\,.\,AB}=0.$$

9. Forces act along the sides of a skew quadrilateral $ABCD$, taken in order, and are proportional to those sides. Prove that they are equivalent to a couple whose plane is parallel to that of the parallelogram formed by joining the middle points of these sides, and whose moment is represented by four times the area of the parallelogram.

10. A system of forces is represented completely by the sides of a skew polygon taken in order. Prove that the moment of the equivalent couple about any axis is proportional to the area of the orthogonal projection of the polygon on a plane perpendicular to that axis.

11. Couples act in the faces of any polyhedron. Their moments are proportional to the areas of the respective faces; and their axes are directed all inwards, or all outwards. Prove that they are in equilibrium. (Möbius.)

12. Prove from first principles that a three-dimensional system of forces can be reduced, in an infinite number of ways, to two forces in perpendicular skew lines. (Monge.)

13. Prove that a wrench can be reduced to two forces, one of which acts through a given point, whilst the other lies in a given plane.

14. If G be the mean centre of n points $P_1, P_2, \ldots, P_n$, and H the mean centre of n points $Q_1, Q_2, \ldots, Q_n$, prove that the central axis of the n forces represented completely by $P_1Q_1, P_2Q_2, \ldots, P_nQ_n$ is parallel to GH.

15. Prove from statical considerations that if three straight lines are parallel to the same plane, any straight line which meets them is parallel to a fixed plane.

16. Prove that a system of forces is in equilibrium if the sum of their moments about each of the six edges of a tetrahedron is zero.

17. If $P_1, P_2, \ldots, P_n$ be a system of forces in equilibrium, and $M_1, M_2, \ldots, M_n$ the moments of any other force (or system of forces) about the respective lines of action, prove that

$$P_1M_1 + P_2M_2 + \ldots + P_nM_n = 0.$$

EXAMPLES. VI.

(Analytical.)

1. Verify by means of the analytical formula of Art. 19 that a uniform hydrostatic pressure over any closed surface constitutes a system of forces in equilibrium.

2. Prove that a uniform hydrostatic pressure over an octant of an ellipsoidal surface is equivalent to a single force; and find where the line of action meets the principal planes.

3. Prove that the analytical conditions of equilibrium (Art. 19) retain the same form if the axes of coordinates are oblique.

4. A rigid body is free to turn about a fixed point O under the attraction of a distant particle at P. Prove that if the principal moments of inertia at O are unequal, the only stable positions of equilibrium are those in which the principal axis of least moment is along OP.

5. Assuming that in the case of the earth $C - A = {\cdot}00109\,Ea^2$, where E is the earth's mass and a its mean radius, prove that the line of the moon's attraction can never deviate from the earth's centre of mass by more than ·0000225 of the radius, or about 475 feet. (Assume that the moon's distance is 60 times the earth's radius, and that its greatest declination is 28°.)

6. Forces X, Y, Z act along alternate edges of a rectangular parallelepiped, the lengths of these edges being a, b, c, respectively. Prove that they have a single resultant if

$$\frac{a}{X} + \frac{b}{Y} + \frac{c}{Z} = 0.$$

7. If couples L, M, N act in the (oblique) coordinate planes the plane of the resultant couple is parallel to

$$\frac{Lx}{\sin\alpha} + \frac{My}{\sin\beta} + \frac{Nz}{\sin\gamma} = 0,$$

where α, β, γ are the angles between the coordinate axes.

8. If the axis of a wrench of pitch ϖ be taken as axis of z, the null-plane of the point (x_1, y_1, z_1) is

$$xy_1 - yx_1 = \varpi(z - z_1);$$

and the null-point of the plane $z = mx + ny + c$ is $(-n\varpi, m\varpi, c)$.

Hence shew that the orthogonal projections on the plane $z = 0$ of the intersection of any two planes, and of the line joining their null-points, are parallel.

9. The moment of a wrench (P, Q, R, L, M, N) about an axis whose line-coordinates are (l, m, n, p, q, r) is

$$(Lp + Mq + Nr + Pl + Qm + Rn)/\omega,$$

where

$$\omega = \surd(p^2 + q^2 + r^2).$$

10. Prove that the null-plane of the point (x_1, y_1, z_1) with respect to the wrench (P, Q, R, L, M, N) is

$$L(x - x_1) + M(y - y_1) + N(z - z_1) = P(yz_1 - zy_1) + Q(zx_1 - xz_1) + R(xy_1 - yx_1).$$

11. Also that the null-point of the plane $Ax + By + Cz = 1$ has the coordinates

$$\frac{P + BN - CM,\ Q + CL - AN,\ R + AM - BL}{AP + BQ + CR}.$$

12. If l, m, n, p, q, r be the six coordinates of a line, find the coordinates of its conjugate with respect to the wrench (P, Q, R, L, M, N); and prove that when the wrench is replaced by two forces acting in the given line and its conjugate the force in the given line will be

$$\frac{(LP + MQ + NR)\,\omega}{Lp + Mq + Nr + Pl + Qm + Rn},$$

where

$$\omega = \surd(p^2 + q^2 + r^2).$$

13. Prove that the condition that the axes of two wrenches should intersect is

$$L'P + M'Q + N'R + LP' + MQ' + NR' = (\varpi + \varpi')(PP' + QQ' + RR'),$$

where ϖ, ϖ' are the two pitches.

14. From a given polyhedron another is constructed whose edges are respectively conjugate to the edges of the former with respect to the paraboloid

$$2cz = x^2 + y^2.$$

Prove that the projections of these figures on the plane $z = 0$ are reciprocal, except that corresponding lines are perpendicular to one another, instead of parallel. (Maxwell.)

15. Prove that the lines which pass through a given point and are perpendicular to their conjugates lie on a cone of the second degree.

16. Verify by transformation of rectangular coordinates that the expression

$$lp' + mq' + nr' + l'p + m'q + n'r$$

for the virtual coefficient of two screws is an absolute invariant.

17. Prove that two screws on the same cylindroid are reciprocal if they are parallel to conjugate diameters of the pitch conic.

18. When a body has three degrees of freedom, the axes of any three screws which are mutually reciprocal are parallel to conjugate diameters of the pitch quadric.

19. Prove that a wrench whose force-component is S, and pitch ϖ, can in general be resolved into six wrenches about six given screws.

If these screws are mutually reciprocal the components are given by formulæ of the type

$$S_r = \frac{\varpi_{0r} S}{2\varpi_r},$$

where ϖ_r is the pitch of the rth given screw, and ϖ_{0r} its virtual coefficient with respect to the given wrench.

Hence shew that

$$\varpi S^2 = \varpi_1 S_1^2 + \varpi_2 S_2^2 + \ldots + \varpi_6 S_6^2.$$

CHAPTER IV

MOMENTS OF INERTIA

25. The Momental Ellipsoid.

If $m_1, m_2, \ldots, m_n$ be the masses of the various particles of a system, and $p_1, p_2, \ldots, p_n$ their respective distances from any given axis, the sum $\Sigma(mp^2)$ is called the 'quadratic moment,' or the 'moment of inertia' of the system with respect to that axis. If we put

$$\frac{\Sigma(mp^2)}{\Sigma(m)} = k^2, \qquad \ldots\ldots\ldots\ldots\ldots\ldots\ldots\ldots\ldots(1)$$

the linear magnitude k is called the 'mean square of the distances' of the particles from the axis, or the 'radius of gyration' about the axis.

The calculation of moments of inertia of bodies of given shape and distribution of density is a matter of integration [S. 71, 72]. It is found, for instance, that the radius of gyration of a homogeneous circular cylinder, or a uniform circular disk, of radius a, about its axis is given by $k^2 = \frac{1}{2}a^2$. The corresponding results for a thin spherical shell and for a solid sphere are $k^2 = \frac{2}{3}a^2$, and $k^2 = \frac{2}{5}a^2$, respectively. For a homogeneous ellipsoid of semiaxes a, b, c, the radius of gyration about the axis a is given by $k^2 = \frac{1}{5}(b^2 + c^2)$.

To find the relation between the moments of inertia with respect to different axes through the same point O, we take this point as origin of rectangular coordinates. Let (λ, μ, ν) be the direction-cosines of any axis OH, and let (x, y, z) be the coordinates of the position P of any particle of the system. If ON be the orthogonal projection of OP on OH, we have

$$\begin{aligned} PN^2 &= OP^2 - ON^2 \\ &= (x^2 + y^2 + z^2)(\lambda^2 + \mu^2 + \nu^2) - (\lambda x + \mu y + \nu z)^2 \\ &= (y^2 + z^2)\lambda^2 + (z^2 + x^2)\mu^2 + (x^2 + y^2)\nu^2 \\ &\quad - 2yz\mu\nu - 2zx\nu\lambda - 2xy\lambda\mu. \qquad \ldots\ldots(2) \end{aligned}$$

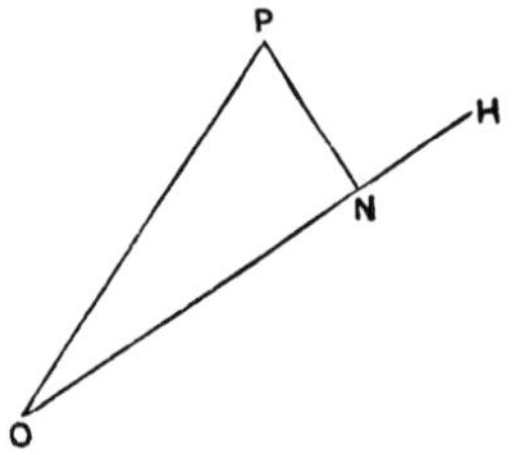

Fig. 26.

The moment of inertia about OH is therefore

$$I = \Sigma(m\,.\,PN^2)$$
$$= A\lambda^2 + B\mu^2 + C\nu^2 - 2F\mu\nu - 2G\nu\lambda - 2H\lambda\mu, \quad\ldots\ldots\ldots(3)$$

where

$$\left.\begin{array}{c} A = \Sigma m\,(y^2 + z^2),\quad B = \Sigma m\,(z^2 + x^2),\quad C = \Sigma m\,(x^2 + y^2), \\ F = \Sigma\,(myz),\quad G = \Sigma\,(mzx),\quad H = \Sigma\,(mxy). \end{array}\right\}\ldots(4)$$

The quantities A, B, C are the moments of inertia about Ox, Oy, Oz, respectively, and F, G, H are called the 'products' of inertia with respect to this system of axes.

A geometrical representation of the formula (3) is obtained as follows. We construct the quadric whose equation is

$$Ax^2 + By^2 + Cz^2 - 2Fyz - 2Gzx - 2Hxy = M\epsilon^4, \quad\ldots\ldots(5)$$

where ϵ is any linear magnitude, and M is the total mass $\Sigma(m)$. If ρ be the radius of this quadric in the direction OH, we have, putting $x, y, z = \lambda\rho, \mu\rho, \nu\rho$, respectively,

$$I = \frac{M\epsilon^4}{\rho^2}. \quad\ldots\ldots\ldots\ldots\ldots\ldots\ldots\ldots\ldots(6)$$

The moment of inertia about any radius therefore varies inversely as the square of its length. Since I is necessarily positive, every radius meets the surface in real points, and the quadric is therefore an ellipsoid. It is called the 'momental ellipsoid' of O*.

If this surface be referred to its principal axes, its equation takes the form

$$Ax^2 + By^2 + Cz^2 = M\epsilon^4. \quad\ldots\ldots\ldots\ldots\ldots\ldots(7)$$

These special axes are called the 'principal axes of inertia' at O, and the moments A, B, C about them are called the 'principal moments of inertia' at O. The products of inertia, $\Sigma(myz)$, $\Sigma(mzx)$, $\Sigma(mxy)$, with respect to the principal axes all vanish.

It is to be noticed that there is a limitation to the possible forms which can be assumed by the momental ellipsoid. It appears from (4) that we must have in general

$$B + C > A,\quad C + A > B,\quad A + B > C. \quad\ldots\ldots\ldots\ldots(8)$$

The greatest principal moment must therefore be less than the sum of the other two. An exception occurs in the case of a

* It appears to have been first employed by Cauchy (1827).

distribution of matter concentrated in a plane. If this be the plane $z=0$ we have $A+B=C$.

If $A=B=C$, the momental ellipsoid is a sphere. All axes through O are then principal axes, and the moment of inertia is the same with respect to each. The system is accordingly said to be 'kinetically symmetrical' about O.

If two of the principal moments are equal, say $A=B$, the ellipsoid is of revolution about Oz, and the system is said to have kinetic symmetry with respect to this axis. All perpendicular diameters are then principal axes*.

Special importance attaches to the momental ellipsoid at the mass-centre G of the system; for when the moment of inertia about an axis through G is known, that about any parallel axis is obtained by adding the quantity Mf^2, where f is the distance between the two axes [S. 73]. The dynamical importance of this 'central ellipsoid,' as it was called by Poinsot, will appear later (Arts. 44, 46).

If α, β, γ be the radii of gyration about the principal axes at any point O, so that

$$A=M\alpha^2,\quad B=M\beta^2,\quad C=M\gamma^2, \qquad \text{(9)}$$

the moment of inertia about an axis through O in the direction (λ, μ, ν) is

$$\begin{aligned} I &= A\lambda^2+B\mu^2+C\nu^2 \\ &= M(\alpha^2\lambda^2+\beta^2\mu^2+\gamma^2\nu^2), \qquad \text{(10)} \end{aligned}$$

or

$$I=Mp^2, \qquad \text{(11)}$$

where p is the perpendicular drawn from O in the direction (λ, μ, ν) to a tangent plane of the ellipsoid

$$\frac{x^2}{\alpha^2}+\frac{y^2}{\beta^2}+\frac{z^2}{\gamma^2}=1. \qquad \text{(12)}$$

This is called the 'ellipsoid of gyration.' It is the reciprocal polar of the momental ellipsoid with respect to a concentric sphere†.

26. Plane Quadratic Moments‡.

Further theorems relating to the quadratic moments of a material system are of mainly geometrical interest; but one or two of the simpler results may be noticed.

* The term 'uniaxal' is used by Routh to designate bodies of this special type.

† It was introduced by J. MacCullagh (1844).

‡ The theory is due to J. Binet (1811).

If the mass m of each particle be multiplied by the square of its distance h from a given *plane*, the sum $\Sigma\,(mh^2)$ is called the quadratic moment with respect to the plane [S. 70].

If we take the principal axes of inertia at any point O as axes of coordinates, the quadratic moment with respect to the plane

$$\lambda x + \mu y + \nu z = 0, \quad\text{.........................(1)}$$

where λ, μ, ν are direction-cosines, is

$$\Sigma m\,(\lambda x + \mu y + \nu z)^2.$$

This reduces to the form

$$A'\lambda^2 + B'\mu^2 + C'\nu^2, \quad\text{.........................(2)}$$

where $\quad A' = \Sigma\,(mx^2),\ \ B' = \Sigma\,(my^2),\ \ C' = \Sigma\,(mz^2).$ ....(3)

These quantities A', B', C' are the quadratic moments with respect to the coordinate planes. They are connected with the principal moments about the axes by the relations

$$B' + C' = A,\ \ C' + A' = B,\ \ A' + B' = C. \quad\text{.........(4)}$$

If, further, we write

$$A' = Ma^2,\ \ B' = Mb^2,\ \ C' = Mc^2, \quad\text{..............(5)}$$

so that a^2, b^2, c^2 are the mean squares of the distances of the particles from the coordinate planes, the quadratic moment with respect to the plane (1) is

$$M\,(a^2\lambda^2 + b^2\mu^2 + c^2\nu^2) = M\varpi^2, \quad\text{..................(6)}$$

where ϖ is the perpendicular drawn from the origin in the direction (λ, μ, ν) to a tangent plane of the ellipsoid

$$\frac{x^2}{a^2} + \frac{y^2}{b^2} + \frac{z^2}{c^2} = 1. \quad\text{.........................(7)}$$

The mean square of the distances from any diametral plane is therefore equal to the square of the distance of a parallel tangent plane from the centre.

27. The Configuration of the Principal Axes at various Points.

Let us now suppose that the origin is taken at the mass-centre G, and that the quantities a^2, b^2, c^2 denote the mean squares of the distances of the particles from the principal planes at that point. The ellipsoid (7) of Art. 26 may then, for the sake of a name, be called 'Binet's central ellipsoid.' Since the mean square of the distances from any plane exceeds the mean square of the distances from a parallel plane through G by the square of the distance between the two planes [S. 73], the quadratic moment with respect to a plane

$$\lambda x + \mu y + \nu z - p = 0 \quad \text{.....................(1)}$$

will be $M(\varpi^2 + p^2)$, where ϖ^2 is given by Art. 26 (6). Hence denoting this quadratic moment by Mf^2, we have

$$f^2 = p^2 + a^2\lambda^2 + b^2\mu^2 + c^2\nu^2. \quad \text{..................(2)}$$

This may be written

$$p^2 = (f^2 - a^2)\lambda^2 + (f^2 - b^2)\mu^2 + (f^2 - c^2)\nu^2, \quad \text{......(3)}$$

shewing that the planes of given quadratic moment Mf^2 are the tangent planes of the quadric

$$\frac{x^2}{f^2 - a^2} + \frac{y^2}{f^2 - b^2} + \frac{z^2}{f^2 - c^2} = 1. \quad \text{...............(4)}$$

The quadrics corresponding to different values of f^2 are confocal. Moreover, writing

$$\theta = f^2 - a^2 - b^2 - c^2, \quad \text{.....................(5)}$$

the equation becomes

$$\frac{x^2}{\alpha^2 + \theta} + \frac{y^2}{\beta^2 + \theta} + \frac{z^2}{\gamma^2 + \theta} = 1, \quad \text{..............(6)}$$

where α, β, γ are the radii of gyration about the principal axes at G. The various quadrics are therefore confocal with MacCullagh's ellipsoid of gyration at the mass-centre (Art. 25).

It is known from Solid Geometry that three real confocals of the system pass through any point P, and that their tangent planes at P are mutually perpendicular. These planes are, in fact, the principal planes of inertia at P. For, consider the tangent plane at P to any one of the three quadrics. Any plane through P which makes an infinitely small angle with this plane will be a tangent plane to a consecutive quadric whose dimensions differ by small quantities of the *second* order. The tangent plane at P is therefore one of 'stationary' quadratic moment, as regards different planes through P. This property is characteristic of a principal plane.

The principal axes of inertia at P are therefore the normals to the three quadrics confocal with the ellipsoid of gyration which pass through P*. If (x, y, z) be the coordinates of P, the corresponding values of θ are given by (6), which, when cleared of fractions, is a cubic. Denoting the roots by $\theta_1, \theta_2, \theta_3$, we have

$$\theta_1+\theta_2+\theta_3 = x^2+y^2+z^2-(\alpha^2+\beta^2+\gamma^2)$$
$$= r^2-(\alpha^2+\beta^2+\gamma^2), \quad \text{......(7)}$$

if $r = GP$. Hence, by (5),

$$f_2^2+f_3^2 = \theta_2+\theta_3+2(a^2+b^2+c^2) = r^2-\theta_1. \quad \text{......(8)}$$

The squares of the radii of gyration about the principal axes at P are therefore

$$r^2-\theta_1, \quad r^2-\theta_2, \quad r^2-\theta_3. \quad \text{.................(9)}$$

EXAMPLES. VII.

1. Prove that the momental ellipsoid of a uniform thin hemispherical shell at its pole is a sphere.

2. Find the shape of a uniform solid right circular cone in order that the momental ellipsoid at the vertex may be a sphere.

[The radius of the base must be double the altitude.]

3. Find the shape of the momental ellipsoid at the centre of a uniform rectangular plate.

Also at the centre of an elliptic plate.

4. Find the shape of the momental ellipsoid at a corner of a uniform cube.

Also at a vertex of a uniform regular tetrahedron.

5. The thickness of a thin ellipsoidal shell having the form

$$\frac{x^2}{a^2}+\frac{y^2}{b^2}+\frac{z^2}{c^2}=1$$

varies as the perpendicular from the centre on the tangent plane; prove that the square of its radius of gyration about the axis of x is $\frac{1}{3}(b^2+c^2)$.

6. A uniform solid rectangular parallelepiped has edges $2a$, $2b$, $2c$; prove that the square of its radius of gyration about a diagonal is

$$\frac{2}{3}\,\frac{b^2c^2+c^2a^2+a^2b^2}{a^2+b^2+c^2}.$$

7. The radius of the edge of a plano-convex lens is a, and the thickness at the centre is t; if k be the radius of gyration about the axis, and k' that about a diameter of the plane face, prove that

$$k^2=\frac{10a^4+5a^2t^2+t^4}{10(3a^2+t^2)}, \quad k'^2=\frac{10a^4+15a^2t^2+7t^4}{20(3a^2+t^2)}.$$

* This result was given independently by W. Thomson and R. Townsend in 1846.

8. If the density of a globe of radius a at a distance r from the centre be

$$\rho=\rho_0\left(1-\beta\frac{r^2}{a^2}\right),$$

prove that the radius of gyration about a diameter is given by

$$\kappa^2=\frac{14-10\beta}{35-21\beta}a^2.$$

If the mean density be twice the surface density, prove that

$$\kappa^2=\frac{24}{70}a^2.$$

9. The density of a sphere of radius a at a distance r from the centre is given by the formula

$$\rho=\rho_0\frac{\sin\beta r}{\beta r};$$

prove that the square of the radius of gyration about a diameter is

$$\frac{(12\beta a-2\beta^3a^3)\cos\beta a+(6\beta^2a^2-12)\sin\beta a}{3\beta^2(\sin\beta a-\beta a\cos\beta a)}.$$

Evaluate this when βa is infinitesimal.

10. A plane area having a line of symmetry revolves about a parallel axis at a distance a from it in the same plane. Prove that the square of the radius of gyration, about this axis, of the annular solid generated is

$$a^2+3\kappa^2,$$

where κ^2 is the mean square of the distances of points of the area from the line of symmetry.

11. A hollow ring of small uniform thickness is generated by the revolution of a circle of radius b about an axis at a distance a from its centre. Prove that the square of its radius of gyration about this axis is

$$a^2+\tfrac{3}{2}b^2.$$

Also that the square of the radius of gyration about a diameter of the circle through the centres of the cross-sections is

$$\tfrac{1}{2}a^2+\tfrac{5}{4}b^2.$$

12. If the axes of coordinates be the principal axes at the mass-centre of a body M, the equation of the momental ellipsoid at the point (f, g, h) is

$$\{A+M(g^2+h^2)\}x^2+\{B+M(h^2+f^2)\}y^2+\{C+M(f^2+g^2)\}z^2$$
$$-2Mghyz-2Mhfzx-2Mfgxy=\text{const.}$$

13. Find at what points, and under what condition, the momental ellipsoid of a given body is a sphere.

14. If Ox, Oy, Oz are the principal axes at the origin, and (l_1, m_1, n_1), (l_2, m_2, n_2), (l_3, m_3, n_3) are the direction-cosines, with respect to these, of another set of rectangular axes Ox', Oy', Oz', prove that the products of inertia with respect to the latter are

$$-(Al_2l_3+Bm_2m_3+Cn_2n_3),\ \text{etc., etc.}$$

15. The principal moments of a body at the origin are A, B, C. If a small mass be added to it whose coefficients of inertia with respect to the principal axes at O are A', B', C', F', G', H', prove that the new principal moments are $A+A'$, $B+B'$, $C+C'$, approximately.

Also that the direction-cosines of that principal axis which is nearly coincident with OX are

$$1, \quad -\frac{H'}{A-B}, \quad -\frac{G'}{A-C},$$

approximately.

16. The principal moments of a uniaxal body are A, A, C, and the direction-cosines of its axis with respect to any rectangular axes through its mass-centre are l, m, n; prove that its moments and products of inertia with respect to these axes are

$$A+(C-A)\,l^2, \quad A+(C-A)\,m^2, \quad A+(C-A)\,n^2,$$

and

$$(A-C)\,mn, \quad (A-C)\,nl, \quad (A-C)\,lm,$$

respectively.

17. Prove that the axes of given moment of inertia which pass through a given point lie on a cone of the second degree.

18. Prove that the principal axes at the various points of a rigid body form a complex of the second order, whose equation in line-coordinates (Art. 9) is

$$Alp+Bmq+Cnr=0,$$

the axes of Cartesian coordinates being the principal axes at the mass-centre.

19. If Ox', Oy', Oz' be conjugate diameters of the ellipsoid of Art. 26, prove that

$$\Sigma(my'z')=0, \quad \Sigma(mz'x')=0, \quad \Sigma(mx'y')=0,$$

$$\frac{\Sigma(mx'^2)}{\Sigma(m)}=a'^2, \qquad \frac{\Sigma(my'^2)}{\Sigma(m)}=b'^2, \qquad \frac{\Sigma(mz'^2)}{\Sigma(m)}=c'^2,$$

where a', b', c' are the lengths of the conjugate semi-diameters.

20. Prove that the lines through a point O of a body which are principal axes at some point of their length lie on a cone whose equation referred to the principal axes at O is

$$(B-C)\,fyz+(C-A)\,gzx+(A-B)\,hxy=0,$$

where f, g, h are the coordinates of the mass-centre.

CHAPTER V

INSTANTANEOUS MOTION OF A BODY (KINEMATICS)

28. Rotation about a Fixed Point. The Instantaneous Axis.

This Chapter is devoted mainly to the investigation of various formulæ relating to the instantaneous state of motion of a rigid body, without any reference to the changes which may be taking place in this state, whether (as it were) spontaneously, or in consequence of the action of external forces.

Suppose, first, that one point O of the body is fixed. The change of position which takes place in any interval δt of time is equivalent (Art. 2) to a rotation about some axis OJ. If P, P' be the initial and final positions of any point of the body, and $\delta\theta$ the angle of rotation, we have

$$PP' = 2PN \,.\, \sin \tfrac{1}{2}\delta\theta, \qquad\qquad (1)$$

where PN is the perpendicular drawn from P to OJ. The mean velocity of the point in question is therefore at right angles to the plane through OJ bisecting PP', and is equal to

$$\frac{\sin \frac{1}{2}\delta\theta}{\frac{1}{2}\delta t} \,.\, PN. \qquad\qquad (2)$$

The limiting position of the axis of rotation when the interval δt is infinitely small is called the 'instantaneous axis,' and the limiting value of the ratio $\delta\theta/\delta t$ is called the 'instantaneous angular velocity.' Denoting this by ω, we see that the velocity of any point P of the body is at right angles to the plane containing P and the instantaneous axis, and equal to $\omega \,.\, PN$, where PN is the perpendicular distance from that axis. Points in the instantaneous axis itself are momentarily at rest.

If we take rectangular axes through O, the rotation $\omega\delta t$ which takes place in the infinitesimal interval δt may be resolved as in Art. 9 into three components about these axes. If with a change of notation we now denote these components by $p\delta t$, $q\delta t$, $r\delta t$, the quantities p, q, r are called the 'components of angular velocity.'

If l, m, n be the direction-cosines of the instantaneous axis, we have

$$p = \omega l, \quad q = \omega m, \quad r = \omega n, \qquad \text{(3)}$$

and

$$p^2 + q^2 + r^2 = \omega^2. \qquad \text{(4)}$$

Allowing for the change in the notation, the formulæ (3) of Art. 9 give the following expressions for the component velocities of that point of the body whose instantaneous position is (x, y, z):

$$\dot{x} = qz - ry, \quad \dot{y} = rx - pz, \quad \dot{z} = py - qx. \qquad \text{(5)}$$

29. Polhode and Herpolhode. Precessional Motion.

The instantaneous state of motion of the body may be completely specified by a vector OJ along the instantaneous axis, of length proportional to the angular velocity, the sense being such that the rotation shall be right-handed in relation to the direction from O to J.

As t varies, the point J will describe a certain curve in the body, and a certain curve in space. These curves are called the 'polhode' (i.e. path of the pole) and 'herpolhode,' respectively. The corresponding *cones* traced out by the instantaneous axis in the body and in space are accordingly termed the cones of the polhode and herpolhode.

In any continuous motion of the body about O, the former of these cones rolls (without sliding) on the latter. To see this, we may imagine two spherical surfaces of the same radius to be described about O as centre, and regard one of these as fixed in the body and therefore moveable with it, whilst the other is fixed in space. The intersection of OJ with these surfaces will trace out two spherical curves. The argument by which the corresponding theorem in Plane Kinematics [S. 16] is inferred can be applied without change to shew that in the continuous motion of the body the former of these curves rolls in contact with the latter.

The case where both cones are of revolution, and the angular velocity is constant, presents itself in several important questions. This type of motion is called 'precessional,' the astronomical phenomenon of the precession of the equinoxes being the most important instance.

In the annexed figures, OC is the axis of the rolling cone, OZ that of the fixed cone, and OJ is the line of contact. We denote by β the semi-angle of the rolling cone, and by α the angle which its axis OC makes with the fixed axis OZ. If ψ be the angle which the plane ZOC makes with some fixed plane through OZ, the angular velocity with which this plane revolves about OZ is $d\psi/dt$, or $\dot{\psi}$. This measures the 'rate of precession.' To find the

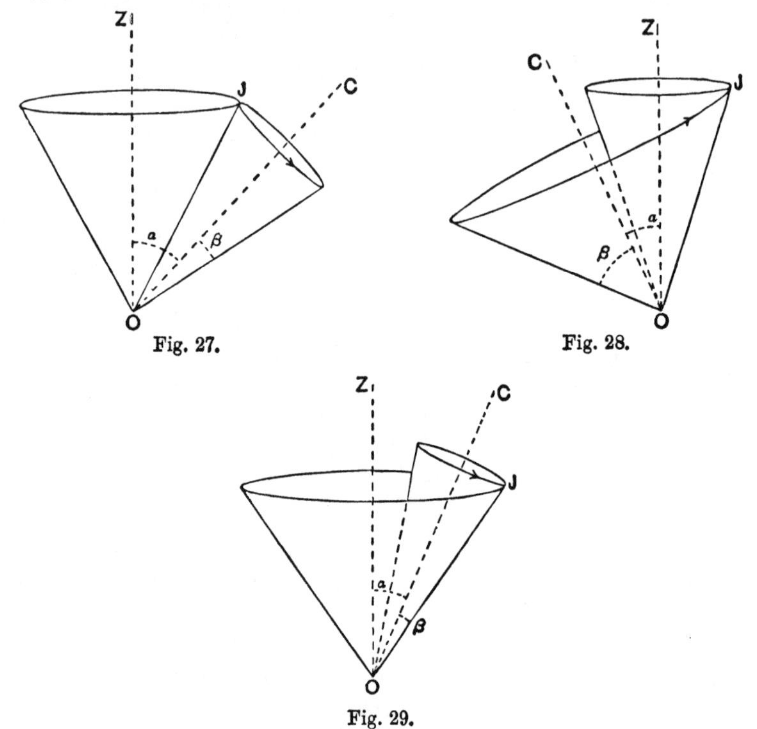

Fig. 27.

Fig. 28.

Fig. 29.

relation between this rate and the angular velocity ω about OJ, we consider the velocity of a point of OC at unit distance from O. This velocity is expressed by $\dot{\psi}\sin\alpha$, or by $\omega\sin\beta$, according as it is considered as due to a rotation $\dot{\psi}$ about OZ, or a rotation ω about OJ. Hence if $\dot{\psi}$ be reckoned positive or negative according as the rotation to which it refers is right-handed or left-handed about OZ, we have

$$\dot{\psi} = \omega \frac{\sin\beta}{\sin\alpha}, \quad \text{.......................(1)}$$

in the cases of Figs. 27 and 28, and

$$\dot{\psi} = -\omega \frac{\sin \beta}{\sin \alpha}, \quad \text{.....................(2)}$$

in the case of Fig. 29. The formula (1) may be held to include all cases if we reckon β to be negative when the rolling cone is inside the other.

We shall require, later, a formula for the rate at which the instantaneous axis describes its cone in the body. Let χ denote the angle which the plane JOC makes with some plane through OC which is fixed in the body. Expressing the equality of elementary arcs of the two pole-curves (now circles) we have, in Figs. 27 and 28,

$$\sin(\alpha - \beta)\,\delta\psi = -\sin\beta\,\delta\chi, \quad \text{..................(3)}$$

if, as we assume, the sense in which χ increases is right-handed with respect to OC. Hence

$$\dot{\chi} = -\frac{\sin(\alpha - \beta)}{\sin\beta}\dot{\psi} = -\omega\frac{\sin(\alpha-\beta)}{\sin\alpha}. \quad \text{............(4)}$$

This formula will apply also in Fig. 29 if we reverse the sign of β.

Fig. 29 corresponds, except as to the proportions, to the astronomical case. OZ then represents the axis of the ecliptic, and OC that of the earth. The obliquity (α) is $23° 28'$, and the annual precession is about $50''$. We have, from (1), in seconds of arc,

$$\beta = \frac{50 \sin 23° 28'}{2\pi \times 366{\cdot}25} = {\cdot}0086.$$

The earth's radius being $6{\cdot}37 \times 10^8$ cm., the polhode cone intersects the earth's surface in a circle of about 27 cm. radius.

Ex. A wheel of radius a rolls on a horizontal plane at a constant inclination α, its centre describing a circle of radius c with the angular velocity $\dot{\psi}$.

In the annexed figure, C is the centre of the wheel, CO is its axis, A the point of contact, N the centre of the circle described by C. Since O is a fixed

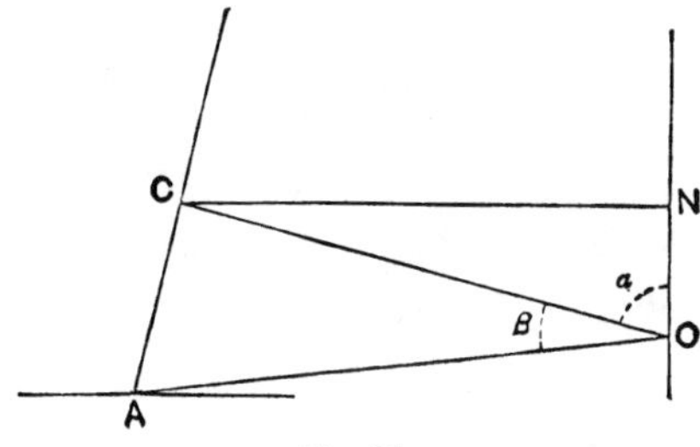

Fig. 30.

point relative to the wheel, OA is the instantaneous axis. The polhode cone is that traced by OA about OC, and its semi-angle β is therefore given by

$$\tan\beta = \frac{a}{c}\sin\alpha. \qquad \text{.............................(5)}$$

The herpolhode cone is that traced by OA about ON. If ω be the angular velocity of the wheel we have, considering the motion of C, as due to rotation about OA,

$$\omega\sin\beta = -\dot{\psi}\sin\alpha, \qquad \text{.................................(6)}$$

as in (2). This gives the component angular velocity about the diameter through the point of contact. The velocity of C may also be regarded as due to a rotation $\omega\cos\beta$ about an axis through A normal to the plane of the wheel. Hence

$$\omega\cos\beta = -\frac{c}{a}\dot{\psi}, \qquad \text{.................................(7)}$$

as may also be derived from (5) and (6). This gives the angular velocity of the wheel about its axis.

The rate $\dot{\chi}$ at which the point of contact travels round the wheel is given by

$$a\dot{\chi} = (c + a\cos\alpha)\dot{\psi}. \qquad \text{.............................(8)}$$

This is easily seen to agree with (4), if allowance be made for the altered sign of β.

30. Momentum.

The momenta of the various particles of any assemblage form a system of localized vectors, which may be replaced in various ways by other systems having the same geometric sum, and the same moment about any arbitrary axis.

For example, we may reduce the system to a 'linear momentum' represented by a localized vector (ξ, η, ζ) through any given point O, and a 'momentum couple,' or 'angular momentum,' represented by a free vector (λ, μ, ν) normal to its plane. The linear momentum is given, as to magnitude and direction, by the geometric sum of the momenta of the several particles, and is therefore independent of the position of O. The angular momentum, on the other hand, varies in general with this position.

The analytical process of reduction is exactly the same as in the case of a system of forces (Art. 19). If we adopt rectangular axes through O, the components of linear momentum are

$$\xi = \Sigma(m\dot{x}), \quad \eta = (\Sigma m\dot{y}), \quad \zeta = \Sigma(m\dot{z}), \qquad \text{........(1)}$$

where (x, y, z) are the coordinates of any particle m, and the summations include all the particles.

The components of angular momentum, referred to O, are simply the total moments of momenta of the particles about the coordinate axes, viz.

$$\lambda = \Sigma \{m(y\dot{z} - z\dot{y})\}, \quad \mu = \Sigma \{m(z\dot{x} - x\dot{z})\}, \quad \nu = \Sigma \{m(x\dot{y} - y\dot{x})\}. \quad \text{.........(2)}$$

This process of reduction is convenient in the case of a rigid body moveable about a fixed point. We are then concerned chiefly with the angular momentum. Taking this point as origin, and substituting the values of $\dot{x}$, $\dot{y}$, $\dot{z}$ from Art. 28 (5), we have, for example,

$$\Sigma \{m(y\dot{z} - z\dot{y})\} = \Sigma \{my(py - qx) - mz(rx - pz)\}$$
$$= \Sigma \{m(y^2 + z^2)\} p - \Sigma (mxy) q - \Sigma (mxz) r.$$

The moments of momentum about the coordinate axes are accordingly

$$\lambda = Ap - Hq - Gr, \quad \mu = -Hp + Bq - Fr, \quad \nu = -Gp - Fq + Cr, \quad \text{.........(3)}$$

where A, B, C, F, G, H are the moments and products of inertia with respect to the axes (Art. 25).

When the axes coincide with the principal axes of inertia at O, the expressions reduce to

$$\lambda = Ap, \quad \mu = Bq, \quad \nu = Cr. \quad \text{..................(4)}$$

The normal to the plane of the momentum couple may be called the axis of angular momentum (referred to O). There is an important geometrical relation between the direction of this axis and that of the instantaneous axis of rotation. For simplicity, let the coordinate axes be the principal axes at O. Since the direction-cosines of the instantaneous axis are, p/ω, q/ω, r/ω, the vector (Ap, Bq, Cr) is normal to that diametral plane of the momental ellipsoid

$$Ax^2 + By^2 + Cz^2 = M\epsilon^4 \quad \text{.....................(5)}$$

which is conjugate to the instantaneous axis. In other words, the plane of the momentum couple is conjugate to the axis of rotation. It is only in the case of complete kinetic symmetry about O, when $A = B = C$, that the axis of the couple always coincides with that of the rotation.

In any case whatever the linear momentum is the same, as to magnitude and direction, as if the whole mass were concentrated at the mass-centre G, and endowed with the velocity of that point.

For, writing

$$x=\bar{x}+a,\quad y=\bar{y}+\beta,\quad z=\bar{z}+\gamma, \quad \text{......................(6)}$$

where $(\bar{x}, \bar{y}, \bar{z})$ are the coordinates of G, we have

$$\Sigma(m\dot{x})=\frac{d}{dt}\Sigma(mx)=\frac{d}{dt}\Sigma\{m(\bar{x}+a)\}$$

$$=\Sigma(m)\,.\,\frac{d\bar{x}}{dt}, \quad \text{..............................(7)}$$

since $\Sigma(ma)=0$. And similarly for the other components*.

Hence if a rigid body rotates about the mass-centre as a fixed point, the linear momentum vanishes.

31. Angular Momentum referred to Mass-Centre.

In the more general type of motion it is convenient to adopt the instantaneous position of the mass-centre as the point of reference.

If P be the position of any particle m of a system, its momentum $(m\mathbf{v})$ may be resolved into two components in lines through P by the vector formula

$$m\mathbf{v}=m\bar{\mathbf{v}}+m\boldsymbol{v}, \quad \text{...........(1)}$$

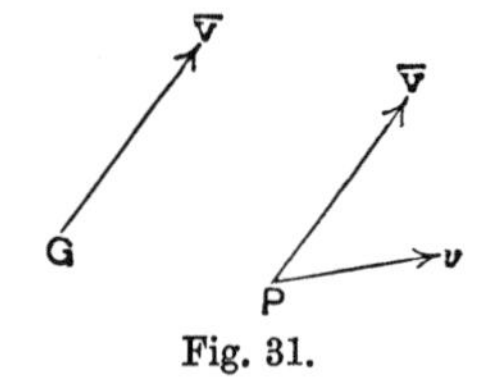

Fig. 31.

where $\bar{\mathbf{v}}$ is the velocity of the mass-centre (G), and $\boldsymbol{v}$ is the velocity of m relative to G. The components $m\bar{\mathbf{v}}$ form a series of parallel vectors proportional to the respective masses m, and are therefore equivalent to a single localized vector $\Sigma(m)\,.\,\bar{\mathbf{v}}$ through G. Since $\Sigma(m\boldsymbol{v})=0$, the aggregate of the remaining components has the formal properties of a 'couple' in Statics.

It follows that in calculating the moment of momentum of a system about any axis through the mass-centre, we may ignore the motion of G itself, and take account only of the motion relative to G.

In symbols, taking any fixed axes, we have with our previous notation

$$\Sigma\{m(y\dot{z}-z\dot{y})\}=\Sigma\left\{m(\bar{y}+\beta)\left(\frac{d\bar{z}}{dt}+\dot{\gamma}\right)-m(\bar{z}+\gamma)\left(\frac{d\bar{y}}{dt}+\dot{\beta}\right)\right\}$$

$$=\Sigma(m)\,.\left(\bar{y}\frac{d\bar{z}}{dt}-\bar{z}\frac{d\bar{y}}{dt}\right)+\Sigma\{m(\beta\dot{\gamma}-\gamma\dot{\beta})\}, \quad \text{..............(2)}$$

* A vector proof of this theorem is given in the author's *Dynamics*, Art. 45.

since $\Sigma(m\beta)=0$, $\Sigma(m\gamma)=0$, and therefore also $\Sigma(m\dot{\beta})=0$, $\Sigma(m\dot{\gamma})=0$. If the origin be at the instantaneous position of G, we have $\bar{y}=0$, $\bar{z}=0$, and therefore

$$\Sigma\{m(y\dot{z}-z\dot{y})\}=\Sigma\{m(\beta\dot{\gamma}-\gamma\dot{\beta})\}. \quad\text{.....................(3)}$$

Similarly for the other components.

Hence in the case of a rigid body, the components of angular momentum with respect to axes through the mass-centre will be given by expressions identical in form with those of Art. 30, viz.

$$Ap-Hq-Gr, \quad -Hp+Bq-Fr, \quad -Gp-Fq+Cr. \text{...(4)}$$

If the axes be the principal axes of inertia at G, we have, simply

$$Ap, \quad Bq, \quad Cr. \quad\text{........................(5)}$$

Again, it appears from the general argument given above, or from equation (2), that the angular momentum of a system about any axis is equal to the angular momentum, about that axis, of the whole mass supposed concentrated at G and moving with this point, together with the angular momentum, with respect to a parallel axis through G, of the motion relative to G.

Hence in the case of a rigid body, if (x, y, z) be the coordinates of the mass-centre with respect to any fixed coordinate axes, and (u, v, w) the velocity of this point, the moments in question are

$$\left.\begin{aligned} &M(yw-zv)+Ap-Hq-Gr,\\ &M(zu-xw)-Hp+Bq-Fr,\\ &M(xv-yu)-Gp-Fq+Cr,\end{aligned}\right\} \quad\text{...............(6)}$$

where A, B, C, F, G, H relate of course to the parallel axes through the mass-centre.

32. Kinetic Energy.

The kinetic energy (T, say) of a rigid body rotating about a fixed point is given by

$$T=\tfrac{1}{2}I\omega^2, \quad\text{..........................(1)}$$

where ω is the angular velocity, and I the moment of inertia with respect to the instantaneous axis. For if PN be the perpendicular drawn to this axis from the position P of any particle m, this particle has a velocity $\omega \,.\, PN$. The energy is therefore

$$T=\tfrac{1}{2}\Sigma(m\,.\,PN^2\,.\,\omega^2)=\tfrac{1}{2}I\omega^2. \quad\text{..............(2)}$$

If (l, m, n) be the direction-cosines of the instantaneous axis with respect to coordinate axes through the fixed point, we have

$$I=Al^2+Bm^2+Cn^2-2Fmn-2Gnl-2Hlm,$$

by Art. 25 (3). Hence

$$2T = Ap^2 + Bq^2 + Cr^2 - 2Fqr - 2Grp - 2Hpq, \quad \ldots\ldots(3)$$

a homogeneous quadratic function (as always) of the symbols which express the velocities of the system. When the axes of coordinates are principal axes, the formula becomes

$$2T = Ap^2 + Bq^2 + Cr^2. \quad \ldots\ldots\ldots\ldots\ldots\ldots\ldots(4)$$

We note, on reference to Art. 30 (3), that the components of angular momentum are the partial derivatives of T with respect to p, q, r, viz.

$$\lambda = \frac{\partial T}{\partial p}, \quad \mu = \frac{\partial T}{\partial q}, \quad \nu = \frac{\partial T}{\partial r}. \quad \ldots\ldots\ldots\ldots\ldots\ldots(5)$$

To obtain formulæ for the kinetic energy in the more general case, we have the theorem that the kinetic energy of any system is equal to the kinetic energy of the whole mass supposed concentrated at the mass-centre and moving with this point, together with the kinetic energy of the motion relative to the mass-centre.

The analytical proof is as follows. In our previous notation

$$\tfrac{1}{2}\Sigma m(\dot{x}^2+\dot{y}^2+\dot{z}^2) = \tfrac{1}{2}\Sigma m\left\{\left(\frac{d\bar{x}}{dt}+\dot{\alpha}\right)^2+\left(\frac{d\bar{y}}{dt}+\dot{\beta}\right)^2+\left(\frac{d\bar{z}}{dt}+\dot{\gamma}\right)^2\right\}$$

$$= \tfrac{1}{2}\Sigma(m)\cdot\left\{\left(\frac{d\bar{x}}{dt}\right)^2+\left(\frac{d\bar{y}}{dt}\right)^2+\left(\frac{d\bar{z}}{dt}\right)^2\right\}$$

$$+\tfrac{1}{2}\Sigma m(\dot{\alpha}^2+\dot{\beta}^2+\dot{\gamma}^2), \quad \ldots\ldots\ldots\ldots\ldots\ldots(6)$$

since $\Sigma(m\dot{\alpha})=0$, $\Sigma(m\dot{\beta})=0$, $\Sigma(m\dot{\gamma})=0$*.

In the case of a rigid body the latter of the two constituents referred to is given by (3), if the inertia coefficients now refer to axes through the mass-centre. Hence, if (u, v, w) be the velocity of the mass-centre, the complete expression is

$$2T = M(u^2+v^2+w^2)$$
$$+ Ap^2 + Bq^2 + Cr^2 - 2Fqr - 2Grp - 2Hpq. \ldots(7)$$

33. Euler's Angular Coordinates.

In order to refer the position of a body which is free to turn about a fixed point O to some standard position, we require (Art. 2) three coordinates, which may be chosen in various ways. The system most frequently used in practical questions was introduced by Euler (1758), and is as follows:

* Other proofs are given in the author's *Dynamics*, Art. 46.

The position of some axis OC fixed in the body is defined by the angle θ which it makes with an axis OZ fixed in space, and the angle ψ which the plane ZOC makes with some fixed plane through OZ. The angle ϕ which some plane through OC, fixed in the body, makes with the plane ZOC completes the determination of the position. The annexed figure, which represents the intersections of the various lines and planes with a unit sphere having O as centre, indicates the usual convention as to the senses in which the three angles θ, ψ, ϕ are measured.

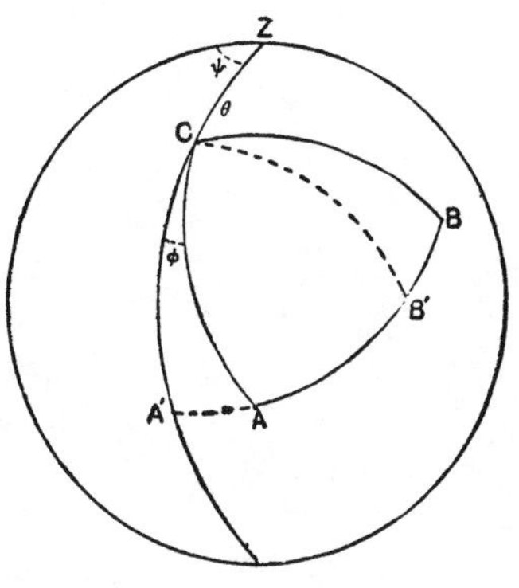

Fig. 32.

The above system of coordinates is exemplified in 'Cardan's suspension' as applied to the marine compass, and in the ordinary gyroscope. In the latter apparatus we have a heavy flywheel free to rotate about a diameter of a circular frame, which is itself free to turn about a horizontal axis at right angles to the former, being pivoted to a vertical ring which again is free to turn about its vertical diameter*.

The coordinate θ here represents the inclination of the axis of the flywheel to the vertical, whilst ψ determines the azimuth of the vertical ring. The remaining angle ϕ fixes the position of the wheel relatively to its circular frame.

The same system of coordinates is applied, in Astronomy, to the earth. If, in Fig. 32, OZ be the normal to the plane of the ecliptic and OC the earth's polar axis, θ will be the 'obliquity,' and ψ will fix the position of the intersection of the equator (AB) with the ecliptic. The slow secular change in ψ (negative on our convention) constitutes the phenomenon of precession. Small periodic fluctuations in the values of θ and ψ are called 'nutations.'

Now let OA' be drawn at right angles to OC in the plane ZOC, and OB' normal to this plane, in such a way that OA', OB', OC shall form a right-handed system of axes. We proceed to calculate the component angular velocities p', q', r about these axes in terms of the rates of variation, $\dot{\theta}$, $\dot{\psi}$, $\dot{\phi}$, of Euler's coordinates.

If θ were constant, the points C and A' would describe circles of radii $\sin\theta$ and $\cos\theta$, respectively, on the unit sphere, in each

* See the frontispiece. The outer vertical ring there shewn is fixed, and serves merely as a support for the pivots of the moveable vertical ring.

case with the angular velocity $\dot{\psi}$. Hence, considering the velocity of C, which moves with the body, in the direction at right angles to the arc ZC, we have

$$p' = -\sin\theta\dot{\psi}. \quad \text{.......................(1)}$$

Since the velocity of C tangential to the same arc is $\dot{\theta}$, we have

$$q' = \dot{\theta}. \quad \text{..............................(2)}$$

The calculation of the third component r is not quite so simple, since A' (unlike C) is not fixed in the body. The point of the body which is momentarily at A' is moving relatively to A' with a velocity whose component perpendicular to the circle ZCA' is $\dot{\phi}$, whilst A' has itself a component velocity $\cos\theta\dot{\psi}$ in the same direction. Hence

$$r = \dot{\phi} + \cos\theta\dot{\psi}. \quad \text{.......................(3)}$$

It is sometimes more convenient to employ axes OA, OB which are fixed in the body. In that case the angle ϕ may be taken to be the angle which the plane COA makes with ZOC. If we resolve the velocity of C in the directions of the tangents to the arcs BC, CA, we find

$$p = \dot{\theta}\sin\phi - \sin\theta\dot{\psi}\cos\phi, \quad q = \dot{\theta}\cos\phi + \sin\theta\dot{\psi}\sin\phi. \quad \text{...(4)}$$

The value of r is given by (3), as before.

From (4), or by direct calculation of the velocity of C along and at right angles to the arc ZC, we have

$$\dot{\theta} = p\sin\phi + q\cos\phi, \quad \sin\theta\dot{\psi} = -p\cos\phi + q\sin\phi. \quad \text{...(5)}$$

The preceding formulæ are specially useful when the body has kinetic symmetry about an axis. If OC be this axis, the principal moments of inertia about perpendicular axes through O are all equal. We have then

$$\begin{aligned} 2T &= A(p'^2 + q'^2) + Cr^2 \\ &= A(\dot{\theta}^2 + \sin^2\theta\dot{\psi}^2) + Cr^2. \quad \text{.............(6)} \end{aligned}$$

Again, the angular momentum about the fixed line OZ is

$$\begin{aligned} &Ap'\cos ZA' + Aq'\cos ZB' + Cr\cos ZC \\ &\qquad = A\sin^2\theta\dot{\psi} + Cr\cos\theta. \quad \text{......(7)} \end{aligned}$$

The angular momentum about an axis at right angles to OZ in the plane ZOC is

$$Ap'\cos\theta + Cr\sin\theta = -A\sin\theta\cos\theta\dot{\psi} + Cr\sin\theta. \quad \text{...(8)}$$

If in (6) and (7) we put $C=0$ we get the case of the spherical pendulum [D. 102, 103]*.

Ex. 1. In the movements of the eye (Art. 4), if OZ be the primary position of the visual axis, and OC any secondary position, Listing's law makes $\phi=-\psi$, if CA be that meridian of the eye which in the primary position coincides with the circle from which ψ is measured. Hence in any small displacement of the visual axis from the position OC, the eyeball has a component rotation about OC of amount

$$-(1-\cos\theta)\,\delta\psi, \quad \text{..............................(9)}$$

by (3). If the displacement of C be δs, in a direction making an angle χ with ZC produced, we have $\sin\theta\,\delta\psi=\delta s\sin\chi$, and the rotation in question is

$$-\tan\tfrac{1}{2}\theta\sin\chi\,\delta s. \quad \text{..............................(10)}$$

Ex. 2. To find the relation between the angular velocities in a Hooke's joint.

This is a contrivance for transmitting rotation from one shaft to another whose axis is in the same plane with that of the first, but makes an angle (θ) with it. It is sufficiently explained by the accompanying sketch. The second

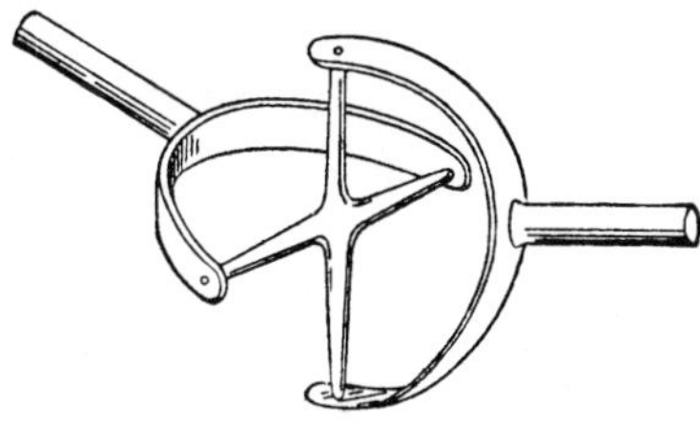

Fig. 33.

figure represents intersections of the various lines with a unit sphere whose centre (O) is at the intersection of the axes of the two shafts. Thus OX, OY

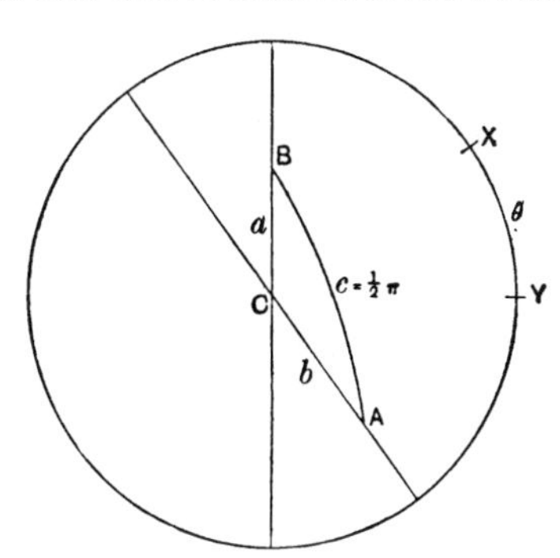

Fig. 34.

* References in this form are to the author's *Dynamics*.

are the axes of the two shafts, and OA, OB those of the two arms of the cross. The great circles which are traced out by A and B intersect in C. Since $AB=\frac{1}{2}\pi$, we have in the spherical triangle ABC

$$\sin a \sin b - \cos a \cos b \cos \theta = 0,$$

or

$$\tan a \tan b = \cos \theta. \quad \text{..............................(11)}$$

Since θ is constant, the ratio of the angular velocity of the second shaft to that of the first is

$$\frac{-\dot{a}}{\dot{b}} = \frac{\cos^2 a \cos \theta}{\sin^2 b} = \frac{\cos \theta}{1 - \cos^2 b \sin^2 \theta}. \quad \text{..................(12)}$$

This varies, in the course of a revolution of the first shaft about OX, between the values $\cos \theta$ and $\sec \theta$. If θ be small the ratio never differs from unity by more than a small quantity of the second order.

A true 'universal flexure-joint' would be one which transmits rotation accurately from one shaft to another independently of the inclination of the axes. It was pointed out by Thomson and Tait* that this condition is satisfied with great exactness by a short length of wire whose ends are fitted axially into the two shafts.

Ex. 3. Let us suppose that we have a pendulum hanging from a vertical spindle by an accurate universal flexure-joint; and let it be required to find the several expressions for the component angular velocities. We refer to Fig. 32 where OZ is now assumed to represent the downward vertical.

If the inclination θ of the axis of the pendulum is alone varied, the component rotations about OA', OB', OC are evidently

$$0, \quad \delta\theta, \quad 0,$$

respectively. If the spindle and the pendulum move in azimuth as one body through an angle $\delta\psi$, we have $\delta\phi=0$, and the component rotations are

$$-\sin\theta\,\delta\psi, \quad 0, \quad \cos\theta\,\delta\psi.$$

Finally, the axis of the pendulum being unchanged in direction, let the spindle be turned back to its former position. This involves rotations

$$0, \quad 0, \quad -\delta\psi.$$

Hence if the spindle be *fixed*, the component angular velocities are

$$p' = -\sin\theta\,\dot{\psi}, \quad q' = \dot{\theta}, \quad r = -(1-\cos\theta)\,\dot{\psi}. \quad \text{............(13)}$$

If, on the other hand, the spindle be rotating with angular velocity ω, we have

$$p' = -\sin\theta\,\dot{\psi}, \quad q' = \dot{\theta}, \quad r = \omega - (1-\cos\theta)\,\dot{\psi}. \quad \text{...........(14)}$$

* *Natural Philosophy*, Art. 109.

EXAMPLES. VIII.

1. Prove that if the instantaneous axis is fixed in space it is fixed in the body, and *vice versâ*.

2. Deduce the formula

$$2T = Ap^2 + Bq^2 + Cr^2 - 2Fqr - 2Grp - 2Hpq$$

for the kinetic energy of a body rotating about a fixed point from the formulæ (5) of Art. 28.

3. A sphere is pressed between parallel boards which are made to rotate with constant angular velocities ω, ω' about fixed axes normal to their planes. Prove that the centre of the sphere describes a circle about an axis which is in the same plane with the former axes, and whose distances from them are as $\omega' : \omega$.

4. Two concentric spherical surfaces of radii a, b rotate with angular velocities ω_1, ω_2, respectively, about fixed diameters inclined to one another at an angle a. A sphere placed between them rolls in contact with both. Prove that the centre of the sphere describes a circle with the angular velocity

$$\frac{\sqrt{(a^2\omega_1{}^2 + 2ab\omega_1\omega_2 \cos a + b^2\omega_2{}^2)}}{2(a+b)}.$$

5. A cone of semi-angle a rolls on a horizontal plane with the angular velocity ω. Prove that its axis revolves about the vertical through the apex with the angular velocity $\omega \tan a$.

6. If (l_1, m_1, n_1), (l_2, m_2, n_2), (l_3, m_3, n_3) be the direction-cosines of a moving system of rectangular axes Ox', Oy', Oz' relative to fixed axes Ox, Oy, Oz, prove that the component angular velocities of the moving system about Ox, Oy, Oz are

$$p = m_1 \frac{dn_1}{dt} + m_2 \frac{dn_2}{dt} + m_3 \frac{dn_3}{dt} = -\left(n_1 \frac{dm_1}{dt} + n_2 \frac{dm_2}{dt} + n_3 \frac{dm_3}{dt}\right), \text{ etc., etc.}$$

Also that the angular velocities about Ox', Oy', Oz' are

$$p' = l_3 \frac{dl_2}{dt} + m_3 \frac{dm_2}{dt} + n_3 \frac{dn_2}{dt} = -\left(l_2 \frac{dl_3}{dt} + m_2 \frac{dm_3}{dt} + n_2 \frac{dn_3}{dt}\right), \text{ etc., etc.}$$

7. The plane $Ax + By + Cz = 1$ is fixed in space, but the rectangular axes to which it is referred are rotating with the angular velocity (p, q, r); prove that

$$\frac{dA}{dt} = Br - Cq, \text{ etc., etc.}$$

8. A rigid body is rotating about the origin with the angular velocity (p, q, r). If A, B, C, F, G, H denote the instantaneous values of the moments and products of inertia with respect to the (fixed) coordinate axes, prove that

$$\frac{dA}{dt} = 2(Hr - Gq), \text{ etc., etc.}$$

$$\frac{dF}{dt} = (C - B)p - Hq + Gr, \text{ etc., etc.}$$

9. A quadric surface whose equation relative to fixed axes is at any instant

$$ax^2+by^2+cz^2+2fyz+2gzx+2hxy=1$$

is of invariable form, but is rotating about the origin with the angular velocity (p, q, r). Prove that

$$\frac{da}{dt}=2(gq-hr), \text{ etc., etc.,}$$

$$\frac{df}{dt}=(b-c)p+gr-hq, \text{ etc., etc.}$$

Deduce the conditions that the surface should be of revolution.

[If f, g, h are none of them zero, the conditions are

$$a-\frac{gh}{f}=b-\frac{hf}{g}=c-\frac{fg}{h}.$$

If one of the coefficients f, g, h vanishes, another must vanish. If $g=0$, $h=0$, the condition is

$$(a-b)(a-c)=f^2.]$$

10. A solid is rolling in contact with a fixed plane. If the angular velocities about the tangents to the lines of curvature and about the normal, at the point of contact, be p, q, r, respectively, and if ρ_1, ρ_2 be the principal radii of curvature, the component accelerations of that point of the solid which is in contact with the plane are

$$-pr\rho_2, \quad -qr\rho_1, \quad p^2\rho_2+q^2\rho_1.$$

11. A wheel, whose mass is M, radius a, and radius of gyration about its axis κ, rolls along the ground at a constant inclination α, so that its centre describes a circle of radius c with velocity v; prove that its kinetic energy is

$$\frac{1}{2}Mv^2\left(1+\frac{\kappa^2}{a^2}+\frac{1}{2}\frac{\kappa^2}{c^2}\sin^2\alpha\right).$$

12. Two equal wheels of radius a can rotate independently about an axle of length l at right angles to both. The mass of each wheel is M; its moments of inertia are C about its axis, and A about a diameter. The combination rolls on a horizontal plane. If ω_1, ω_2 be the angular velocities of the wheels, the kinetic energy is

$$\tfrac{1}{2}C(\omega_1{}^2+\omega_2{}^2)+\tfrac{1}{2}A\frac{a^2}{l^2}(\omega_1-\omega_2)^2+\tfrac{1}{2}M(\omega_1{}^2+\omega_2{}^2)a^2.$$

(The inertia of the axle is neglected.)

13. Prove that in Fig. 32, p. 83, the angular velocities of the body about axes OX, OY, OZ, of which OX is drawn perpendicular to OZ in the plane $\psi=0$, and OY in the plane $\psi=\frac{1}{2}\pi$, are

$$\left.\begin{aligned} -\dot\theta\sin\psi+\dot\phi\sin\theta\cos\psi,\\ \dot\theta\cos\psi+\dot\phi\sin\theta\sin\psi,\\ \dot\psi+\cos\theta\dot\phi. \end{aligned}\right\}$$

Explain the relation of these formulæ to (3) and (4) of Art. **33.**

CHAPTER VI

DYNAMICAL EQUATIONS

34. Kinetics of a Particle.

A few notes on the Kinematics and Kinetics of a particle will be of service later.

The component velocities, referred to rectangular axes, of a particle moving in three dimensions are $\dot{x}, \dot{y}, \dot{z}$; and the component accelerations are $\ddot{x}, \ddot{y}, \ddot{z}$.

In 'cylindrical' coordinates the position of a point P is specified by its distance ρ from a fixed axis OZ, the angle ψ which the plane ZOP makes with some fixed plane through OZ, and the distance z of P from a fixed plane through O, perpendicular to OZ. These are connected with a certain system of rectangular coordinates by the relations

$$x = \rho \cos \psi, \quad y = \rho \sin \psi, \quad z = z. \qquad \text{.........(1)}$$

As in two dimensions, the two velocities $\dot{x}, \dot{y}$ are equivalent to velocities $\dot{\rho}$ and $\rho\dot{\psi}$, in the direction of ρ, and at right angles to the

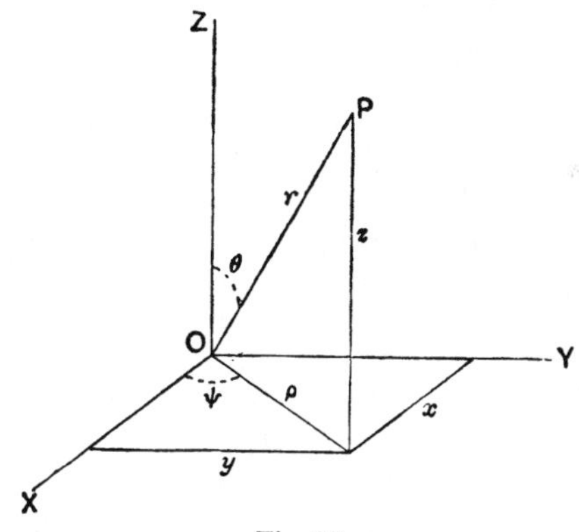

Fig. 35.

plane ZOP, respectively. Again, the accelerations $\ddot{x}, \ddot{y}$ are equivalent to

$$\ddot{\rho} - \rho\dot{\psi}^2 \text{ and } \frac{1}{\rho}\frac{d}{dt}(\rho^2\dot{\psi}) \qquad \text{...............(2)}$$

in the same two directions. The third component is of course $\ddot{z}$.

In 'spherical polar' coordinates the position is specified by the distance r from a fixed origin O, the angle θ which OP makes with a fixed line OZ, and the angle ψ which the plane ZOP makes with a fixed plane through OZ. These are connected with the corresponding cylindrical coordinates by the relations

$$\rho = r \sin \theta, \quad z = r \cos \theta. \quad \text{..................(3)}$$

The velocities $\dot{z}$ and $\dot{\rho}$ are equivalent to $\dot{r}$ and $r\dot{\theta}$ in the direction of r, and at right angles to r in the plane of θ, in the direction of θ increasing. We have thus the three components

$$\dot{r}, \quad r\dot{\theta}, \quad r \sin \theta \dot{\psi} \quad \text{........................(4)}$$

along OP, perpendicular to OP in the plane ZOP, and normal to this plane, respectively.

Again, referring to (2), we note that an acceleration $\ddot{\rho}$ in the direction of ρ, combined with $\ddot{z}$, is equivalent to

$$\ddot{r} - r\dot{\theta}^2 \quad \text{and} \quad \frac{1}{r}\frac{d}{dt}(r^2\dot{\theta}) \quad \text{.....................(5)}$$

along OP, and at right angles to OP in the plane of θ, respectively. Hence, resolving the component $-\rho\dot{\psi}^2$ along and at right angles to OP, we obtain the formulæ

$$\left.\begin{aligned} \alpha &= \ddot{r} - r\dot{\theta}^2 - r \sin^2 \theta \dot{\psi}^2, \\ \beta &= \frac{1}{r}\frac{d}{dt}(r^2\dot{\theta}) - r \sin \theta \cos \theta \dot{\psi}^2, \\ \gamma &= \frac{1}{r \sin \theta}\frac{d}{dt}(r^2 \sin^2 \theta \dot{\psi}), \end{aligned}\right\} \quad \text{...........(6)}$$

where α is the radial acceleration, β the transverse acceleration in the plane of θ, and γ the acceleration normal to the latter plane.

35. Intrinsic Formulæ.

To obtain expressions having a more intrinsic relation to the path, we take two adjacent positions P, P', and draw the corresponding tangents PT, $P'T'$ in the direction of motion. Let δs be the arc PP', δt the corresponding element of time, and $\delta\epsilon$ the angle between the directions of PT, PT'. If from a fixed point O we draw vectors OV, OV′ to represent the velocities at P and P', respectively, the vector VV′ will represent the change of velocity in the interval δt, and VV′/δt will accordingly be the mean acceleration in this interval. The limiting value of this latter vector when P' is taken infinitely near to P is (by definition) the acceleration of the moving point when at P.

The osculating plane of the path at P is the limiting position of the plane through PT parallel to $P'T'$. Since OV, OV′ are parallel to PT, $P'T'$, respectively, the vector $VV'/\delta t$ will in the limit

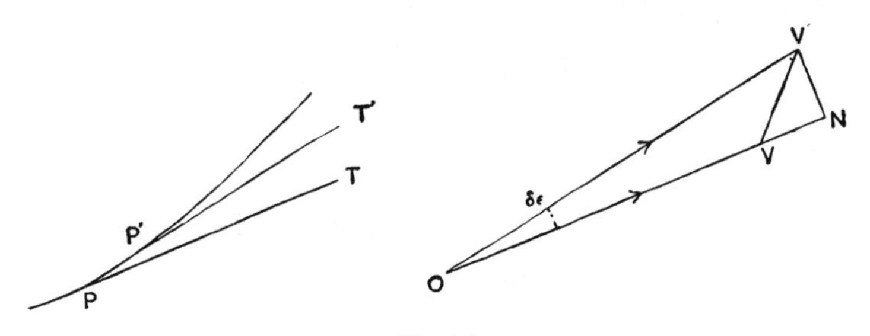

Fig. 36.

be parallel to the osculating plane. It may be resolved into two rectangular components, in the direction of the tangent PT, and of the principal normal, i.e. the normal to the path in the osculating plane, respectively. If we draw $V'N$ perpendicular to OV, the tangential component will be the limit of $VN/\delta t$. If v denote the magnitude of the velocity, we have $VN = \delta v$, ultimately, and the tangential acceleration is accordingly dv/dt. An alternative form is

$$\frac{dv}{ds}\cdot\frac{ds}{dt} \quad \text{or} \quad v\frac{dv}{ds}. \quad \text{......................(1)}$$

The component along the principal normal will be the limit of $NV'/\delta t$, and its magnitude will be the limit of $OV\,\delta\epsilon/\delta t$, or $v d\epsilon/dt$. The limit of $d\epsilon/ds$ is, by definition, the curvature of the path at P, usually denoted by $1/\rho$. We have, then,

$$v\frac{d\epsilon}{dt} = v\frac{d\epsilon}{ds}\frac{ds}{dt} = \frac{v^2}{\rho}, \quad \text{......................(2)}$$

since $ds/dt = v$. The two components are accordingly dv/dt (or vdv/ds) and v^2/ρ.

These results may be obtained also as follows. Using accents to indicate differentiations with respect to the arc s of the path, we have

$$\dot{x} = \frac{dx}{ds}\frac{ds}{dt} = x'\dot{s}, \quad \ddot{x} = x'\ddot{s} + x''\dot{s}^2, \quad \text{etc.} \quad \text{..................(3)}$$

The component velocities are accordingly

$$x'\dot{s}, \quad y'\dot{s}, \quad z'\dot{s}, \quad \text{..................................(4)}$$

shewing that the resultant velocity is $\dot{s}$, along the tangent to the path, whose direction-cosines are x', y', z'. The component accelerations may be written

$$x'\ddot{s}+\rho x''\frac{\dot{s}^2}{\rho},\quad y'\ddot{s}+\rho y''\frac{\dot{s}^2}{\rho},\quad z'\ddot{s}+\rho z''\frac{\dot{s}^2}{\rho}. \quad \ldots\ldots\ldots\ldots\ldots\ldots(5)$$

By a formula of Solid Geometry, the direction-cosines of the principal normal are $\rho x''$, $\rho y''$, $\rho z''$. The acceleration may therefore be resolved into a tangential component $\ddot{s}$, or dv/dt, and a normal component $\dot{s}^2/\rho$.

If F be the force resolved always in the direction of motion we have

$$mv\frac{dv}{ds}=F, \quad \ldots\ldots\ldots\ldots\ldots\ldots\ldots\ldots(6)$$

and therefore

$$\tfrac{1}{2}mv^2=\int F ds+\text{const.} \quad \ldots\ldots\ldots\ldots\ldots\ldots(7)$$

If the particle is subject only to a constant field of force, and if V be its potential energy in relation to this field, we have $F=-\partial V/\partial s$, whence

$$\tfrac{1}{2}mv^2+V=\text{const.}, \quad \ldots\ldots\ldots\ldots\ldots\ldots(8)$$

the equation of energy.

36. Motion on a Surface.

One immediate consequence of the preceding investigation is that if a particle be restricted to lie in a given smooth surface, and be subject to no force except the normal reaction (R') of this surface, it will describe a geodesic or 'straightest' line on the surface with constant velocity. For, since the acceleration must be wholly normal, the osculating plane at any point P of the path must contain the normal to the surface This is the known condition for a geodesic. Also, since the tangential acceleration vanishes, we have $dv/dt=0$, or $v=\text{const.}$ The reaction is given by

$$R'=\frac{mv^2}{\rho}, \quad \ldots\ldots\ldots\ldots\ldots\ldots\ldots\ldots(1)$$

where m is the mass, and therefore varies as the (principal) curvature of the path.

In the general case of a particle moving on a given surface it is convenient to take as directions of reference (i) the tangent PT to the path at P, (ii) a perpendicular to PT in the tangent plane of the surface, and (iii) the normal to the surface. For definiteness we will suppose that the positive directions of these lines, taken in the above order, form a right-handed system. Let the tangent $P'T'$ at an adjacent point P' of the path be projected orthogonally on the normal plane to the surface through PT, and on the tangent plane; and let the angles which these projections make with PT be $\delta\epsilon'$ and $\delta\chi$, respectively. The increments in time δt of the component velocities in our present standard directions are δv, $v\delta\chi$, $v\delta\epsilon$, respectively. The component accelerations are accordingly

$$\frac{dv}{dt}, \quad v\frac{d\chi}{dt}, \quad v\frac{d\epsilon'}{dt}. \qquad \text{.....................(2)}$$

If ρ', σ be the radii of curvature at P of the projections of the path on the normal plane and the tangent plane*, respectively, we have

$$\frac{1}{\rho'} = \frac{d\epsilon'}{ds}, \quad \frac{1}{\sigma} = \frac{d\chi}{ds}. \qquad \text{.....................(3)}$$

The expressions (2) may accordingly be written

$$\frac{dv}{dt}, \quad \frac{v^2}{\sigma}, \quad \frac{v^2}{\rho'}, \qquad \text{.......................(4)}$$

respectively. There is also the alternative form vdv/ds for the first component.

These results may also be derived from those of Art. 35. If ϕ be the angle which the osculating plane of the path makes with the normal to the surface at P, the component accelerations will be

$$\frac{dv}{dt}, \quad \frac{v^2}{\rho}\sin\phi, \quad \frac{v^2}{\rho}\cos\phi. \qquad \text{..........................(5)}$$

Since

$$\rho = \rho'\cos\phi, \quad \sigma = \rho/\sin\phi, \qquad \text{..........................(6)}$$

by known formulæ of Solid Geometry, this agrees with (4).

To obtain the equations of motion of a particle m, we must suppose the forces acting on it (other than the normal reaction R') to be resolved into components P, Q, R acting in the tangential, lateral, and normal directions, respectively. Thus

$$mv\frac{dv}{ds} = P, \quad \frac{mv^2}{\sigma} = Q, \quad \frac{mv^2}{\rho'} = R + R'. \qquad \text{...........(7)}$$

From the first of these we deduce, as in two dimensions,

$$\tfrac{1}{2}mv^2 = \int Pds + \text{const.}, \qquad \text{.....................(8)}$$

which is the equation of energy, since P is the only component of force which does work. When v has been found from this equation, the second of equations (7) gives σ, and the third R'.

* σ is the so-called 'geodesic' radius of curvature. Its reciprocal $1/\sigma$ actually measures the rate at which the path is *deviating from* a geodesic. From the present point of view it might be called the 'lateral curvature' of the path.

37. The Principles of Linear and Angular Momentum.

There are two general principles which are held to apply to any material system, whatever the nature of its internal forces and connections, viz. the principles of linear and angular momentum.

The first of these asserts that if the system be free from *external* force, its total momentum, resolved in any fixed direction, is constant. It follows from Art. 30 that the mass-centre describes a straight line with constant velocity. When external forces act, the rate of increase of the total momentum, resolved in any direction, is equal to the total external force resolved in that direction. The motion of the mass-centre is therefore the same as if the whole mass were concentrated at that point and subject to all the external forces acting parallel to their original directions.

The second principle is to the effect that if the system is free from external force, the total moment of momentum of its particles about any fixed axis is constant. Further, when external forces act, the rate of increase of this moment of momentum is equal to the total moment of the external forces about this axis.

If x, y, z be the coordinates of any particle m of the system relative to fixed rectangular axes, the analytical expression of these principles is

$$\frac{d}{dt}\Sigma m\dot{x} = X, \quad \frac{d}{dt}\Sigma m\dot{y} = Y, \quad \frac{d}{dt}\Sigma m\dot{z} = Z, \quad \text{.........(1)}$$

$$\frac{d}{dt}\Sigma m(y\dot{z} - z\dot{y}) = L, \quad \frac{d}{dt}\Sigma m(z\dot{x} - x\dot{z}) = M, \quad \Sigma m(x\dot{y} - y\dot{x}) = N. \quad \text{......(2)}$$

The sign Σ denotes summation with respect to all the particles of the system. The symbols X, Y, Z represent the total components of the external forces resolved parallel to the axes of coordinates, whilst L, M, N are the moments of these forces about the same axes.

The above principles are established without difficulty in the case of a system of discrete particles subject to mutual forces which act in the lines joining them, and obey the law of equality of action and reaction. If we wish to avoid making any hypothesis as to the structure of an apparently continuous body such as a solid or a fluid, we may either formulate the principles as physical postulates, which is perhaps the best plan, or we may fall back on 'd'Alembert's principle,' which historically formed the basis of the first systematic treatment of the Dynamics of extended bodies [D. 53].

If we consider any particle m, the resultant force upon it, due both to internal and external influences, must have as components $m\ddot{x}, m\ddot{y}, m\ddot{z}$. This is called by d'Alembert the 'effective force' on m; and his assumption is that the whole assemblage of effective forces on the various particles of the system must be statically equivalent to the actual *external* forces; i.e. they must have the same total component in any direction, and the same moment about any axis. This gives

$$\Sigma m\ddot{x} = X, \quad \Sigma m\ddot{y} = Y, \quad \Sigma m\ddot{z} = Z, \qquad (3)$$

$$\Sigma m(y\ddot{z} - z\ddot{y}) = L, \quad \Sigma m(z\ddot{x} - x\ddot{z}) = M, \quad \Sigma m(x\ddot{y} - y\ddot{x}) = N, \qquad (4)$$

which are seen to be equivalent to (1) and (2).

It has already been pointed out (Art. 31) that in calculating the angular momentum about any axis which passes through the instantaneous position of the mass-centre (G) we may ignore the motion of G itself, and take account only of the motion relative to G.

A further very important remark is that in calculating the *rate of change* of the angular momentum about a *fixed* axis drawn through the instantaneous position of G we may have regard only to the relative motion.

It was shewn in the Art. referred to that the momentum of the system at time t may be completely summed up, for all purposes of resolution and taking moments, in a linear momentum $\Sigma(m).\bar{\mathbf{v}}$ resident in the tangent line to the path of G, and an angular momentum $\mathbf{H}$ (say), the latter being of the nature of a free vector. At time $t + \delta t$ these are replaced by a linear momentum $\Sigma(m).(\bar{\mathbf{v}} + \delta\bar{\mathbf{v}})$ in the tangent line at G', the new position of G, and a free vector $\mathbf{H} + \delta\mathbf{H}$. Since the distance of G from the tangent line at G' is a small quantity of the second order, the terms corresponding to the linear momentum at time $t + \delta t$ will disappear when we take moments about G. Hence in calculating the changes in the interval δt we are only concerned with $\delta\mathbf{H}$. And we have seen that $\mathbf{H}$ depends only on the relative motion.

Analytically, the angular momentum about Ox is

$$\Sigma(m)\left(\bar{y}\frac{d\bar{z}}{dt} - \bar{z}\frac{d\bar{y}}{dt}\right) + \Sigma\{m(\beta\dot{\gamma} - \dot{\beta}\gamma)\}, \qquad (5)$$

by Art. 31 (2). Differentiating with respect to t, we obtain

$$\Sigma(m)\left(\bar{y}\frac{d^2\bar{z}}{dt^2} - \bar{z}\frac{d^2\bar{y}}{dt^2}\right) + \frac{d}{dt}\Sigma\{m(\beta\dot{\gamma} - \dot{\beta}\gamma)\}. \qquad (6)$$

This reduces, when $\bar{y} = 0$, $\bar{z} = 0$, to the second term.

The equations (4) above may therefore be replaced by

$$\frac{d}{dt}\Sigma\{m(\beta\dot{\gamma}-\dot{\beta}\gamma)\}=L,\quad \frac{d}{dt}\Sigma\{m(\gamma\dot{\alpha}-\alpha\dot{\gamma})\}=M,\quad \frac{d}{dt}\Sigma\{m(\alpha\dot{\beta}-\dot{\alpha}\beta)\}=N, \text{ ...(7)}$$

provided L, M, N be now understood to denote the moments of the external forces about axes through G parallel to the coordinate axes.

Ex. 1. A uniform bar rotates like a conical pendulum about the vertical through its upper end, making a constant angle θ with the vertical; to find the requisite angular velocity (ω).

This comes most easily from d'Alembert's principle. If μ be the line-density, the effective force on an element at a distance x from the upper end (O) is $\mu\delta x\,.\,\omega^2 x\sin\theta$ towards the vertical through O; and the moment of this about O is $\mu\delta x\,.\,\omega^2 x\sin\theta\,.\,x\cos\theta$. If l be the length, the moment of gravity is $\mu lg\,.\,\frac{1}{2}l\sin\theta$. Hence

$$\mu\omega^2\sin\theta\cos\theta\int_0^l x^2\,dx=\tfrac{1}{2}\mu g l^2\sin\theta,$$

or

$$\omega^2=\frac{3}{2}\frac{g}{l\cos\theta}. \text{ ....(8)}$$

Ex. 2. A body of any form rotates about a fixed axis through its mass-centre G, with a constant angular velocity ω whose components about the principal axes at G are p, q, r; to find the pressures on the axis.

If PN (Fig. 26, p. 66) be the perpendicular from a particle m at (x, y, z) to the axis of rotation, the effective force on m is $m\omega^2$. PN. Since

$$ON=(px+qy+rz)/\omega, \text{ ....(9)}$$

the components of this effective force are

$$m\{p(px+qy+rz)-\omega^2x\},\quad m\{q(px+qy+rz)-\omega^2y\},$$
$$m\{r(px+qy+rz)-\omega^2z\}. \text{ ......(10)}$$

Since $$\Sigma(mx)=0,\quad \Sigma(my)=0,\quad \Sigma(mz)=0,$$

the whole system of effective forces reduces to a couple. The moment of this couple about Ox, for example, is

$$\Sigma\{m(ry-qz)(px+qy+rz)\}=\Sigma\{m(y^2-z^2)\}qr=(C-B)qr, \text{ ......(11)}$$

in the usual notation, since

$$\Sigma(myz)=0,\quad \Sigma(mzx)=0,\quad \Sigma(mxy)=0,$$

by the property of principal axes. Hence, the body exerts on its bearings a 'centrifugal couple' whose components are

$$(B-C)qr,\quad (C-A)rp,\quad (A-B)pq. \text{ ....(12)}$$

This does not vanish unless the axis of rotation be a principal axis.

The plane of the couple is that of the instantaneous axis and the axis of angular momentum (Art. 30). This plane is fixed in the body, and revolves with it, so that there is a periodic stress on the bearings. The effect, sometimes dangerous, is felt in machinery which is not properly 'balanced.'

38. Equations of Motion of a Rigid Body.

We proceed to apply the foregoing general principles to the case of a rigid body. The motion of the body may be expressed in terms of the velocity (u, v, w) of its mass-centre G, and the angular velocity (p, q, r). If M_0 be the total mass*, the components of linear momentum are M_0u, M_0v, M_0w, whilst those of angular momentum are given by the formulæ (4) of Art. 31. Hence, in virtue of the principles explained in Art. 37 we have

$$M_0\frac{du}{dt}=X, \quad M_0\frac{dv}{dt}=Y, \quad M_0\frac{dw}{dt}=Z, \qquad (1)$$

and

$$\left.\begin{aligned}\frac{d}{dt}(Ap-Hq-Gr)&=L,\\ \frac{d}{dt}(-Hp+Bq-Fr)&=M,\\ \frac{d}{dt}(-Gp-Fq+Cr)&=N,\end{aligned}\right\} \qquad (2)$$

where L, M, N now denote the moments of the external forces about lines through G parallel to the coordinate axes.

It appears that the motion of G and the motion relative to G are independent of one another.

The quantities A, B, C, F, G, H are the moments and products of inertia with respect to lines through G which are fixed in direction. They are therefore not in general constants, but are continually changing in consequence of the rotation. For this reason the equations (2) are not often convenient for direct application.

39. Motion of a Sphere on a Plane.

There is one case, however, which is specially simple, viz. that of a homogeneous sphere, or (more generally) of any body which is kinetically symmetrical with respect to the mass-centre. We then have

$$A=B=C=I, \quad F=G=H=0,$$

where I is the moment of inertia about any diameter.

* The zero suffix is introduced for the sake of distinction from the M of equations (2). It is omitted afterwards whenever there is no risk of confusion.

As an example, take the case of a sphere rolling on a plane, and suppose that the external forces, other than the reactions of the plane, reduce to a force (X, Y, Z) at the centre. The axes of x and y being taken parallel to the plane, let the components of

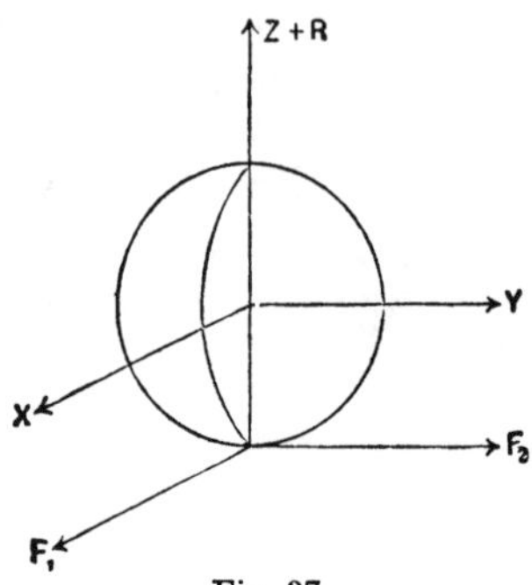

Fig. 37.

the reaction at the point of contact be F_1, F_2, R. If the coordinates x, y refer to the centre of the sphere, we have

$$M\ddot{x} = X + F_1, \quad M\ddot{y} = Y + F_2, \quad 0 = Z + R. \quad \text{.........(1)}$$

Again, if a be the radius,

$$I\dot{p} = F_2 a, \quad I\dot{q} = -F_1 a, \quad I\dot{r} = 0. \quad \text{..............(2)}$$

The angular velocity r about the normal to the plane is therefore constant. Since the velocity of the point of the sphere which is in contact with the plane is zero, we have in addition the kinematical relations

$$\dot{x} = qa, \quad \dot{y} = -pa, \quad \text{.......................(3)}$$

as is seen by considering the effect of the rotations about lines through the point of contact parallel to x and y respectively.

Eliminating p and q between these equations, we find

$$(M + I/a^2)\,\ddot{x} = X, \quad (M + I/a^2)\,\ddot{y} = Y. \quad \text{...........(4)}$$

The acceleration of the centre is therefore the same as if the plane were perfectly smooth, and the inertia of the sphere were increased by the amount I/a^2, or $M\kappa^2/a^2$, if κ be the radius of gyration about a diameter.

Thus a sphere rolling, under gravity, on a plane whose inclination is α will have a constant acceleration

$$\frac{a^2}{\kappa^2 + a^2} g \sin \alpha, \quad \text{........................(5)}$$

parallel to a line of greatest slope. The path of the centre is therefore in general a parabola. In the case of a homogeneous sphere $\kappa^2 = \frac{2}{5}a^2$, and the acceleration is $\frac{5}{7}g \sin \alpha$.

If the sphere slides, as well as rotates, the equations (1) and (2) will still hold, but the geometrical relations (3) are replaced by

$$\dot{x} - aq = V\cos\theta, \quad \dot{y} + ap = V\sin\theta, \quad \text{.....................(6)}$$

where V is the velocity of sliding, i.e. the velocity of that point of the sphere which is in contact with the plane, and θ the angle which the direction of V makes with the axis of x.

Suppose, for simplicity, that the plane is horizontal, and that there are no forces except those of gravity and friction. We have, then, $X = Y = 0$, $Z = -Mg$, and therefore $R = Mg$. The dynamical equations thus reduce to

$$M\ddot{x} = F_1, \quad M\ddot{y} = F_2, \quad \text{and} \quad I\dot{p} = F_2 a, \quad I\dot{q} = -F_1 a, \quad \text{.........(7)}$$

whence

$$\ddot{x} - a\dot{q} = \left(1 + \frac{a^2}{\kappa^2}\right)\frac{F_1}{M}, \quad \ddot{y} + a\dot{p} = \left(1 + \frac{a^2}{\kappa^2}\right)\frac{F_2}{M}. \quad \text{...............(8)}$$

If we denote the resultant of the frictional forces by S, and assume that this acts in a direction opposite to the sliding, we have

$$F_1 = -S\cos\theta, \quad F_2 = -S\sin\theta. \quad \text{.........................(9)}$$

Hence, from (6) and (8),

$$\left.\begin{aligned} \dot{V}\cos\theta - V\sin\theta\,\dot{\theta} &= -\left(1 + \frac{a^2}{\kappa^2}\right)\frac{S}{M}\cos\theta, \\ \dot{V}\sin\theta + V\cos\theta\,\dot{\theta} &= -\left(1 + \frac{a^2}{\kappa^2}\right)\frac{S}{M}\sin\theta. \end{aligned}\right\} \quad \text{...............(10)}$$

Hence $\dot{\theta} = 0$, i.e. the direction of sliding is constant, and

$$\dot{V} = -\left(1 + \frac{a^2}{\kappa^2}\right)\frac{S}{M}. \quad \text{..............................(11)}$$

The motion of the centre is given by

$$\ddot{x} = -\frac{S}{M}\cos\theta, \quad \ddot{y} = -\frac{S}{M}\sin\theta. \quad \text{.........................(12)}$$

Hence if S is constant the acceleration of the centre is constant in magnitude and direction, and the path is therefore a parabola. On the usual law of friction we have $S = \mu Mg$, and therefore

$$\dot{V} = -\mu\left(1 + \frac{a^2}{\kappa^2}\right)g. \quad \text{..............................(13)}$$

Sliding will only take place so long as V is positive. If V_0 be the initial value, V will vanish after a time

$$t = \frac{\kappa^2}{\mu\,(\kappa^2 + a^2)} \cdot \frac{V_0}{g}. \quad \text{..............................(14)}$$

With such values of the constants, and of V_0, as occur in the game of billiards, the time of sliding is very short.

40. Sphere rolling on a Fixed Sphere.

As a further example, bringing in some interesting points, we take the case of a sphere rolling on a fixed spherical surface, under no forces except the reaction at the point of contact.

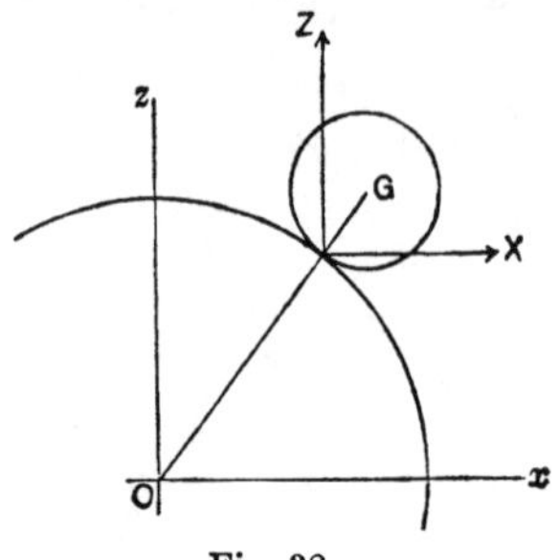

Fig. 38.

Let a be the radius of the rolling sphere, and x, y, z the coordinates of its centre G relative to axes through O, the centre of the fixed surface. Let c be the radius of the spherical surface which is the locus of G. We take as the standard case that of a sphere rolling on the *outside* of another; the opposite case is included in the results if we change the sign of a. If X, Y, Z be the components of the reaction we have

$$M\ddot{x}=X, \quad M\ddot{y}=Y, \quad M\ddot{z}=Z, \qquad \text{(1)}$$

$$I\dot{p}=-\frac{a}{c}(yZ-zY), \quad I\dot{q}=-\frac{a}{c}(zX-xZ), \quad I\dot{r}=-\frac{a}{c}(xY-yX), \qquad \text{(2)}$$

since the coordinates of the point of contact, relative to the centre of the rolling sphere, are $-ax/c$, $-ay/c$, $-az/c$. We have also the kinematical relations

$$\dot{x}=\frac{a}{c}(qz-ry), \quad \dot{y}=\frac{a}{c}(rx-pz), \quad \dot{z}=\frac{a}{c}(py-qx). \qquad \text{(3)}$$

Thus we have, in all, nine equations involving the nine quantities $x, y, z, p, q, r, X, Y, Z$. Eliminating X, Y, Z between (1) and (2), we have

$$I\dot{p}=-\frac{Ma}{c}(y\ddot{z}-z\ddot{y}), \text{ etc.} \qquad \text{(4)}$$

Hence

$$\left.\begin{aligned} p=-\frac{Ma}{Ic}(y\dot{z}-z\dot{y})+\alpha, \quad q=-\frac{Ma}{Ic}(z\dot{x}-x\dot{z})+\beta,\\ r=-\frac{Ma}{Ic}(x\dot{y}-y\dot{x})+\gamma, \end{aligned}\right\} \qquad \text{(5)}$$

where α, β, γ are arbitrary constants. Substituting these values of p, q, r in (3), we find after reduction

$$\left.\begin{aligned} \left(1+\frac{Ma^2}{I}\right)\dot{x}=\frac{a}{c}(\beta z-\gamma y), \quad \left(1+\frac{Ma^2}{I}\right)\dot{y}=\frac{a}{c}(\gamma x-\alpha z),\\ \left(1+\frac{Ma^2}{I}\right)\dot{z}=\frac{a}{c}(\alpha y-\beta x). \end{aligned}\right\} \qquad \text{(6)}$$

Hence $$\alpha\dot{x}+\beta\dot{y}+\gamma\dot{z}=0, \qquad \text{(7)}$$

and therefore $$\alpha x+\beta y+\gamma z=\text{const.} \qquad \text{(8)}$$

The path of G is accordingly the circle in which the plane (8) meets the spherical surface of radius c.

If we take the axis of z normal to the plane of the path we have $\alpha=0$, $\beta=0$, and

$$\left(1+\frac{a^2}{\kappa^2}\right)\ddot{x}=-\frac{\gamma a}{c}\dot{y}, \quad \left(1+\frac{a^2}{\kappa^2}\right)\ddot{y}=\frac{\gamma a}{c}\dot{x}. \quad \text{..............(9)}$$

These may be treated by the ordinary methods applicable to simultaneous linear equations; but it is neater to take advantage of the special form, and to unite them in the equation

$$\left(1+\frac{a^2}{\kappa^2}\right)\ddot{\zeta}=\frac{i\gamma a}{c}\dot{\zeta}, \quad \text{..............................(10)}$$

where $i=\sqrt{(-1)}$, and

$$\zeta=x+iy. \quad \text{....................................(11)}$$

The solution is

$$\zeta=He^{i(\lambda t+\epsilon)}, \quad \text{...............................(12)}$$

where $He^{i\epsilon}$ is the (complex) arbitrary constant, and

$$\lambda=\frac{\gamma\kappa^2 a}{(\kappa^2+a^2)c}. \quad \text{..................................(13)}$$

Separating the real and 'imaginary' parts, we have

$$x=H\cos(\lambda t+\epsilon), \quad y=H\sin(\lambda t+\epsilon). \quad \text{.................(14)}$$

The circular path of G is therefore described with the constant angular velocity λ, which depends on the initial circumstances.

41. Sphere Spinning on a Fixed Sphere, under Gravity.

If in the preceding question gravity be taken into account, the difficulties become even greater than in the problem of the spherical pendulum [D. 103]. An approximate solution may however be given of the case where the moveable sphere is supposed to have been slightly disturbed from an equilibrium position at the highest point of a convex (or the lowest point of a concave) spherical surface, where it was spinning about a vertical axis with a given angular velocity n.

We take the axis of z vertically upwards. In the earlier phases, at all events, of the disturbed motion, the quantities x, y, p, q, X, Y will be small, whilst

$$z=c, \quad r=n, \quad Z=-Mg, \quad \text{.........................(1)}$$

approximately. Hence, neglecting small quantities of the second order, the equations (1) and (2) of Art. 40, with the addition of terms representing the effect of gravity, give

$$M\ddot{x}=X, \quad M\ddot{y}=Y, \quad \text{..............................(2)}$$

$$I\dot{p}=aY-\frac{Mga}{c}y, \quad I\dot{q}=-aX+\frac{Mga}{c}x. \quad \text{.................(3)}$$

The kinematical relations reduce to

$$\dot{x}=aq-\frac{na}{c}y, \quad \dot{y}=-ap+\frac{na}{c}x. \quad \text{......................(4)}$$

Eliminating X, Y, p, q, we find

$$\left.\begin{aligned}(\kappa^2+a^2)\ddot{x}+\frac{\kappa^2 a}{c}n\dot{y}-\frac{ga^2}{c}x=0,\\ (\kappa^2+a^2)\ddot{y}-\frac{\kappa^2 a}{c}n\dot{x}-\frac{ga^2}{c}y=0.\end{aligned}\right\} \quad\text{.....................(5)}$$

These combine into the equation

$$(\kappa^2+a^2)\ddot{\zeta}-i\frac{\kappa^2 a}{c}n\dot{\zeta}-\frac{ga^2}{c}\zeta=0, \quad\text{.....................(6)}$$

where $\zeta=x+iy$, as before.

Assuming that ζ varies as $e^{i\lambda t}$, we have

$$(\kappa^2+a^2)\lambda^2-\frac{\kappa^2 a}{c}n\lambda+\frac{ga^2}{c}=0. \quad\text{.....................(7)}$$

For periodic solutions, the values of λ must be real. This requires

$$n^2>\frac{4gc(\kappa^2+a^2)}{\kappa^4}. \quad\text{.....................(8)}$$

If the angular velocity of a sphere spinning about a vertical axis on the top of a fixed sphere exceed the limit thus indicated, the position is in a sense stable. In the case of a sphere spinning in a spherical bowl, at the lowest point, the sign of c must be reversed, and the condition of stability is therefore always fulfilled.

If the roots of (7), supposed real, are λ_1, λ_2, the complete solution of (6) has the form

$$\left.\begin{aligned}x=H_1\cos(\lambda_1 t+\epsilon_1)+H_2\cos(\lambda_2 t+\epsilon_2),\\ y=H_1\sin(\lambda_1 t+\epsilon_1)+H_2\sin(\lambda_2 t+\epsilon_2),\end{aligned}\right\} \quad\text{.................(9)}$$

where the constants H_1, H_2, ϵ_1, ϵ_2 are arbitrary. The disturbed path of G, approximately plane, is therefore an epicyclic curve. If c be positive in (7), λ_1 and λ_2 have the same sign, and the epicyclic is 'direct.' In the opposite case λ_1 and λ_2 have contrary signs, and the epicyclic is 'retrograde' [D. 23].

When the roots of (7) are imaginary they will be of the forms $\mu\pm i\nu$. The values of x and y will accordingly be made up of terms of the types $e^{\pm\nu t}\cos\mu t$, $e^{\pm\nu t}\sin\mu t$. The terms with the upper sign in the exponentials will increase continually in importance until the approximations on which the investigation is based cease to be valid. Imaginary values of λ thus indicate instability.

It is to be remarked, however, that the stability which has been attributed, under the condition (8), to a sphere spinning on the top of a spherical surface is of a qualified kind. Unlike statical stability, it is impaired by the operation of dissipative forces, however slight. The influence of friction in gradually diminishing the spin until it falls below the limit indicated by (8) is obvious enough, but there is a practical instability independent of this cause.

Suppose, for instance, that there are frictional forces, proportional to the velocity, acting in lines through the centre of the sphere, and therefore not directly affecting the rotation. The equations corresponding to (5) will be of the type

$$\left.\begin{aligned}\ddot{x}+k\dot{x}+\beta\dot{y}-m^2x=0,\\ \ddot{y}+k\dot{y}-\beta\dot{x}-m^2y=0,\end{aligned}\right\} \quad\text{.....................(10)}$$

combining into the single equation

$$\ddot{\zeta}+k\dot{\zeta}-i\beta\dot{\zeta}-m^2\zeta=0. \quad \text{..........................(11)}$$

Assuming that ζ varies as $e^{i\lambda t}$ we have

$$\lambda^2-ik\lambda-\beta\lambda+m^2=0, \quad \text{..........................(12)}$$

or, as we may write it,

$$(\lambda-\lambda_1)(\lambda-\lambda_2)=ik\lambda, \quad \text{..........................(13)}$$

where λ_1, λ_2 are the roots of

$$\lambda^2-\beta\lambda+m^2=0, \quad \text{..........................(14)}$$

assumed to be real.

It will be sufficient to consider the case where the frictional coefficient k is small compared with λ_1 or λ_2. The approximate roots of (13) are then

$$\lambda_1+\frac{ik\lambda_1}{\lambda_1-\lambda_2}, \quad \lambda_2+\frac{ik\lambda_2}{\lambda_2-\lambda_1}. \quad \text{..........................(15)}$$

In the question in view, λ_1 and λ_2 have the same sign, so that the imaginary part of *one* of the roots is negative. Hence when the values of x and y are expressed in real form one term in each will involve an exponential factor increasing indefinitely with t. Hence, one of the two superposed circular vibrations represented by (9) gradually dies out, whilst the other continually increases in amplitude. The sphere therefore tends to fall further and further away from its equilibrium position in an ever widening spiral path.

In the other problem, of a sphere spinning near the bottom of a spherical bowl, it will be found on examination that *both* circular vibrations gradually dwindle.

The distinction here illustrated between 'temporary' and 'practical' stability of motion will present itself in various problems to be considered later (Art. 98).

42. Equation of Energy.

If we were to assume that a rigid body may be regarded as a system of discrete particles whose mutual distances are invariable, the theorem that the increase of the kinetic energy in any interval of time is equal to the total work done by the *external* forces would follow at once, since the internal forces do no work [S. 50]. If on the other hand we adopt d'Alembert's principle as our basis, or merely postulate the principles of linear and angular momentum (Art. 37), a formal proof is logically necessary.

The expression for the kinetic energy may be written

$$T=\tfrac{1}{2}M_0(u^2+v^2+w^2)+\tfrac{1}{2}\Sigma\{m(\dot{\alpha}^2+\dot{\beta}^2+\dot{\gamma}^2)\}, \quad \text{......(1)}$$

where (u, v, w) is the velocity of the mass-centre, and α, β, γ are coordinates relative to this point. The first part represents the energy of the whole mass M_0 supposed concentrated at the mass-centre, and moving with this point, whilst the second part represents the energy of the relative motion (Art. 32). The equations of motion may be written

$$\left.\begin{array}{lll} M_0\dot{u} = X, & M_0\dot{v} = Y, & M_0\dot{w} = Z, \\ \Sigma\{m(\beta\ddot{\gamma} - \gamma\ddot{\beta})\} = L, & \Sigma\{m(\gamma\ddot{\alpha} - \alpha\ddot{\gamma})\} = M, & \Sigma\{m(\alpha\ddot{\beta} - \beta\ddot{\alpha})\} = N. \end{array}\right\} \quad \text{......(2)}$$

If we multiply these by u, v, w, p, q, r, respectively, (p, q, r) denoting the angular velocity, we get

$$M_0(u\dot{u} + v\dot{v} + w\dot{w}) + \Sigma[m\{(q\gamma - r\beta)\ddot{\alpha} + (r\alpha - p\gamma)\ddot{\beta} + (p\beta - q\alpha)\}\ddot{\gamma}]$$
$$= Xu + Yv + Zw + Lp + Mq + Nr. \quad \text{......(3)}$$

Since

$$q\gamma - r\beta = \dot{\alpha},\ (r\alpha - p\gamma) = \dot{\beta},\ p\beta - q\alpha = \dot{\gamma}, \quad \text{.........(4)}$$

by Art. 28, this is equivalent to

$$\frac{dT}{dt} = Xu + Yv + Zw + Lp + Mq + Nr. \quad \text{.........(5)}$$

The right-hand member, when multiplied by δt, gives the work done by the external forces in the time δt. Hence, integrating over any finite interval, we obtain the proposed theorem.

If there are no external forces, $dT/dt = 0$, and the kinetic energy is constant.

We notice further that the two parts of the kinetic energy are affected independently. Thus

$$\frac{d}{dt}\cdot\tfrac{1}{2}M(u^2 + v^2 + w^2) = Xu + Yv + Zw, \quad \text{.........(6)}$$

i.e. the energy of translation (as we may term it) is affected only by the external forces imagined transferred to the mass-centre. Again

$$\frac{d}{dt}\cdot\tfrac{1}{2}\Sigma\{m(\dot{\alpha}^2 + \dot{\beta}^2 + \dot{\gamma}^2)\} = Lp + Mq + Nr, \quad \text{.........(7)}$$

i.e. the energy of rotation is affected only by such external forces as have moment about the mass-centre. If there are no such forces, the energy of rotation is constant.

43. Time-Integrals.

The following statements follow immediately from the principles of momentum as enunciated in Art. 37.

1°. In any interval of time the total momentum of a system, resolved in any fixed direction, is increased by an amount equal to the time-integral of all the external forces similarly resolved;

2°. The total angular momentum of the system about any fixed axis is increased by an amount equal to the time-integral of the moment of the external forces about that axis.

More formally, if we integrate each member of the equations (1) and (2) of Art. 37, with respect to t, between limits t_0 and t_1, we obtain

$$\Big[\Sigma(m\dot{x})\Big]_{t_0}^{t_1} = \int_{t_0}^{t_1} X dt,\quad \Big[\Sigma(m\dot{y})\Big]_{t_0}^{t_1} = \int_{t_0}^{t_1} Y dt,\quad \Big[\Sigma(m\dot{z})\Big]_{t_0}^{t_1} = \int_{t_0}^{t_1} Z dt, \quad\ldots\ldots(1)$$

$$\left.\begin{aligned}\Big[\Sigma\{m(y\dot{z}-z\dot{y})\}\Big]_{t_0}^{t_1} = \int_{t_0}^{t_1} L dt,\quad \Big[\Sigma\{m(z\dot{x}-x\dot{z})\}\Big]_{t_0}^{t_1} = \int_{t_0}^{t_1} M dt,\\ \Big[\Sigma\{m(x\dot{y}-y\dot{x})\}\Big]_{t_0}^{t_1} = \int_{t_0}^{t_1} N dt.\end{aligned}\right\}\quad(2)$$

Again, the angular momentum of the motion relative to the mass-centre (G), about any axis through G which is fixed in direction, is increased by an amount equal to the time-integral of the moment of the external forces about this (moving) axis.

This follows, analytically, by integration of the equations (7) of Art. 37; thus

$$\left.\begin{aligned}\Big[\Sigma\{m(\beta\dot{\gamma}-\gamma\dot{\beta})\}\Big]_{t_0}^{t_1} = \int_{t_0}^{t_1} L dt,\quad \Big[\Sigma\{m(\gamma\dot{\alpha}-\alpha\dot{\gamma})\}\Big]_{t_0}^{t_1} = \int_{t_0}^{t_1} M dt,\\ \Big[\Sigma\{m(\alpha\dot{\beta}-\beta\dot{\alpha})\}\Big]_{t_0}^{t_1} = \int_{t_0}^{t_1} N dt.\end{aligned}\right\}\quad(3)$$

In particular, if the system is free from external force, the vector which represents the angular momentum of the relative motion is constant in direction and magnitude.

44. Motion due to an Instantaneous Impulse.

We may apply the preceding results to the case of instantaneous impulses acting on a rigid body. The external forces are here supposed, by a mathematical idealization, to be infinitely great, and their time of action to be infinitely short, in such a way that their time-integrals are finite.

We write, then, for brevity,

$$\left.\begin{aligned}\xi=\int_{t_0}^{t_1}X\,dt,\quad \eta=\int_{t_0}^{t_1}Y\,dt,\quad \zeta=\int_{t_0}^{t_1}Z\,dt,\\ \lambda=\int_{t_0}^{t_1}L\,dt,\quad \mu=\int_{t_0}^{t_1}M\,dt,\quad \nu=\int_{t_0}^{t_1}N\,dt,\end{aligned}\right\}\quad\text{.........(1)}$$

where the interval t_1-t_0 is supposed in the limit to be infinitely small. Consequently, we neglect the change of position of the body during the impulse, since the velocities remain finite. The six quantities ξ, η, ζ, λ, μ, ν may be called the components of the 'impulse' on the body.

Referring to Art. 31 for the components of momentum we have

$$M(u_1-u_0)=\xi,\quad M(v_1-v_0)=\eta,\quad M(w_1-w_0)=\zeta,\quad\text{...(2)}$$

$$\left.\begin{aligned}\Big[Ap-Hq-Gr\Big]_{t_0}^{t_1}=\lambda,\quad \Big[-Hp+Bq-Fr\Big]_{t_0}^{t_1}=\mu,\\ \Big[-Gp-Fq+Cr\Big]_{t_0}^{t_1}=\nu.\end{aligned}\right\}\text{...(3)}$$

If we adopt the principal axes of inertia at the mass-centre as axes of reference, the latter equations reduce to

$$A(p_1-p_0)=\lambda,\quad B(q_1-q_0)=\mu,\quad C(r_1-r_0)=\nu.\quad\text{...(4)}$$

For example, considering any given state of motion of a rigid body, as defined by the six quantities u, v, w, p, q, r, the components of the impulsive wrench by which this state would be generated instantaneously from rest are

$$\xi=Mu,\quad \eta=Mv,\quad \zeta=Mw,\quad\text{..................(5)}$$

$$\lambda=Ap,\quad \mu=Bq,\quad \nu=Cr.\quad\text{..................(6)}$$

The equations (6) shew that when a body is started by an impulsive couple (λ, μ, ν), the initial axis of rotation is that diameter of the momental ellipsoid which is conjugate to the plane of the couple (Art. 30). If H be the magnitude of the couple, and therefore of the resulting angular momentum, the component of angular momentum about the instantaneous axis will be $\varpi/\rho\,.\,H$, where ρ is the radius of the momental ellipsoid in the direction of the instantaneous axis, and ϖ the perpendicular from the centre on the tangent plane at its extremity. This component must be equal to $I\omega$, where ω is the initial angular velocity, and I the moment of inertia about the instantaneous axis. Hence

$$\omega = \frac{\varpi}{\rho} \cdot \frac{H}{I}, \quad \text{.........................(7)}$$

or, since $I = M\epsilon^4/\rho^2$ in the notation of Art. 25,

$$\omega = \frac{H\varpi\rho}{M\epsilon^4}. \quad \text{.........................(8)}$$

The kinetic energy generated by the impulse is

$$T = \tfrac{1}{2}I\omega^2 = \tfrac{1}{2}\frac{H^2\varpi^2}{I\rho^2} = \tfrac{1}{2}\frac{H^2\varpi^2}{M\epsilon^4} \quad \text{.........................(9)}$$

For a given value of H this is least when the axis of the couple is the shortest axis of the momental ellipsoid, i.e. the axis of greatest principal moment.

Ex. A body which is free only to rotate about a fixed axis through the mass-centre, whose direction-cosines relative to the principal axes at this point are (l, m, n), is struck by a given impulsive couple (λ, μ, ν).

Taking moments about the axis in question, we have

$$I\omega = \lambda l + \mu m + \nu n, \quad \text{.............................(10)}$$

where

$$I = Al^2 + Bm^2 + Cn^2. \quad \text{.........................(11)}$$

Thus

$$\omega = \frac{\lambda l + \mu m + \nu n}{Al^2 + Bm^2 + Cn^2}. \quad \text{.........................(12)}$$

The kinetic energy generated is given by

$$2T = I\omega^2 = \frac{(\lambda l + \mu m + \nu n)^2}{Al^2 + Bm^2 + Cn^2}. \quad \text{.........................(13)}$$

The constraining forces exerted by the fixed axis form an impulsive couple whose components are

$$Al\omega - \lambda, \quad Bm\omega - \mu, \quad Cn\omega - \nu. \quad \text{.........................(14)}$$

If the body had been perfectly free, the component angular velocities would have been determined by (6). In terms of these, the kinetic energy would have been given by

$$2T' = Ap^2 + Bq^2 + Cr^2. \quad \text{.........................(15)}$$

Hence

$$2(T' - T) = Ap^2 + Bq^2 + Cr^2 - \frac{(Alp + Bmq + Cnr)^2}{Al^2 + Bm^2 + Cn^2}$$

$$= \frac{BC(nq - mr)^2 + CA(lr - np)^2 + AB(mp - lq)^2}{Al^2 + Bm^2 + Cn^2}. \quad \text{......(16)}$$

This vanishes, as it ought, when $l : m : n = p : q : r$; but is in all other cases positive. The constraint has therefore the effect of diminishing the energy due to a given impulse. This example is due to Euler. It is a particular case of a very general theorem, to be noticed later (Arts. 75, 77).

45. Energy due to an Impulse.

In the general case the kinetic energy generated by an impulse is

$$\begin{aligned} T_1 - T_0 &= \tfrac{1}{2}M\{(u_1^2 - u_0^2) + (v_1^2 - v_0^2) + (w_1^2 - w_0^2)\} \\ &\quad + \tfrac{1}{2}\{A(p_1^2 - p_0^2) + B(q_1^2 - q_0^2) + C(r_1^2 - r_0^2)\} \\ &= \xi \,.\, \tfrac{1}{2}(u_1 + u_0) + \eta \,.\, \tfrac{1}{2}(v_1 + v_0) + \zeta \,.\, \tfrac{1}{2}(w_1 + w_0) \\ &\quad + \lambda \,.\, \tfrac{1}{2}(p_1 + p_0) + \mu \,.\, \tfrac{1}{2}(q_1 + q_0) + \nu \,.\, \tfrac{1}{2}(r_1 + r_0), \ldots(1) \end{aligned}$$

by Art. 44 (5), (6). As in the case of two dimensions [D. 73], we can give to this result a simple interpretation. If F be any one of the forces constituting the impulse, and v_0, v_1 the initial and final velocities of any point on its line of action, resolved in the direction of this line, the expression in the last member of (1) is equivalent to

$$\Sigma F \,.\, \tfrac{1}{2}(v_0 + v_1), \quad \ldots\ldots\ldots\ldots\ldots\ldots\ldots\ldots(2)$$

where the summation embraces all the impulsive forces. For a system of *finite* forces proportional to F would be equivalent to a wrench whose components are proportional (on the same scale) to ξ, η, ζ, λ, μ, ν. And an infinitesimal twist whose components are proportional to

$$\tfrac{1}{2}(u_1 + u_0),\ \tfrac{1}{2}(v_1 + v_0),\ \tfrac{1}{2}(w_1 + w_0),$$
$$\tfrac{1}{2}(p_1 + p_0),\ \tfrac{1}{2}(q_1 + q_0),\ \tfrac{1}{2}(r_1 + r_0)$$

would involve a displacement of the point of application of F, the resolved part of which in the direction of F would be $\tfrac{1}{2}(v_0 + v_1)$, since the relations between the velocities of different points of the body are linear. The equivalence of (2) to the last member of (1) then follows from Art. 21.

EXAMPLES. IX.

1. Prove that if a particle move under gravity on a smooth cylindrical surface (of any form of section) whose generators are vertical, the path when developed into a plane is a parabola. (Euler.)

2. The bob of a simple pendulum of length l is projected horizontally with velocity v at right angles to the string, which makes an angle α with the vertical. Prove that the bob will rise or fall according as

$$v^2 \gtrless gl \sin^2 \alpha / \cos \alpha.$$

3. A particle is constrained to move on a smooth surface having the form of the anchor ring

$$x = \rho \cos \psi, \quad y = \rho \sin \psi, \quad z = b \sin \theta,$$

where
$$\rho = a + b \cos \theta.$$

Prove that the component accelerations in the tangent plane, tangential and perpendicular to the cross-section, are

$$b\ddot{\theta}+\rho\sin\theta\dot{\psi}^2, \quad \text{and} \quad \frac{1}{\rho}\frac{d}{dt}(\rho^2\dot{\psi}),$$

respectively, and that the acceleration in the direction of the inward normal is

$$b\dot{\theta}^2+\rho\cos\theta\dot{\psi}^2.$$

4. Prove that, if there are no forces except the reaction of the surface, motion along the outer equatorial circle is stable, and that if this motion is slightly disturbed the path will intersect this circle at intervals

$$\pi\sqrt{\{b(a+b)\}}.$$

Prove also that motion along the inner equatorial circle is unstable, and that if this motion is slightly disturbed the path will intersect the outer equatorial circle at an angle

$$\cos^{-1}\left(\frac{a-b}{a+b}\right).$$

5. Prove that if a uniform string of line-density μ be pulled with constant velocity v through a smooth twisted tube under a tension μv^2 the pressure on the tube will vanish; and thence that a wave of any form can be propagated along an otherwise straight string with the velocity $\sqrt{(T/\mu)}$, where T is the tension.

6. A circular plate is spun by means of a stick supporting it at an eccentric point which is made to describe a horizontal circle. Prove that when a state of steady motion has been attained the inclination θ of the plate to the horizontal is given by

$$\sin\theta=\frac{Mga}{A\omega^2},$$

where a is the distance of the point of support from the centre, ω is the angular velocity about the vertical, and A is the moment of inertia about a diameter. (It is assumed that the centre of the disk is at rest.)

How is the problem modified if the centre of the disk describes a horizontal circle of given radius?

7. A sphere is set rolling on a horizontal plane which is made to rotate about a fixed vertical axis with the constant angular velocity ω. Prove that the path of the centre of the sphere is a circle described with the angular velocity

$$\frac{\kappa^2\omega}{a^2+\kappa^2},$$

where a is the radius, and κ the radius of gyration about a diameter.

8. A sphere rolls on an inclined plane which rotates about an axis normal to it with constant angular velocity ω. Prove that the centre of the sphere describes a trochoid, the average velocity being $ga^2\sin\alpha/\kappa^2\omega$ in a horizontal direction, where a is the radius of the sphere, κ its radius of gyration about a diameter, and α the inclination of the plane to the horizontal.

9. A rigid body is free to rotate about a fixed point O; the coordinates of its mass-centre relative to the principal axes at O are a, b, c; and the direction-cosines of the vertical are l, m, n. Prove that it can rotate in steady motion about the vertical provided

$$(B-C)\frac{a}{l}+(C-A)\frac{b}{m}+(A-B)\frac{c}{n}=0.$$

Find the requisite angular velocity. (Staude.)

10. A solid of mass M is held so as to be in contact at one point with a smooth horizontal plane, and is then released. Prove that the initial pressure (R) on the plane is given by

$$\frac{Mg}{R}=1+\frac{M(ny-mz)^2}{A}+\frac{M(lz-nx)^2}{B}+\frac{M(mx-ly)^2}{C},$$

where (x, y, z) are the coordinates of the points of contact, and (l, m, n) is the direction of gravity, relative to the principal axes at the mass-centre.

11. Under what condition can the motion of a free rigid body be arrested by a single impulsive *force*?

[The velocity of the mass-centre G must be at right angles to the axis of angular momentum at G.]

12. If two bodies rotating independently about the same fixed axis with angular velocities ω_1, ω_2 be suddenly connected rigidly together, the loss of kinetic energy is

$$\frac{1}{2}\frac{I_1 I_2}{I_1+I_2}(\omega_1-\omega_2)^2,$$

where I_1, I_2 are the respective moments of inertia about the axis.

13. A chain of equal uniform links is laid out in a straight line, and is set in motion by given impulses in one plane, at right angles to the length. If u_n be the initial velocity of the nth joint, prove that in a part of the chain not subject to the external impulses

$$u_{n+1}+4u_n+u_{n-1}=0.$$

14. A uniform rectangular plate receives a blow normal to its plane at a corner. Prove that it will begin to rotate about the line whose equation, relative to axes through the centre parallel to the edges $(2a, 2b)$, is

$$\frac{3x}{a}+\frac{3y}{b}+1=0,$$

the blow being given at the point (a, b).

15. A uniform triangular plate ABC is free to turn about C as a fixed point. A blow is struck at B normal to the plane of the triangle. Prove that the initial axis of rotation is that trisector of the side AB which is furthest from B.

16. A rectangular block at rest receives a blow along an edge ($2a$). Prove that the initial motion is a rotation about an axis whose equations, referred to axes through the centre parallel to the edges ($2a$, $2b$, $2c$), are

$$x=0, \quad \frac{3by}{a^2+b^2}+\frac{3cz}{a^2+c^2}+1=0,$$

the line of the blow being $y=b$, $z=c$.

17. Two wheels of radii a_1, a_2, and moments of inertia I_1, I_2, are rotating freely about parallel axes with angular velocities ω_1, ω_2. If they be suddenly put in gear with one another, find the subsequent angular velocities, and prove that the loss of energy is

$$\frac{I_1 I_2 (\omega_1 a_1+\omega_2 a_2)^2}{2(I_1 a_2{}^2+I_2 a_1{}^2)}.$$

18. A uniform solid cube is spinning freely about a diagonal, when one of the edges which does not meet that diagonal suddenly becomes fixed. Prove that the energy is diminished in the ratio 1 : 12.

19. An impulse of unit amount is given to a free rigid body at the point (x, y, z) in the direction (l, m, n); prove that the velocity generated at the point (x', y', z') in the direction (l', m', n') is

$$\frac{ll'+mm'+nn'}{M}+\frac{1}{ABC}\{(All'+Bmm'+Cnn')(Axx'+Byy'+Czz')$$
$$-(Al'x+Bm'y+Cn'z)(Alx'+Bmy'+Cnz')\},$$

the axes of coordinates being the principal axes at the centre of mass.

20. Three particles subject to any laws of mutual attraction start from rest; prove that at any subsequent instant the tangents to their paths are concurrent.

21. A solid is immersed in a homogeneous incompressible fluid which is bounded externally by fixed rigid walls. Everything being at rest, the solid is suddenly set in motion with velocity V. Prove that the momentum acquired by the fluid is $-M'V$, where M' is the mass of fluid displaced by the solid.

CHAPTER VII

FREE ROTATION OF A RIGID BODY

46. Poinsot's Construction.

We have seen that in any dynamical system which is free from external force the vector which represents the angular momentum with respect to the mass-centre (G) is fixed in direction and magnitude. A line drawn through G in this direction is called the 'invariable line' of the system, and a plane through G at right angles to it is called the 'invariable plane.'

Thus, the configuration and motion of all the members of the solar system, if given at any instant, determine a fixed plane which will stand in the same dynamical relation to the system for all time, or at all events so long as the attractions of the distant stars can be neglected. This invariable plane has been proposed by Laplace and others as a fixed plane of reference for 'secular' purposes, in preference to the plane of the ecliptic (i.e. of the earth's orbit), which is subject to continual slight perturbations.

The same conclusion as to the invariability in all respects of the angular momentum relative to the mass-centre will hold even if there are external forces, provided these have zero moment about G. It would apply, for instance, to the case of a swarm of particles subject to gravity, if the resistance of the air could be neglected.

We proceed to consider the motion of a rigid body relative to the mass-centre, under the condition just stated. The same investigation would apply also to the case of a body rotating about a fixed point O other than the mass-centre, if the external forces (including the constraining forces) had zero moment about O. This case is however hardly a practical one, since gravity would have to be ignored.

It has been pointed out in Art. 42 that on the present assumption the energy of rotation is constant. Denoting it by T, we have

$$2T = I\omega^2 = M\epsilon^4 \cdot \frac{\omega^2}{\rho^2}, \qquad \text{.....................(1)}$$

where ω is the angular velocity, I the moment of inertia about the instantaneous axis, ρ the corresponding radius (OJ) of the momental ellipsoid

$$Ax^2 + By^2 + Cz^2 = M\epsilon^4. \quad \text{.................}(2)$$

Hence ω varies as ρ.

Again, the tangent plane to the momental ellipsoid at J is perpendicular to the invariable line, by Art. 30, and is therefore fixed in direction. Also, if ϖ be its distance from the centre, and H the resultant angular momentum, we have, taking moments of momentum about OJ,

$$H\frac{\varpi}{\rho} = I\omega = \frac{M\epsilon^4\omega}{\rho^2}, \quad \text{....................}(3)$$

or

$$\varpi = \frac{M\epsilon^4}{H}\cdot\frac{\omega}{\rho}. \quad \text{.........................}(4)$$

Since H is constant, whilst ω varies as ρ, ϖ is constant. The tangent plane at J is therefore fixed.

Hence we obtain a complete mental representation of the successive phases of the motion if we imagine the momental ellipsoid, carrying the body with it, to roll (with its centre fixed) in contact with a fixed plane, with an angular velocity proportional at each instant to the radius OJ drawn to the point of contact. This remarkable theorem is due to Poinsot*.

Since the vector OJ represents on a certain scale the angular velocity of the body, the curve traced out by J on the ellipsoid is the 'polhode' curve of Art. 29, and that traced out by J on the fixed plane is the 'herpolhode'†. The cones described by OJ in the solid and in space have similar designations.

47. Case of Kinetic Symmetry.

Before proceeding further with the general problem, we consider the very important case where two of the principal moments of inertia at O are equal, say $A = B$, and the momental ellipsoid is accordingly of revolution. The cones of the polhode and herpolhode are therefore circular, and the angular velocity ω is constant. The motion is therefore of the type called 'precessional' (Art. 29).

* *Théorie nouvelle de la rotation des corps*, Paris, 1834.

† Since OJ has evidently one maximum and one minimum value, the herpolhode curve touches alternately two fixed circles between which it lies. It has been shewn by de Sparre, Darboux, and others that it has no points of inflexion, contrary to an impression of Poinsot. The proof depends on the fact that the sum of any two of the principal moments of inertia is greater than the third (Art. 25).

We denote by β the semi-angle of the polhode cone, and by α the angle which the axis of kinetic symmetry (OC) makes with the invariable line (OZ). Taking components of angular momentum about OC, and about an axis at right angles to it in the plane ZOC, we have

$$C\omega \cos \beta = H \cos \alpha, \quad A\omega \sin \beta = H \sin \alpha, \quad \text{.........(1)}$$

whence

$$\tan \alpha = \frac{A}{C} \tan \beta. \quad \text{.........................(2)}$$

Hence $\alpha \gtrless \beta$ according as $A \gtrless C$. The two cases are represented in the annexed figures. The relations between the two cones are of the types shewn in Figs. 27 and 28, respectively (p. 76).

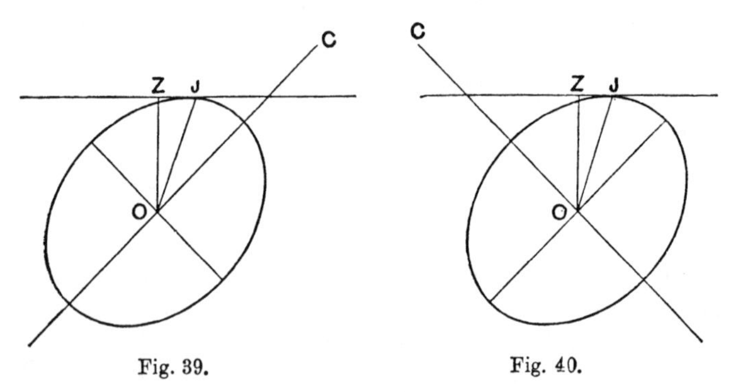

Fig. 39. Fig. 40.

If $\dot{\psi}$ be the rate at which the plane ZOC revolves about OZ, and $\dot{\chi}$ the rate at which J describes the polhode circle in the body, we have, by Art. 29,

$$\dot{\psi} = \frac{\sin \beta}{\sin \alpha}\, \omega, \quad \dot{\chi} = -\frac{\sin(\alpha - \beta)}{\sin \alpha}\, \omega. \quad \text{...........(3)}$$

If we denote by n the constant angular velocity of the body about the axis of kinetic symmetry, viz.

$$n = \omega \cos \beta, \quad \text{.........................(4)}$$

we have from (1)

$$\dot{\psi} = \frac{H}{A} = \frac{Cn}{A \cos \alpha}, \quad \text{.......................(5)}$$

and

$$\dot{\chi} = \frac{\tan \beta - \tan \alpha}{\tan \alpha}\, n = \frac{C - A}{A}\, n. \quad \text{..............(6)}$$

The point J completes the circuit of the polhode in the period $2\pi/\dot{\chi}$.

If the instantaneous axis deviates only slightly from the axis of symmetry the angles α, β are small, so that $\alpha/\beta = A/C$, and

$$\dot{\psi} = \frac{C}{A}\omega, \quad \dot{\chi} = \frac{C-A}{A}\omega, \quad \text{.....................(7)}$$

approximately.

In the case of the earth, it is inferred from the independent phenomenon of the Precession of the Equinoxes that

$$(C-A)/A = \cdot 00327,$$

about. Hence if the axis of rotation of the earth (supposed absolutely rigid) were not quite coincident with the axis of figure, it would describe a cone round the latter in about 306 sidereal days*. Since $C > A$, the circumstances are those of Figs. 28 and 40, except that the angle $\beta - \alpha$ of the herpolhode cone is now very small compared with that of the polhode cone (β). The annexed Fig. 41 is intended

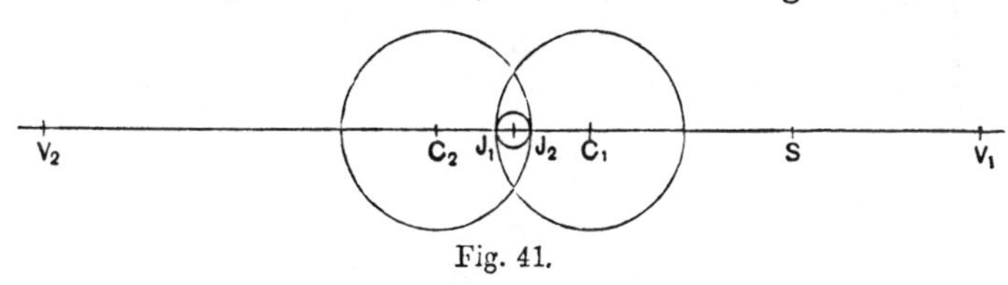

Fig. 41.

to shew the intersections of the polhode cone with the celestial sphere, in two opposite positions, but it is of course impossible to reproduce the true proportions. The point S is supposed to represent a fixed circumpolar star, and V_1, V_2 are the two positions of the zenith of an observer, when at its least and greatest distances from S. The mean of the two zenith distances V_1S, V_2S is usually taken as the colatitude of the place, but we see that in strictness it gives the distance of the zenith from the *invariable line* (OZ). Since this line describes in the body a cone about the axis of figure in the aforementioned period, the result of the want of coincidence between the axis of rotation and the axis of symmetry would be a periodic variation in the observational latitudes of all places on the earth's surface.

There is in fact some evidence of a periodic variation of latitude of very minute amount (one or two-tenths of a second), but the period which fits best with the observations appears to be about

* This is known as the 'Eulerian nutation.' The determination of its actual amount in the case of the earth, by observation of the effect on latitude, was suggested by Maxwell (1857).

427 days*. The discrepancy is attributed to a defect of rigidity in the earth (see Art. 71).

48. The Invariable and Polhode Cones.

We return to the general case where the principal moments at O are unequal. We will suppose that $A > B > C$.

The invariable line describes relatively to the body a certain cone. To find its equation, referred to the principal axes at O, we note that

$$Ap^2 + Bq^2 + Cr^2 = 2T, \qquad (1)$$

$$A^2p^2 + B^2q^2 + C^2r^2 = H^2. \qquad (2)$$

Hence
$$A^2p^2 + B^2q^2 + C^2r^2 = \frac{H^2}{2T}(Ap^2 + Bq^2 + Cr^2). \qquad (3)$$

At a point on the 'invariable cone,' as it is called, we have

$$x : y : z = Ap : Bq : Cr, \qquad (4)$$

and therefore

$$\left(1 - \frac{H^2}{2AT}\right) x^2 + \left(1 - \frac{H^2}{2BT}\right) y^2 + \left(1 - \frac{H^2}{2CT}\right) z^2 = 0, \qquad (5)$$

which is the required equation.

From (1) and (2) we have

$$\left.\begin{aligned} 2AT - H^2 &= B(A - B)q^2 + C(A - C)r^2, \\ 2BT - H^2 &= C(B - C)r^2 + A(B - A)p^2, \\ 2CT - H^2 &= A(C - A)p^2 + B(C - B)q^2. \end{aligned}\right\} \qquad (6)$$

Hence, of the three coefficients in (5), the first is always positive, and the third is always negative, whilst the remaining coefficient may have either sign. If $H^2 = 2BT$, this coefficient vanishes, and the cone degenerates into a pair of planes through the mean axis, viz. that of B.

The configuration of the various possible forms of the invariable cone may be understood from their intersections with a unit sphere having O as centre. When viewed from a distant point on the axis of x these curves present the appearance of a system of similar ellipses The appearance from a distant point on Oz has the same character. But when viewed from a distant point on the mean axis Oy the appearance is that of two conjugate systems of similar hyperbolas, the asymptotes representing the two planes into which the cone breaks up in the special case

* S. C. Chandler (1891).

referred to. This latter aspect is illustrated in the figure, which is drawn for the case of

$$A : B : C = 45 : 40 : 32.$$

The lines correspond to equidistant values of the ratio $2BT/H^2$.

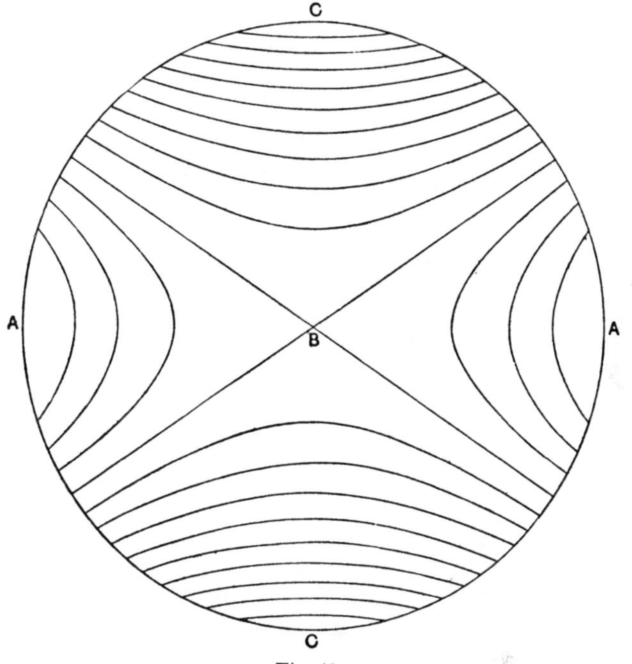

Fig. 42.

The figure enables us to judge the effect of a slight disturbance from a state of steady rotation about a principal axis. If the axis in question is that of x the invariable cone, in the disturbed motion, will closely surround it, and the axis accordingly never deviates far from the invariable line, which is fixed in space. A rotation about the axis of x is therefore reckoned as stable. The same conclusion holds with regard to rotation about Oz. But if the body be slightly disturbed from rotation about the mean axis Oy, the invariable line will in general begin by deviating further and further from this axis. A rotation about the mean axis is therefore unstable.

At a point of the *polhode* cone we have

$$x : y : z = p : q : r. \qquad \text{.....................(7)}$$

Substituting in (3) we have the equation

$$A^2\left(1-\frac{H^2}{2AT}\right)x^2+B^2\left(1-\frac{H^2}{2BT}\right)y^2+C^2\left(1-\frac{H^2}{2CT}\right)z^2=0. \quad \text{...(8)}$$

The signs of the coefficients are the same as for the invariable cone, and the configuration of the possible forms is therefore analogous.

There is another conical surface which is of some interest, viz. that described in the body by the perpendicular OK drawn to the invariable line OZ in the plane containing this line and the instantaneous axis. Since

$$I\omega = H\varpi/\rho, \quad \text{...(9)}$$

by Art. 46 (3), the component angular velocity of the body about the invariable line is

$$\omega\varpi/\rho = I\omega^2/H = 2T/H, \quad \text{...(10)}$$

and is therefore constant. The points of the body which lie in OK are therefore moving at right angles to OZ with the above constant angular velocity. This is to be distinguished of course from the variable rate at which the geometrical plane ZOJ revolves about OZ. Since OZ is perpendicular to the consecutive position of OK in the body, the locus of OK in the body is the reciprocal of the invariable cone; and its equation is therefore obtained by inverting the coefficients in (5), thus

$$\frac{A}{2AT-H^2}x^2+\frac{B}{2BT-H^2}y^2+\frac{C}{2CT-H^2}z^2=0. \quad \text{...(11)}$$

Since this cone touches the invariable plane through O, whilst the line of contact, considered as a line of the body, is revolving about O, it was called by Poinsot the 'rolling and sliding cone.' If we imagine a massless disk, free to turn about O in the invariable plane, to remain always in contact with the cone without slipping, the revolution of the disk will give a measure of the time.

49. Euler's Equations.

Many investigations on the rotation of a body about a fixed point, whether under the influence of external forces or not, are based on a remarkable system of equations due to Euler (1758), and known by his name. It has been remarked (Art. 38) that equations relative to fixed axes have the inconvenience that the coefficients of inertia are as a rule continually changing. For this reason Euler conceived the plan of using axes fixed in the body and moving with it. For greater simplicity they are taken to be the principal axes of inertia OA, OB, OC at the fixed point O.

Let Ox, Oy, Oz be a system of axes which are fixed in space, but are chosen so as to coincide with OA, OB, OC, respectively, at the instant t under consideration. The position of the latter axes after an interval δt is obtained by rotations $p\,\delta t$, $q\,\delta t$, $r\,\delta t$ about Ox, Oy, Oz, respectively; and if we neglect squares and products of small quantities, it is immaterial in what order we suppose these rotations to be effected. The rotation about Oy does not affect the position of OB, but that about Oz moves OB away from Ox through an angle $r\,\delta t$. The rotation about Ox does not affect the inclination to Ox. Hence the cosine of the angle between Ox and OB is now $-r\,\delta t$. Again, the rotation about Oz does not affect OC, whilst the rotation about Oy moves OC towards Ox through an angle $q\,\delta t$. The cosine of the angle between Ox and OC is now $q\,\delta t$. Further, the angle between OA and Ox is infinitely small. Hence the cosines of the angles which OA, OB, OC respectively make with Ox are

$$1, \quad -r\,\delta t, \quad q\,\delta t,$$

to the first order of small quantities. Since the components of angular momentum about the new positions of OA, OB, OC at the instant $t + \delta t$ are

$$A\,(p + \delta p), \quad B\,(q + \delta q), \quad C\,(r + \delta r),$$

the component about Ox is now

$$A\,(p + \delta p) + B\,(q + \delta q)(-r\,\delta t) + C\,(r + \delta r)\,q\,\delta t.$$

If L, M, N be the moments of external force about Ox, Oy, Oz, this expression must be equal to $Ap + L\,\delta t$. Hence, neglecting small quantities of the second order, and passing to the limit, we obtain the first of the following system of equations:

$$\left.\begin{aligned} A\frac{dp}{dt} - (B - C)\,qr &= L, \\ B\frac{dq}{dt} - (C - A)\,rp &= M, \\ C\frac{dr}{dt} - (A - B)\,pq &= N. \end{aligned}\right\} \qquad \text{(1)}$$

These are Euler's equations, above referred to*.

* A slightly different derivation of the equations is given in Art. 62.

If we multiply the equations by p, q, r, in order, and add, we get

$$\frac{d}{dt}\cdot\tfrac{1}{2}(Ap^2+Bq^2+Cr^2)=Lp+Mq+Nr, \quad\text{.........(2)}$$

which is the equation of energy.

An interpretation of the equations (1) has been given by Poinsot. Regarding Adp/dt, Bdq/dt, Cdr/dt as the *apparent* rates of change of angular momentum about the principal axes, and transferring the terms

$$(B-C)\,qr,\quad (C-A)\,rp,\quad (A-B)\,pq \quad\text{.....................(3)}$$

to the right-hand sides of the equations, these terms appear as the components of a fictitious 'centrifugal couple' acting on the body. They are really the components of the couple which the body would exert on its bearings if constrained to rotate with the angular velocity (p, q, r) about a fixed axis (cf. Art. 37, Ex. 2).

50. Application to Free Rotation.

When the moments of the external forces vanish, we have

$$\left.\begin{aligned} A\frac{dp}{dt} &= (B-C)\,qr,\\ B\frac{dq}{dt} &= (C-A)\,rp,\\ C\frac{dr}{dt} &= (A-B)\,pq. \end{aligned}\right\}\quad\text{.....................(1)}$$

If we multiply these by p, q, r, respectively, and add, we obtain

$$Ap\frac{dp}{dt}+Bq\frac{dq}{dt}+Cr\frac{dr}{dt}=0. \quad\text{..............(2)}$$

Again, multiplying by Ap, Bq, Cr, respectively, and adding, we find

$$A^2p\frac{dp}{dt}+B^2q\frac{dq}{dt}+C^2r\frac{dr}{dt}=0. \quad\text{..............(3)}$$

Since $$Ap^2+Bq^2+Cr^2=2T, \quad\text{....................(4)}$$

and $$A^2p^2+B^2q^2+C^2r^2=H^2, \quad\text{....................(5)}$$

in our previous notation, we verify that the kinetic energy T, and the resultant angular momentum H, are constant. It is on these facts, and on the invariability of the direction of H, that Poinsot's representation of the motion was based (Art. 46).

The equations (4) and (5) may be regarded as two integrals of the equations, with T and H^2 as arbitrary constants. To connect p, q, r with the time t, a third relation is necessary. In the general case this involves the use of elliptic functions (Art. 52).

In the case of symmetry about an axis, however, the process is simple. If we put $A = B$, the third of equations (1) shews that the angular velocity r about the axis of symmetry is constant. Denoting it by n, the remaining equations become

$$A \frac{dp}{dt} = -(C - A)\, nq, \quad A \frac{dq}{dt} = (C - A)\, np. \qquad \text{(6)}$$

These combine into

$$A \frac{d}{dt}(p + iq) = in\,(C - A)(p + iq). \qquad \text{(7)}$$

Hence, writing

$$\lambda = \frac{C - A}{A}\, n, \qquad \text{(8)}$$

we have

$$p + iq = Ke^{i(\lambda t + \epsilon)}, \qquad \text{(9)}$$

where $Ke^{i\epsilon}$ is the arbitrary constant of integration. In real form

$$p = K \cos(\lambda t + \epsilon), \quad q = K \sin(\lambda t + \epsilon). \qquad \text{(10)}$$

The polhode cone is therefore right circular, and is described in the period

$$\frac{2\pi}{\lambda} = \frac{A}{C - A} \cdot \frac{2\pi}{n}, \qquad \text{(11)}$$

as already found in Art. 47.

51. Sylvester's Theorem.

The theory of free rotation has engaged the attention of many distinguished mathematicians, from the time of Euler onwards. Their efforts have been directed for the most part to the analytical solution of the problem, rather than to the development of geometrical or dynamical conceptions which should facilitate a mental grasp of the phenomena. In this respect Poinsot's work stands alone, if we except some remarkable extensions of his results by Sylvester*.

The simplest and perhaps the most interesting of the theorems here referred to is to the effect that a homogeneous material ellipsoid, identical in size and shape with the momental ellipsoid of a given body, set rolling (with its centre fixed) on a rigid material plane in the position of Poinsot's fixed plane, would if properly started of itself keep step with the given body. That is, the directions of the principal axes, and the angular velocities about them, would continue to be identical in the two cases.

* *Phil. Trans.*, 1866; *Collected Papers*, vol. ii, p. 577.

To prove this, we inquire in the first place what forces (other than the reaction at the centre) must act on the material ellipsoid in order that it may keep step with the given body, and then proceed to shew that these forces are such as can be supplied by the reaction at the point of contact.

If A', B', C' be the principal moments of the material ellipsoid, the required forces are given by

$$\left.\begin{aligned} L &= A'\frac{dp}{dt} - (B' - C')\,qr, \\ M &= B'\frac{dq}{dt} - (C' - A')\,rp, \\ N &= C'\frac{dr}{dt} - (A' - B')\,pq, \end{aligned}\right\} \qquad \text{...........(1)}$$

where the values of p, q, r are supposed to be the same as for the given body. Substituting from Art. 50 (1), we have

$$\left.\begin{aligned} L &= \left\{\frac{A'}{A}(B - C) - (B' - C')\right\} qr, \\ M &= \left\{\frac{B'}{B}(C - A) - (C' - A')\right\} rp, \\ N &= \left\{\frac{C'}{C}(A - B) - (A' - B')\right\} pq. \end{aligned}\right\} \qquad \text{...........(2)}$$

The moment of the constraining forces about the diameter through the point of contact will vanish, provided

$$Lp + Mq + Nr = 0, \qquad \text{...................(3)}$$

or

$$\frac{A'}{A}(B - C) + \frac{B'}{B}(C - A) + \frac{C'}{C}(A - B) = 0. \qquad \text{........(4)}$$

Now the squares of the principal semi-axes of the momental ellipsoid are proportional to $1/A$, $1/B$, $1/C$, respectively, so that

$$\begin{aligned} A' : B' : C' &= \frac{1}{B} + \frac{1}{C} : \frac{1}{C} + \frac{1}{A} : \frac{1}{A} + \frac{1}{B} \\ &= A(B + C) : B(C + A) : C(A + B). \qquad \text{.........(5)} \end{aligned}$$

The condition (4) is therefore satisfied.

The components X, Y, Z of the reaction at the point of contact (x, y, z) must satisfy the relations

$$yZ - zY = L, \quad zX - xZ = M, \quad xY - yX = N. \qquad \text{.....(6)}$$

Hence,
$$\left.\begin{aligned}\rho^2 X &= zM - yN + x(xX + yY + zZ),\\ \rho^2 Y &= xN - zL + y(xX + yY + zZ),\\ \rho^2 Z &= yL - xM + z(xX + yY + zZ),\end{aligned}\right\}\quad\text{.........(7)}$$

where $\rho^2 = x^2 + y^2 + z^2$, and the values of L, M, N are supposed given by (2).

It appears that X, Y, Z are indeterminate to the extent of additive functions of the types Rx, Ry, Rz, respectively. The indeterminateness is of the kind often met with in statical problems [S. 25], and was to be expected. It is evident that a force at the point of contact in the direction of the centre is without influence on the motion.

52. Solution of Euler's Equations.

The solution of the equations (1) of Art. 50 in the general case is as follows*.

In the usual notation of elliptic functions, if

$$u = \int_0^\chi \frac{d\chi}{\Delta\chi}, \quad\text{.........(1)}$$

where
$$\Delta\chi = \sqrt{(1 - k^2 \sin^2\chi)}, \quad\text{.........(2)}$$

we write
$$\chi = \text{am}\, u. \quad\text{.........(3)}$$

The three functions $\cos\chi$, $\sin\chi$, $\Delta\chi$, or $\cos \text{am}\, u$, $\sin \text{am}\, u$, $\Delta\, \text{am}\, u$, are denoted for shortness by $\text{cn}\, u$, $\text{sn}\, u$, $\text{dn}\, u$. Their differential coefficients are

$$\left.\begin{aligned}\frac{d}{du}\text{cn}\, u &= -\text{sn}\, u\, \text{dn}\, u,\\ \frac{d}{du}\text{sn}\, u &= \quad \text{cn}\, u\, \text{dn}\, u,\\ \frac{d}{du}\text{dn}\, u &= -k^2 \text{sn}\, u\, \text{cn}\, u.\end{aligned}\right\}\quad\text{.........(4)}$$

The resemblance to Euler's equations suggests that a solution of the latter is to be obtained by equating p, q, r to constant multiples of $\text{cn}\,\lambda t$, $\text{sn}\,\lambda t$, $\text{dn}\,\lambda t$.

A reference to Fig. 42, p. 117, which indicates the general arrangement of the polhode cones, shews that the only one of the principal planes which is certainly crossed by the instantaneous

* It appears to be due substantially to Rueb (1834).

axis in all cases is that which is perpendicular to the axis (OB) of mean moment. Also, if the polhode cone surround the axis (OC) of least moment, r cannot vanish. We assume, therefore, for this case

$$p = p_0 \operatorname{cn} \lambda t, \quad q = b \operatorname{sn} \lambda t, \quad r = r_0 \operatorname{dn} \lambda t. \quad \text{.........(5)}$$

The time is here reckoned from the instant when the instantaneous axis crosses the plane $y = 0$. The constants p_0, r_0 are the values of p, r at this instant, and may be considered as arbitrary. The coefficient b, together with λ and k^2, is to be determined.

If we substitute from (5) in Euler's equations, we find that these are satisfied provided

$$\lambda p_0 = -\frac{B-C}{A} b r_0, \quad \lambda b = -\frac{A-C}{B} r_0 p_0, \quad -k^2 \lambda r_0 = \frac{A-B}{C} p_0 b. \quad \text{.........(6)}$$

The first two of these equations give, by division,

$$\frac{b^2}{p_0^2} = \frac{A(A-C)}{B(B-C)}, \quad \text{.........................(7)}$$

and so determine b^2. Again, from the first and third,

$$k^2 = \frac{(A-B) A p_0^2}{(B-C) C r_0^2}. \quad \text{.....................(8)}$$

Finally, by multiplication of the first two equations,

$$\lambda^2 = \frac{(A-C)(B-C)}{AB} \cdot r_0^2. \quad \text{..................(9)}$$

Since it is assumed that $A > B > C$, these expressions are all positive. We may adopt either sign for λ, provided that of b be made to correspond to it in agreement with any one of equations (6). A reversal of both these signs in (6) clearly alters nothing.

It is further necessary, however, that the value of k^2 given by (8) should be less than unity. This requires

$$A^2 p_0^2 + C^2 r_0^2 < B(A p_0^2 + C r_0^2), \quad \text{...............(10)}$$

or

$$H^2 < 2BT, \quad \text{.........................(11)}$$

which is in fact the condition that the polhode cone should surround the axis of least moment.

The solution we have obtained involves three arbitrary constants, viz. p_0, r_0, and a third constant which we may suppose added to t. The solution is therefore complete, subject to the condition (11).

If
$$H^2 > 2BT, \quad \text{.......................(12)}$$
so that the polhode cone surrounds the axis of greatest moment, the proper assumption is
$$p = p_0 \operatorname{dn} \lambda t, \quad q = b \operatorname{sn} \lambda t, \quad r = r_0 \operatorname{cn} \lambda t. \text{.... ....(13)}$$
Obviously this merely requires us to interchange p_0 and r_0, and A and C, respectively, in the formulæ (7), (8), (9).

In the critical case of
$$H^2 = 2BT, \text{....................... ...(14)}$$
where the polhode cone breaks up into two planes, we have $k^2 = 1$, and the elliptic functions assume a simpler form. Thus (1) makes
$$u = \log \tan(\tfrac{1}{4}\pi + \tfrac{1}{2}\chi), \quad \text{.................(15)}$$
and the formulæ (5) become
$$p = p_0 \operatorname{sech} \lambda t, \quad q = b \tanh \lambda t, \quad r = r_0 \operatorname{sech} \lambda t. \text{......(16)}$$
It will be found that
$$b^2 = 2T/B, \quad \text{.................................(17)}$$
$$\lambda^2 = \frac{(A-B)(B-C)}{AC} \cdot \frac{2T}{B}. \text{..................(18)}$$
The instantaneous axis tends asymptotically to coincide with the principal axis of mean moment, and therefore with the invariable line. If the body be originally in rotation about the mean axis, and receive a slight impulse subject to the condition (14), the instantaneous axis will travel along a polhode plane, the asymptotic state being that in which the body again rotates about its mean axis, but with this axis reversed in space, end for end.

When in any problem of free rotation the angular velocities p, q, r about the principal axes have been determined as functions of t, it still remains to ascertain the relation of the body to lines or planes of reference which are fixed in space. For this purpose Euler's coordinates (Art. 33) may be used. We have
$$\dot{\theta} = p \sin\phi + q \cos\phi, \text{....................(19)}$$
$$\sin\theta\,\dot{\psi} = -p\cos\phi + q\sin\phi, \quad \text{..............(20)}$$
$$\dot{\phi} + \cos\theta\,\dot{\psi} = r, \text{.......................(21)}$$
from which θ, ψ, ϕ are theoretically determinate.

The question is simplified if we adopt the invariable line as the axis from which θ is measured. Referring to Fig. 32, p. 83, we have

$$\cos AZ = -\sin\theta\cos\phi, \quad \cos BZ = \sin\theta\sin\phi, \quad \cos CZ = \cos\theta. \quad \text{.........(22)}$$

Hence

$$Ap = -H\sin\theta\cos\phi, \quad Bq = H\sin\theta\sin\phi, \quad Cr = H\cos\theta. \quad \text{.........(23)}$$

Thus
$$\tan\phi = -\frac{Bq}{Ap}, \quad \text{.........................(24)}$$

which determines ϕ. From (19) and (23) we deduce

$$\frac{\dot\theta}{\sin\theta} = \frac{(A-B)H}{AB}\sin\phi\cos\phi, \quad \text{...............(25)}$$

from which θ is to be found. Finally, from (20) and (23),

$$\dot\psi = \frac{H}{AB}(A\sin^2\phi + B\cos^2\phi). \quad \text{..............(26)}$$

When $A=B$, θ is constant, by (25). We found in this case (Art. 50)

$$p = K\cos(\lambda t+\epsilon), \quad q = K\sin(\lambda t+\epsilon), \quad \text{.................(27)}$$

where
$$\lambda = \frac{C-A}{A}n. \quad \text{....................................(28)}$$

Hence, from (24),
$$\phi = -(\lambda t+\epsilon), \quad \text{...............................(29)}$$

and from (26)
$$\dot\psi = H/A. \quad \text{....................................(30)}$$

These agree with previous results.

EXAMPLES. X.

1. The section of the momental ellipsoid by the invariable plane through the centre has a constant area.

2. The length of the normal drawn from any point of the polhode to meet a principal plane is constant.

3. When the polhode cone breaks up into two planes, these are the planes of contact of the two circular cylinders which envelope the momental ellipsoid.

4. The sum of the squares of the distances of the ends of the principal axes of the momental ellipsoid from the invariable line is constant. (Poinsot.)

5. The ellipsoid of gyration passes always through a fixed point on the invariable line; the instantaneous axis is the perpendicular from the centre to the tangent plane at this point; and the angular velocity varies inversely as the length of this perpendicular. (MacCullagh.)

6. If OH be the axis of angular momentum, OJ the instantaneous axis, and OP a line fixed in the body and moving with it, the rate at which the angular momentum about OP is changing is

$$-H\omega\,.\sin HP \sin HJ \sin JHP. \qquad \text{(Hayward.)}$$

7. A uniform thin circular disk is set spinning about an axis through the centre making an angle β with the normal. Prove that the semi-angle α of the cone described by the axis of the disk is given by

$$\tan\alpha = \tfrac{1}{2}\tan\beta.$$

If ω be the angular velocity, prove that the above cone is described in the period

$$\frac{2\pi}{\omega\sqrt{(1+3\cos^2\beta)}}.$$

8. Prove that in the case of a body symmetrical about its axis of greatest moment the semi-angle of the herpolhode cone cannot exceed 19° 28′.

9. Prove that a solid having kinetic symmetry about an axis can be made to execute a steady precessional motion, in which the angles α and β of Art. 29 have prescribed values, by the continual application of a couple

$$(C\sin\alpha\cos\beta - A\sin\beta\cos\alpha)\,\omega^2\frac{\sin\beta}{\sin\alpha}$$

in the plane of the axes of the two cones.

10. If a body having kinetic symmetry about the axis of greatest moment be subject to a retarding couple about the instantaneous axis, whose moment varies as the angular velocity, the instantaneous axis will approach asymptotically the axis of symmetry.

11. Obtain the equation of the rolling and sliding cone (Art. 48) as the envelope of the invariable plane.

12. A wheel is carried by a rotating shaft; its centre of gravity is in the axis of the shaft; but the axis of the wheel is inclined at an angle α. Prove that if ω be the angular velocity of the shaft, the centrifugal couple on the bearings is

$$(C-A)\,\omega^2\sin\alpha\cos\alpha,$$

in the usual notation.

Work this out in ft.-lbs., for the case of a circular disk of iron (sp. gr. $=8$), whose diameter is 1 ft., and thickness 1 in., assuming $\alpha=1°$, and that the speed is 100 revolutions per second. [7·3.]

13. Apply Euler's equations to examine the stability of rotation about a principal axis, when the principal moments are unequal. Also find the period of oscillation in the case of stability.

14. If a body moveable about a fixed point is subject to forces whose moment about the instantaneous axis is always zero, the angular velocity varies as the length of the radius of the momental ellipsoid in the direction of the instantaneous axis.

15. A uniform rectangular plate is set spinning about a diagonal with the angular velocity ω. Prove that it will be spinning about the other diagonal after a time

$$\frac{2}{\omega\sqrt{(\cos 2a)}} F_1(\sin a),$$

if $2a$ be the angle between the diagonals.

16. Prove that

$$\omega \frac{d\omega}{dt} = \sqrt{\{(\omega_1^2 - \omega^2)(\omega_2^2 - \omega^2)(\omega_3^2 - \omega^2)\}},$$

where

$$\omega_1^2 = \frac{2T(B+C) - H^2}{BC}, \text{ etc.}$$

If $A > B > C$, prove that the maximum value of ω is ω_2, and that the minimum is ω_1 or ω_3 according as the polhode surrounds the axis of least or greatest moment of inertia.

17. Prove that when $2BT = H^2$ the polar equation of the herpolhode curve is of the form

$$\frac{a}{r} = \cosh m\theta. \qquad \text{[See } \textit{Dynamics}\text{, Fig. 88.]}$$

18. If the invariable line meet the unit sphere whose centre is O in Q, the moment about the principal axis OA of the velocity of Q relative to the body is

$$\left(\frac{2AT}{H^2} - 1\right) p.$$

If τ be the time which the invariable line takes to perform a complete circuit in the body, prove that

$$\int_0^\tau p\,dt = 2\pi \sqrt{\left\{\frac{BC}{(A-B)(A-C)}\right\}}.$$

19. If the instantaneous axis meet the unit sphere in P, the moment of the velocity of P relative to the body about the axis OA is

$$\frac{2AT - H^2}{BC\omega^2} p.$$

20. Prove that

$$A^3p^2 + B^3q^2 + C^3r^2 = ABC\omega^2 - 2(BC + CA + AB)T + (A + B + C)H^2.$$

21. If θ be the inclination of the instantaneous axis to the invariable line, and $\dot{\psi}$ the rate at which the plane of these two lines revolves in space, prove that

$$\dot{\psi} = \frac{2T}{H} + \frac{(2AT - H^2)(2BT - H^2)(2CT - H^2)}{4ABCT^2H} \cot^2\theta.$$

22. If a be the angle which the principal axis OA makes with the invariable line OZ, the rate $(\dot{\xi})$ at which the plane ZOA is turning about OZ is given by

$$\sin^2 a\,\dot{\xi} = \frac{2T}{H} - \frac{H}{A}\cos^2 a.$$

CHAPTER VIII

GYROSTATIC PROBLEMS

53. Introduction.

We have seen that the problem of the free rotation of a rigid body is greatly simplified in the case of kinetic symmetry about an axis. It is not that the solution (Art. 47) is more complete than in the general case, but that it contains all that is usually of interest. In studying a dynamical problem we are not really concerned, as a rule, to know the position of every part of the system at any given instant. We wish rather to fix our attention on the salient features of the phenomena, and to follow their successive phases, ignoring as far as possible subordinate details. Thus in the case of a solid of revolution, such as a gyroscope, or a projectile, or a planet, we are interested chiefly in the changes of direction of the axis. The dynamical peculiarity which enables us to concentrate on this feature is that the momentary orientation of the body about its axis is without influence. Analytically, the corresponding *coordinate* of the body, viz. the ϕ of Euler's scheme, does not itself enter into the expressions for the energy and momentum (Art. 33 (6), (7), (8)).

It will be anticipated that the hypothesis of kinetic symmetry about an axis will tend also to simplify the more difficult question of rotation under the action of force. It has the further advantage that it covers almost all the cases which are of interest from a practical point of view. It is exemplified in machinery, in the gyroscope with its various applications, in the spinning top, and in the planets. We shall assume then, throughout this Chapter, that the body of which we treat has two of its principal moments at the mass-centre (or at a fixed point) equal to one another.

A further assumption which we shall make is that the external forces are such as to have (on the whole) zero moment about the axis of symmetry. It follows that the component angular velocity about this axis is constant; for it is constant in *free* motion (Art. 47), and it is not affected by an impulsive couple about a perpendicular axis, since this leaves the angular momentum about the former axis unaltered. And the effect of continuous forces can be imitated as closely as we please by a succession of small disconnected impulses.

The same conclusion follows, more artificially, on putting $A = B$ in the third of Euler's equations (Art. 50).

In the spinning top, and in the ordinary use of a gyroscope, frictional forces are of course operative which tend to destroy the rotation, but their effect is often only gradual. In practical applications, as e.g. in the gyrostatic compass, the necessary angular velocity is maintained against friction by a motor.

Whenever in the course of this Chapter we speak of a 'flywheel' or a 'gyrostat' it will be assumed that both the above conditions are fulfilled. The moments of inertia about the axis of symmetry, and about any perpendicular axis through the 'fixed point,' will be denoted by C and A, respectively. The constant angular velocity about the former axis will be denoted by n.

54. Condition for Steady Precession.

We begin with a very simple case, which serves however to illustrate the chief characteristic of gyrostatic phenomena. We

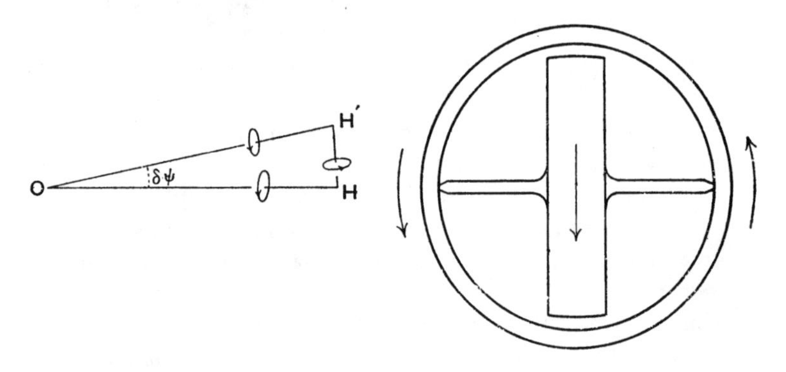

Fig. 43.

inquire what constraining forces are necessary in order that the axis of a rotating flywheel may be made to revolve in one plane with the constant angular velocity $\dot{\psi}$.

Let the plane in question be that of the paper. In the time δt the vector OH, which represents the angular momentum (Cn) of the flywheel about its axis, is turned through an angle $\delta\psi$, into the position OH′. The angular momentum $A\dot{\psi}$ about the normal to the plane of the paper is unaltered. Hence the vector HH′, whose magnitude is $Cn\,\delta\psi$, represents the total change of angular momentum due to the constraining forces, and therefore the 'impulse' of these forces. Dividing by δt we see that the required forces reduce to a couple $Cn\dot{\psi}$ about an axis in the plane of the paper, at right angles to that of the flywheel. In the case of the figure this couple is supplied by the equal and opposite reactions of the supporting frame on the two pivots, normal to the paper.

In the absence of such a constraining couple, any attempt to turn the axis of the flywheel in the manner described (as for instance by rotating the vertical ring of a gyroscope, the other ring being free) will cause the positive end of this axis to approach the positive end of the axis about which rotation is imposed. This forms one of the simplest and most striking experiments which can be made with an ordinary gyroscope.

For a similar reason, the nose of a tractor aeroplane will tend to dip downwards in a turn to the right, and to tilt upwards in a turn to the left, if the rotation of the propeller is right-handed as viewed by the pilot.

The more general case of constrained precessional motion is almost equally simple. Let us suppose that the axis of the flywheel is made to describe a cone of semiangle θ about the line OZ of Fig. 32 (p. 83). We have seen in Art. 33 that the angular momentum has a component

$$A\sin^2\theta\dot{\psi} + Cn\cos\theta \quad\text{..............................(1)}$$

about OZ, and a component

$$-A\sin\theta\cos\theta\dot{\psi} + Cn\sin\theta \quad\text{..........................(2)}$$

about an axis at right angles to OZ in the plane ZOC. Since θ is supposed constant, the remaining component (about OB') is zero. The preceding argument will therefore apply if we suppose OH in Fig. 43 to represent the component (2), which is alone variable. The requisite constraining couple is therefore

$$M = -A\sin\theta\cos\theta\dot{\psi}^2 + Cn\sin\theta\dot{\psi} \quad\text{.....................(3)}$$

about OB'. When this is positive, the tendency of the couple is to resist an approach of OC towards OZ.

A particular case is where the couple is supplied by gravity and an equal and opposite reaction, as in the case of the spinning top, or a weighted gyroscope. If the centre of gravity is in OC at a distance h from O, the couple in question has a moment

$$M = M_0 gh \sin\theta, \quad \text{.................................(4)}$$

where M_0 is the mass of the flywheel, if OZ be now supposed to be the upward vertical. Substituting in (3), and rejecting the trivial cases of $\theta=0$, $\theta=\pi$, we have as the condition for steady precession

$$A\cos\theta\dot{\psi}^2 - Cn\dot{\psi} + M_0 gh = 0. \quad \text{........................(5)}$$

If θ be given, this is a quadratic in $\dot{\psi}$, whose roots will be real if

$$n^2 > \frac{4AM_0 gh\cos\theta}{C^2}. \quad \text{..............................(6)}$$

This condition is always fulfilled if $\theta > \frac{1}{2}\pi$, i.e. if the centre of gravity is below the level of the fixed point, as in the gyroscopic pendulum (Art. 58).

When n is great, the two solutions tend to the values

$$\dot{\psi} = \frac{Cn}{A\cos\theta}, \quad \dot{\psi} = \frac{M_0 gh}{Cn}. \quad \text{..........................(7)}$$

The corresponding types of motion are known as the 'rapid' and the 'slow' precession, respectively. The former is practically identical with the free Eulerian nutation of Arts. 47, 50, gravity having little influence. The slow precession is the type usually realized approximately in a rapidly spinning top. In the case of a gyroscope weighted so that the centre of gravity is below O, the slow precession is retrograde, as is seen most simply by changing the sign of h.

Ex. 1. A solid of revolution is free to turn about a fixed point O on its axis; to find the condition for steady rotation about the upward vertical through O.

With the previous notation we now have

$$n = \dot{\psi}\cos\theta. \quad \text{...................................(8)}$$

Hence, from (3) and (4),

$$M_0 gh\sin\theta = (C-A)\sin\theta\cos\theta\dot{\psi}^2.$$

Omitting the obvious solutions $\theta=0$, $\theta=\pi$, we have

$$\cos\theta = \frac{M_0 gh}{(C-A)\dot{\psi}^2}. \quad \text{.............................(9)}$$

This case may be realized by suspending the body from a transverse axis through O, which is made to rotate in a horizontal plane with the prescribed angular velocity $\dot{\psi}$.

The vertical positions of OG are alone possible unless

$$\dot{\psi}^2 > \frac{M_0 gh}{|C-A|}. \quad \text{.............................(10)}$$

If $A > C$, as in the case of a thin rod, $\cos\theta$ is negative, i.e. G must be below the level of O.

Ex. 2. To find the condition that a wheel may roll round in a circle, as in Art. 29, Ex., under the influence of gravity and the reaction at the point of contact with the ground.

In our previous notation the acceleration of the mass-centre is $c\dot{\psi}^2$ towards the centre of the circle which it describes. The reaction will therefore consist of a horizontal force $M_0 c\dot{\psi}^2$ inwards, and an upward force $M_0 g$. The moment about the horizontal diameter of the wheel is therefore

$$M = M_0 c\dot{\psi}^2 a \sin\theta - M_0 g a \cos\theta. \quad \text{.....................(11)}$$

We have, further, the kinematical relation

$$c\dot{\psi} = -na. \quad \text{.....................................(12)}$$

Substituting in (3), we find

$$\left(C + M_0 a^2 + \frac{Aa}{c}\cos\theta\right)\dot{\psi}^2 = \frac{M_0 g a^2}{c}\cot\theta. \quad \text{..............(13)}$$

55. Intrinsic Equations of Motion of a Gyroscope.

We denote by C one of the two points in which the axis of the flywheel meets a unit sphere having its centre at the fixed point O, and for definiteness we choose that point of the pair which is such that the angular velocity (n) about OC shall be positive (i.e. right-handed). This point C may be called the 'pole' of the gyrostat; it is with its path that we are chiefly concerned.

We draw from C, on the unit sphere, a quadrant CA tangential to the path, in the direction of motion, and a quadrant CB at right angles, so that OA, OB, OC (in this order) shall form a right-handed system. We distinguish the positions of the corresponding lines after an interval δt by OA', OB', OC'*. If v be the velocity of the pole in its path, and $\delta\chi$ the angle between the projections on the tangent plane to the sphere of two consecutive tangent lines to the path, we have

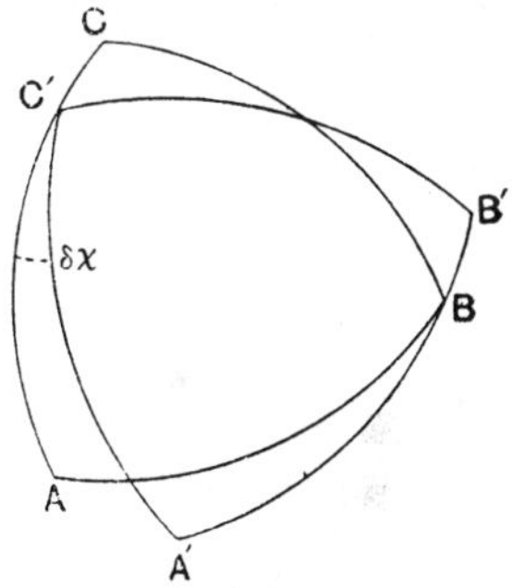

$$CC' = v\,\delta t, \quad AC'A' = \delta\chi. \quad \text{......(1)}$$

Fig. 44.

At the instant t the component rotations about OA, OB, OC are 0, v, n respectively, and the components of angular momentum are accordingly

$$0,\ Av,\ Cn. \quad \text{...........................(2)}$$

In the time δt these are altered to

$$0,\quad A(v + \delta v),\quad C(n + \delta n) \quad \text{..................(3)}$$

* The figure is simplified by the assumption that C' may be taken to lie in CA. The error thus involved, and in the consequent positions of A', B', is of the second order, and so does not affect the final results.

about OA', OB', OC' and therefore to

$$-Av\,\delta\chi + Cnv\,\delta t, \quad A\,(v+\delta v), \quad C\,(n+\delta n) \quad \text{.........(4)}$$

about OA, OB, OC, terms of the second order being neglected. If, as we assume, the external forces have zero moment about the axis of symmetry, they may be replaced by two forces P, Q acting at C along the tangents to the arcs CA, CB respectively, i.e. along and at right angles to the path of C. P is in fact equal to the moment of the external forces about OB, and $-Q$ to that about OA. Hence, equating the increments of angular momentum to $-Q\delta t$, $P\delta t$, 0, respectively, we find

$$Av\frac{d\chi}{dt} = Q + Cnv, \quad A\frac{dv}{dt} = P, \quad \text{..............(5)}$$

with $n = \text{const.}$ These equations may be called 'intrinsic,' as related directly to the path, and not to arbitrary lines or planes of reference.

The expressions dv/dt and $v\,d\chi/dt$ are the accelerations of the pole C along and at right angles to its path on the sphere (Art. 36). If we put $n = 0$ we have as a particular case the equations of motion of a particle on a spherical surface, and we infer that the motion of the pole C in the present case will be exactly the same as that of such a particle, of mass A, under the same forces P, Q, provided we introduce in addition a fictitious 'deviating force' Cnv acting always towards the *left* of the path, as viewed from without the sphere.

For example, in precessional motion the pole describes a circle on the unit sphere with constant angular velocity ($\dot{\psi}$). If θ be the angular radius of this circle the acceleration of C is $v^2/\sin\theta$ towards its centre, and the component at right angles to the path, in the tangent plane, is therefore $v^2\cot\theta$. Hence if there are no external forces

$$Av^2\cot\theta = Cnv, \quad \text{.......................(6)}$$

or, since $v = \dot{\psi}\sin\theta$,

$$\dot{\psi} = \frac{Cn}{A\cos\theta}, \quad \text{.......................(7)}$$

which is the ordinary formula for the Eulerian nutation (Art. 47). As already remarked, the same formula holds approximately for the 'rapid' precession of a top whose rate of spin is very great.

In the case of the 'slow' precession of a top we may ignore the acceleration of C, which varies as v^2, and equate the deviating force $Cn\dot{\psi}\sin\theta$ to the effective component of gravity, viz. $M_0gh\sin\theta$. Thus

$$\dot{\psi}=\frac{M_0gh}{Cn}, \qquad (8)$$

as in Art. 54.

The exact condition for steady motion of a top, including both cases, is

$$Av^2\cot\theta=-M_0gh\sin\theta+Cnv, \qquad (9)$$

or

$$A\dot{\psi}^2\sin\theta\cos\theta=-M_0gh\sin\theta+Cn\dot{\psi}\sin\theta, \qquad (10)$$

in agreement with Art. 54 (5).

56. Further applications of the Intrinsic Equations.

This fiction of a deviating or 'gyrostatic' force, as we may call it, of amount Cnv, acting always at right angles to the path of the poles, enables us to understand the general character of the motion in cases where exact calculation would be difficult.

Take for instance the small oscillations of a rapidly spinning top about a state of precessional motion. As a particular case, suppose that the pole C is initially at rest. It will at once begin to descend under the influence of gravity, but the deviating force which is quickly called into play will deflect it continually to the left, so that it presently turns upwards again, describing a sort of cycloidal curve. If the initial circumstances are of a less special character, the resemblance is to a trochoid.

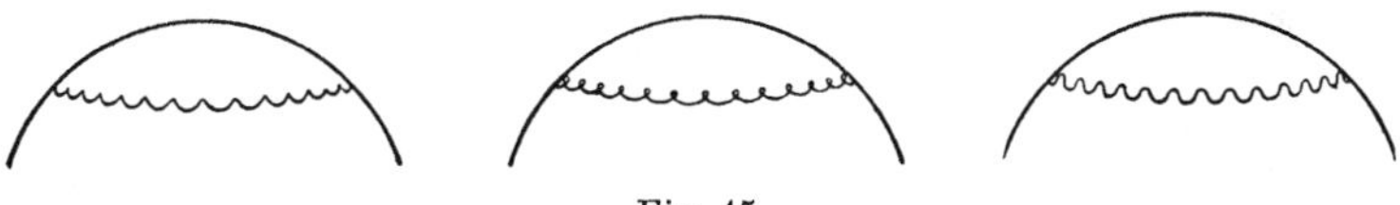

Fig. 45.

If we fix our attention on a small portion of the path, the circumstances, when the undulations are small, are closely analogous to the case of a particle moving in a plane under two forces, one of which is constant in magnitude and direction, whilst the other is at right angles to the path, and varies as the velocity *. The equations of motion under these conditions have the forms

$$\ddot{x}=-\beta\dot{y}, \quad \ddot{y}=f+\beta\dot{x}, \qquad (1)$$

* This case occurs in several important physical problems.

which combine into

$$\ddot{\xi} - i\beta\dot{\xi} = if, \qquad (2)$$

if $\zeta = x + iy$. Hence

$$\zeta = -\frac{ft}{\beta} + a + ib + re^{i(\beta t + \epsilon)}, \qquad (3)$$

where a, b, r, ϵ are arbitrary. Thus

$$\left.\begin{aligned} x &= a - \frac{ft}{\beta} + r\cos(\beta t + \epsilon), \\ y &= b + r\sin(\beta t + \epsilon), \end{aligned}\right\} \qquad (4)$$

which are the equations of a trochoid. For comparison with the case of the top we must put

$$f = -M_0 gh \sin\theta / A, \quad \beta = Cn/A. \qquad (5)$$

The term proportional to t in the expression for x corresponds to the steady precession, on which is superposed a circular vibration of period

$$\frac{2\pi}{\beta} = \frac{A}{C} \cdot \frac{2\pi}{n}. \qquad (6)$$

These results agree exactly with the more elaborate theory, to be given presently (Art. 57).

Again, we may recall the familiar experiment where the projecting cylindrical stem of a top is observed to follow the windings of a metal arc or wire brought into contact with it. The friction of the wire causes the stem to roll along the arc, whilst the deviating force called into play tends to maintain the contact.

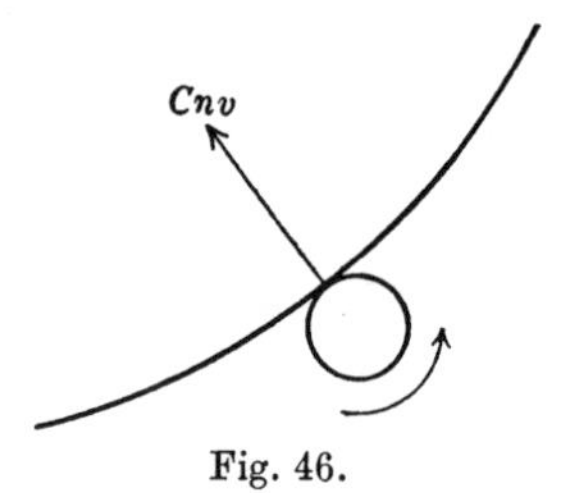

Fig. 46.

It is further evident that any force which tends to accelerate or retard the precessional motion of a top, and so to increase or diminish v, will cause the axis to rise or fall, respectively. This is known as Lord Kelvin's rule, and was applied by him to explain the familiar phenomenon of the 'sleeping' top, where the axis gradually assumes the vertical position. In Fig. 47 the spin is supposed to be right-handed with respect to the axis GC, so that the point of the peg which is in contact with the ground at P is moving from the reader. There is therefore at this point a force (F) of sliding friction acting on the top towards the reader. This may be supposed transferred to G if we introduce a couple of moment $F . GP$. In considering the angular motion we need only attend to the couple, which has a component about an axis at right angles to GC in the plane of the figure, tending to accelerate the precession about the vertical through G, and so causing the top to rise. The remaining (smaller) component about GC tends to check the spin.

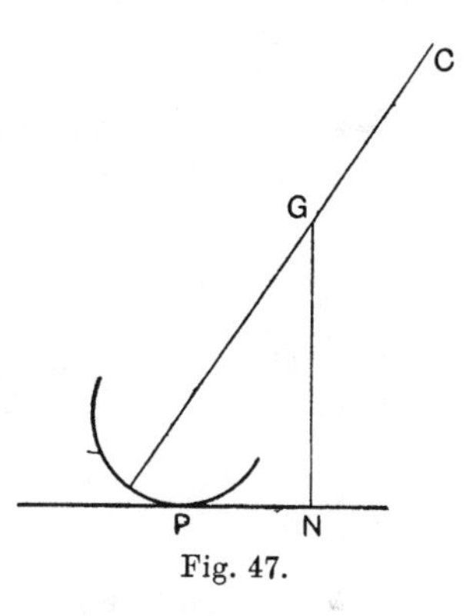

Fig. 47.

57. Investigations in Polar Coordinates.

The deviating force Cnv, being proportional and at right angles to the velocity, is easily resolved into suitable components in any system of coordinates. Thus, to obtain the general equations of motion of a gyrostat in terms of the spherical coordinates θ, ψ of its pole, we note that since the components of the velocity v in and perpendicular to the plane of θ are $\dot{\theta}$ and $\sin\theta\dot{\psi}$, respectively, those of deviating force will be $-Cn\sin\theta\dot{\psi}$ and $Cn\dot{\theta}$. Hence, recalling the expressions for the accelerations of a point in spherical polars (Art. 34), we have at once

$$\left.\begin{aligned} A(\ddot{\theta}-\dot{\psi}^2\sin\theta\cos\theta) &= -Cn\dot{\psi}\sin\theta+P, \\ \frac{A}{\sin\theta}\frac{d}{dt}(\dot{\psi}\sin^2\theta &= Cn\dot{\theta}+Q, \end{aligned}\right\} \quad \text{......(1)}$$

where P, Q are the couples in the plane of θ, and in the perpendicular plane through OC, due to the external forces. In the case of the top we have

$$P = M_0 gh\sin\theta, \quad Q=0, \quad \text{..................(2)}$$

provided the line OZ, from which θ is measured, is supposed drawn vertically upwards, and h is reckoned positive or negative according as the mass-centre G is above or below the fixed point O when $\theta=0$.

Analytical investigations relating to the top are often based on the principles of energy and angular momentum. The former of these gives

$$\tfrac{1}{2}A\,(\dot{\theta}^2+\sin^2\theta\dot{\psi}^2)+\tfrac{1}{2}Cn^2+Mgh\cos\theta=\text{const.},\ \ldots\ldots(3)$$

by Art. 33 (6). Again, if μ denote the constant angular momentum about the vertical through the point of support, we have

$$A\sin^2\theta\dot{\psi}+Cn\cos\theta=\mu,\ \ldots\ldots\ldots\ldots\ldots\ldots(4)$$

by Art. 33 (7). These equations may of course be deduced also from (1)*.

Putting $Cn=\nu$, and eliminating $\dot{\psi}$, we have

$$\left.\begin{aligned}\dot{\theta}^2+\frac{(\mu-\nu\cos\theta)^2}{A^2\sin^2\theta}+\frac{2Mgh}{A}\cos\theta\qquad\qquad\\ -\left\{\sigma^2+\frac{(\mu-\nu\cos\alpha)^2}{A^2\sin^2\alpha}+\frac{2Mgh}{A}\cos\alpha\right\}=0,\end{aligned}\right\}\ \ldots\ldots(5)$$

where σ and α are any two simultaneous values of $\dot{\theta}$ and θ. If we write this equation for shortness in the form

$$\dot{\theta}^2+f(\theta)=0,\ \ldots\ldots\ldots\ldots\ldots\ldots\ldots\ldots(6)$$

the stationary values of θ are given by

$$f(\theta)=0.\ \ldots\ldots\ldots\ldots\ldots\ldots\ldots\ldots\ldots(7)$$

If we put $\sin^2\theta=1-\cos^2\theta$, and clear of fractions, we have a cubic in $\cos\theta$. Now if θ is small, $f(\theta)$ is large and positive; when $\theta=\alpha$ it is negative; and when θ is nearly equal to π it is again positive. Finally, when $\cos\theta=-\infty$, $f(\theta)$ is negative. Hence one root of (7) lies between 0 and α, and a second between α and π. The third root of the cubic in $\cos\theta$ lies between -1 and $-\infty$, and so does not give a real value of θ. The path of the pole on the unit sphere therefore lies between two horizontal circles. As extreme cases, the upper or lower of these circles may contract to a point. When the two circles coincide we have the steady precessional types of motion already investigated. If this motion be slightly disturbed, the two circles will only slightly separate; this shews that the precessional motions are stable.

The above statement as to the character of the path is however obvious without any analysis. There must in any case be points of maximum and minimum altitude of the pole. Any such

* They were given by Lagrange, *Mécanique Analytique*, 1st ed. (1788).

point may be called an 'apse,' and the great-circle arc drawn to it from the highest point Z of the unit sphere may be called an 'apse-line.' The familiar argument employed in the theory of central forces [D. 88] and of the spherical pendulum [D. 103] avails here also to shew that every apse-line divides the orbit symmetrically, and thence that there are two apsidal distances, and a constant apsidal angle *. The limiting cases where the path passes through the highest or lowest point present no difficulty. Cusps, again, may be regarded as infinitely small loops.

The condition for steady precession is obtained by putting $\ddot{\theta}=0$ in (1). Denoting by α and ω the constant values of θ and $\dot{\psi}$, we find

$$A\omega^2 \cos\alpha - Cn\omega + Mgh = 0 \qquad (8)$$

as in Art. 54 (5).

To investigate a small disturbance from this state, we put

$$\theta = \alpha + \xi, \quad \dot{\psi} = \omega + \frac{\dot{\eta}}{\sin\alpha}, \qquad (9)$$

where ξ, η are the component small deviations of the pole, along and at right angles to the meridian, from the position in the steady motion. Making these substitutions in the first of equations (1), and omitting terms of the second order, we find after reduction by means of (8)

$$A\ddot{\xi} + A\omega^2 \sin^2\alpha\xi + (\nu - 2A\omega\cos\alpha)\,\dot{\eta} = 0. \qquad (10)$$

In like manner, from (4) we have

$$(2A\omega\cos\alpha - \nu)\,\xi + A\dot{\eta} = 0. \qquad (11)$$

Hence, eliminating $\dot{\eta}$,

$$A^2\ddot{\xi} + \{A^2\omega^2\sin^2\alpha + (\nu - 2A\omega\cos\alpha)^2\}\,\xi = 0. \qquad (12)$$

Since the coefficient of ξ is positive, we learn, again, that a steady precession is stable. If we put

$$p^2 = \frac{A^2\omega^2\sin^2\alpha + (\nu - 2A\omega\cos\alpha)^2}{A^2}, \qquad (13)$$

the period of a small oscillation about the steady motion is $2\pi/p$.

In the case of the slow precession of a rapidly spinning top

* The argument referred to makes use of the hypothesis of a reversal of the motion. In the present application the reversal must include that of the *spin*. If the motion of the axis of a top, but *not* the spin, were reversed, the pole would not retrace its former path.

$A\omega$ is small compared with ν, so that $p = \nu/A$, approximately. We have then

$$\xi = F\cos\left(\frac{Cnt}{A} + \epsilon\right), \quad \eta = F\sin\left(\frac{Cnt}{A} + \epsilon\right), \quad \text{......(14)}$$

where F, ϵ are arbitrary. An additive constant to the value of η is omitted as immaterial. A small circular vibration is thus superposed on the steady precession, producing the trochoidal type of path already referred to in Art. 56.

In the rapid precession we have $\omega = \nu/A\cos\alpha$, $p = \omega$, approximately. We find

$$\xi = F\cos(\omega t + \epsilon), \quad \eta = -F\cos\alpha\sin(\omega t + \epsilon). \quad \text{......(15)}$$

If x, y be the coordinates of the orthogonal projection of the pole C on a horizontal plane, we have

$$\left.\begin{aligned} x &= \sin\alpha\cos\omega t + \xi\cos\alpha\cos\omega t - \eta\sin\omega t \\ &= \sin\alpha\cos\omega t + F\cos\alpha\cos\epsilon, \\ y &= \sin\alpha\sin\omega t + \xi\cos\alpha\sin\omega t + \eta\cos\omega t \\ &= \sin\alpha\sin\omega t - F\cos\alpha\sin\epsilon. \end{aligned}\right\} \quad \text{......(16)}$$

The projection of the path is therefore a circle. We have, practically, an Eulerian nutation about an invariable line slightly inclined to the vertical (cf. Art. 47).

The case of a bomb, dropped from an aeroplane, comes under the same theory. The bomb is fitted with fins near the tail, set obliquely, so that a spinning moment about the axis is generated by the reaction of the air. When the axis makes a small angle θ with the vertical the force-resultant (R) of the air pressures meets the axis at a point P below the mass-centre G. This has a moment $R \, . \, GP\sin\theta$, in the above notation, about a horizontal line through G perpendicular to the axis of the bomb. When terminal velocities are practically reached, the motion relative to G is determined by (1) with the proper change in the meaning of A, and the substitution of $R \, . \, GP$ for Mgh.

58. Oscillations of a nearly Vertical Top.

The equations (1) of the preceding Art. are not very appropriate when the inclination of the axis to the vertical is small throughout the motion. The question is however easily dealt with on the basis of Art. 55.

If x, y be the projections of the unit vector OC on a horizontal plane, the components of deviating force are $-Cn\dot{y}$, $Cn\dot{x}$, whence

$$A\ddot{x} = -Cn\dot{y} + Mghx, \quad A\ddot{y} = Cn\dot{x} + Mghy, \quad \text{.........(1)}$$

approximately. Putting $\zeta = x + iy$ we have

$$A\ddot{\zeta} - iCn\dot{\zeta} - Mgh\zeta = 0, \quad \text{.................(2)}$$

the solution of which is

$$\zeta e^{-ipt} = He^{i\sigma t} + Ke^{-i\sigma t}, \quad \text{.....................(3)}$$

where

$$\rho = \frac{Cn}{2A}, \quad \sigma = \frac{\sqrt{(C^2n^2 - 4AMgh)}}{2A}, \quad \text{...........(4)}$$

and the (complex) constants H, K are arbitrary. If

$$n^2 > 4AMgh/C^2, \quad \text{.......................(5)}$$

σ is real, and the path of the pole C consists of two superposed circular vibrations of periods $2\pi/(\rho + \sigma)$ and $2\pi/(\rho - \sigma)$. The path is therefore an epicyclic, direct if $|\rho| > \sigma$, as in the case of the nearly upright top.

The path may be otherwise described as an ellipse which revolves about the origin with the angular velocity ρ, the period in the ellipse being $2\pi/\sigma$. For if x', y' be coordinates relative to axes revolving at the rate ρ, we have

$$x' + iy' = (x + iy)\,e^{-i\rho t} = He^{i\sigma t} + Ke^{-i\sigma t}. \quad \text{.........(6)}$$

Expressed in real form, this gives an elliptic orbit relative to the rotating axes.

If we change the sign of h in the above formulæ we get the case of the 'gyroscopic pendulum,' which consists of a thin stem free to turn about its upper end, and carrying at its lower end a rotating flywheel whose axis coincides with that of the stem. The epicyclics are then retrograde.

The variety of forms which the epicyclic curves may assume is of course endless*. The annexed figures shew two examples of the direct and retrograde types, respectively.

Fig. 48. Fig. 49.

* A number of stereoscopic and other diagrams of the path of the pole on the unit sphere are given by Greenhill, *Proc. Lond. Math. Soc.* t. xxvii, and *Report on Gyroscopic Theory*, London, 1914. In the cases there depicted the conditions are adjusted so that the paths are re-entrant.

59. Foucault's Experiments.

An interesting application of the theory is to experiments devised by Foucault* and others for obtaining, by means of a rapidly spinning gyroscope, mechanical demonstrations of the earth's rotation†.

If there were no constraint and no friction, the axis of the fly-wheel (more strictly the axis of angular momentum) would maintain its direction in space. Relatively to the earth there would be a progressive westerly displacement in longitude, and it is this which Foucault in one of his experiments sought to put in evidence.

If, on the other hand, the axis of the flywheel is restricted to motion in one plane, it will tend to approach as closely as possible to the polar axis of the earth, regard being had to the sense of the rotations.

Let us suppose in the first place that the axis of the flywheel is restricted to the plane of the meridian, as by clamping the vertical ring in the E. and W. plane.

In the annexed diagram of the unit sphere P denotes the N. pole of the earth, C the pole of the flywheel, A the E. point of the horizon. Let ω be the earth's angular velocity, θ the angle POC. The velocity of C is made up of $\dot{\theta}$ along the arc PC and $\omega \sin \theta$ parallel to OA, where O is the centre of the sphere.

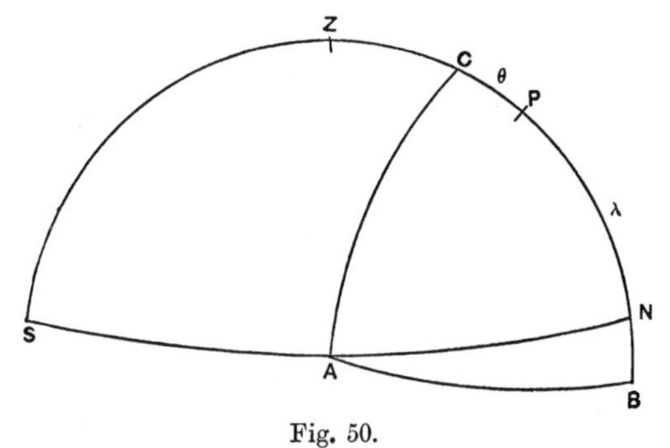

Fig. 50.

* L. Foucault (1819–68), distinguished for his experimental researches in Optics (direct measurement of the velocity of light, etc.).

† The word 'gyroscope' seems to have been first coined by Foucault in this connection, as combining the ideas of *rotation* (e.g. of the earth) and *detection*. It has come to be used indifferently in cases where the rotation of a flywheel plays the leading part in controlling the phenomena. The term 'gyrostat' was applied primarily by Kelvin to a body with a *concealed* flywheel, but has come to be used in the same widened sense as 'gyroscope.'

The principal moments of the flywheel being denoted as usual by A, A, C, and its angular velocity by n, the gyrostatic force will have components $Cn\dot{\theta}$ parallel to OA, and $Cn\omega \sin\theta$ along CP. The former component is resisted by the constraint. The latter leads to the equation

$$A\ddot{\theta} = -Cn\omega \sin\theta + L, \quad \text{.................(1)}$$

where L is the couple exerted on the pivots of the flywheel by the ring which carries it. For the motion of this ring itself about its E. and W. diameter we have

$$A_1 \ddot{\theta} = -L, \quad \text{..........................(2)}$$

where A_1 is the corresponding moment of inertia. Hence, writing $A + A_1 = I$,

$$I\ddot{\theta} + Cn\omega \sin\theta = 0. \quad \text{....................(3)}$$

There are two positions of apparent equilibrium, viz. $\theta = 0$ and $\theta = \pi$. If the rotation n is positive, i.e. right-handed about OC, the former position is stable and the latter unstable. The period of a small oscillation about the stable position is

$$2\pi\sqrt{\left(\frac{I}{Cn\omega}\right)}. \quad \text{........................(4)}$$

It is to be noticed that in the above derivation of (1) we have written $\ddot{\theta}$ for the acceleration of C along the meridian, as if this were a fixed basis of reference. Since it is really turning slowly about OP this is not quite accurate, but it is not difficult to see that the consequent error will depend on ω^2; it is utterly negligible from a practical point of view.

In Gilbert's 'barygyroscope' the ring or frame carrying the flywheel was slightly weighted, so that in the absence of rotation the axis was vertical. The effect of rotation then is to cause a deviation χ towards the N. Instead of (3) we now have

$$I\ddot{\theta} + Cn\omega \sin\theta = Mgh \sin\chi, \quad \text{..........................(5)}$$

where M is the mass of the frame, and h the (small) depth of its centre of gravity below its horizontal line of pivots. Since $\theta + \chi = \frac{1}{2}\pi - \lambda$, where λ is the latitude, the position of apparent equilibrium ($\ddot{\theta} = 0$), which is the subject of observation, is given by

$$Cn\omega \cos(\lambda + \chi) = Mgh \sin\chi,$$

or

$$\tan\chi = \frac{Cn\omega \cos\lambda}{Mgh + Cn\omega \sin\lambda}. \quad \text{..........................(6)}$$

The second term in the denominator was small compared with the first in spite of the smallness of h. Under these conditions the deviation to be expected is greater the more rapid the spin n, and the smaller the distance h.

In another experiment suggested by Foucault the axis of the flywheel is restricted to motion in a *horizontal* plane. In the annexed figure Z is the zenith, X and Y are the E. and N. points of the horizon, and P the north pole. If λ is the latitude YP, the earth's rotation may be resolved into $\omega\cos\lambda$ about OY and $\omega\sin\lambda$

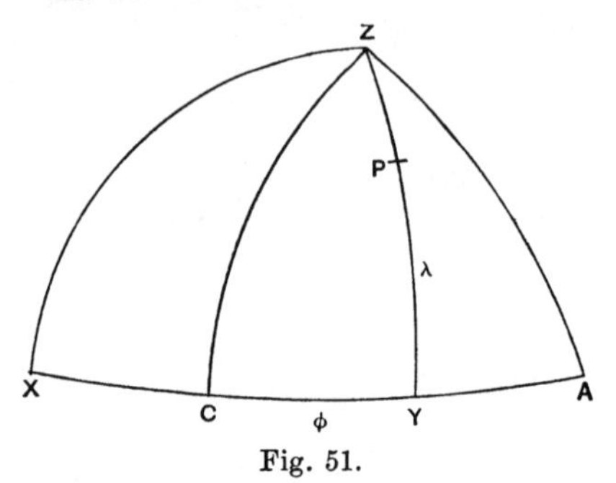

Fig. 51.

about OZ. Hence, if ϕ be the azimuth YC of the axis of the fly-wheel, the velocity of C will consist of $\omega\sin\lambda - \dot{\phi}$ along XY, and $\omega\cos\lambda\sin\phi$ parallel to ZO*. The gyrostatic force along CY is therefore $Cn\omega\cos\lambda\sin\phi$. Hence

$$A\ddot{\phi} + Cn\omega\cos\lambda\sin\phi = N, \quad\text{.................(7)}$$

where N is the horizontal moment exerted on the axis by the ring which carries it. If A_1 is the moment of inertia of the frame-work about the vertical, we have

$$A_1\ddot{\phi} = -N, \quad\text{.........................(8)}$$

whence, writing $A + A_1 = I$,

$$I\ddot{\phi} + Cn\omega\cos\lambda\sin\phi = 0. \quad\text{.................(9)}$$

The pole of the flywheel tends (if n is positive) towards the N. point of the horizon, and oscillates about this in a period

$$2\pi\sqrt{\left(\frac{I}{Cn\omega\cos\lambda}\right)}. \quad\text{.................(10)}$$

60. The Gyrostatic Compass.

This latter arrangement, if it were practicable, would serve as a compass indicating the true as distinguished from the magnetic north. But all these experiments, at least in their primitive forms, are difficult to carry out with definite success. Not to speak of the extreme nicety of construction and accurate centering required, the gyrostatic forces are feeble and liable to be neutralized by the

* There is here, again, an unimportant error of the kind already explained.

inevitable friction at the bearings. They decay, moreover, with the rotation unless, as in the modern gyrostatic compass, this is artificially maintained.

In this apparatus the flywheel turns on an axis which is fixed relatively to a base which itself floats on mercury. The outstanding friction is thus minimized, but the axis OC has now two degrees of directional freedom, so that the theory is modified.

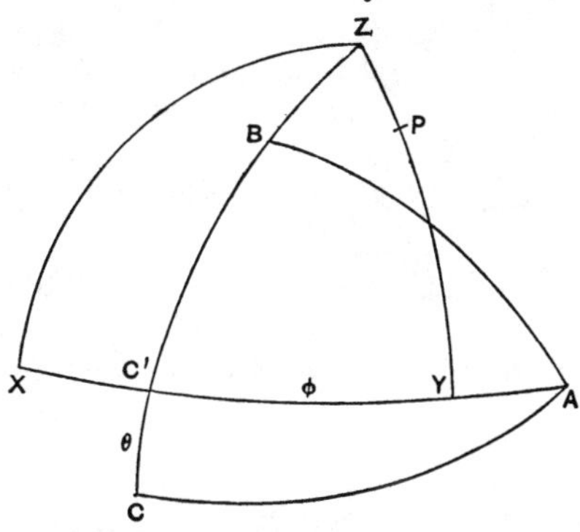

Fig. 52.

Let θ be the inclination of OC below the horizontal, and ϕ the azimuth YZC. The earth's rotation can be resolved into $\omega \sin \lambda$ about OZ and $\omega \cos \lambda$ about OY. The latter component can again be resolved into $\omega \cos \lambda \sin \phi$ about OA and $\omega \cos \lambda \cos \phi$ about OC' (see the figure). Since the arc ZC is turning about OZ at the rate $\omega \sin \lambda - \dot{\phi}$ the total velocity of C at right-angles to the vertical circle ZC is

$$(\omega \sin \lambda - \dot{\phi}) \sin ZC + \omega \cos \lambda \cos \phi \sin C'C$$
$$= (\omega \sin \lambda - \dot{\phi}) \cos \theta + \omega \cos \lambda \cos \phi \sin \theta.$$

The velocity along ZC is $\dot{\theta} + \omega \cos \lambda \sin \phi$.

In the following outline of the theory we assume θ and ϕ to be small, so that these expressions reduce to

$$\omega \sin \lambda - \dot{\phi} + \omega\theta \cos \lambda, \text{ and } \dot{\theta} + \omega\phi \cos \lambda.$$

The equations of motion of the flywheel axis are therefore, with the usual notation for the moments of inertia,

$$\left.\begin{aligned} A(\ddot{\theta} + \omega\dot{\phi} \cos \lambda) &= Cn(\dot{\phi} - \omega \sin \lambda - \omega\theta \cos \lambda) + L, \\ A(\ddot{\phi} - \omega\dot{\theta} \cos \lambda) &= -Cn(\dot{\theta} + \omega\phi \cos \lambda) + M, \end{aligned}\right\} \quad \text{............(1)}$$

where L, M are the couples due to the reactions at the pivots.

The equations of motion of the base of the apparatus, apart from the flywheel, will have the forms

$$\left.\begin{aligned} A_1(\ddot{\theta} + \omega\dot{\phi} \cos \lambda) &= -L - K(\theta - \alpha), \\ A_2(\ddot{\phi} - \omega\dot{\theta} \cos \lambda) &= -M, \end{aligned}\right\} \quad \text{.....................(2)}$$

where $K(\theta - \alpha)$ represents the restoring couple due to the combined action of gravity and buoyancy, the angle α being the value of θ in the equilibrium position when there is no rotation of the flywheel. When the flywheel rotates, the equilibrium position is given by

$$\phi=0, \quad Cn\omega(\sin\lambda+\theta\cos\lambda)+K(\theta-a)=0. \quad \text{......(3)}$$

This gives a certain tilt, depending on the latitude.

Now denoting by θ and ϕ the *deviations* from this position, and eliminating the couples L, M, we have

$$\left.\begin{aligned} I(\ddot\theta+\omega\dot\phi\cos\lambda)&=Cn(\dot\phi-\omega\theta\cos\lambda)-K\theta,\\ J(\ddot\phi-\omega\dot\theta\cos\lambda)&=-Cn(\dot\theta+\omega\phi\cos\lambda),\end{aligned}\right\} \quad \text{......(4)}$$

where

$$I=A+A_1,\ J=A+A_2. \quad \text{......(5)}$$

In the absence of rotation we should have $\ddot\phi=\omega\dot\theta\cos\lambda$. Omitting a possible steady motion which would decay by fluid friction, we should have then

$$I\ddot\theta+(K+I\omega^2\cos^2\lambda)\theta=0. \quad \text{......(6)}$$

This indicates an oscillation of period $2\pi/p$, with

$$p^2=K/I+\omega^2\cos^2\lambda. \quad \text{......(7)}$$

The second term is negligible, and we accordingly write p^2I for K in what follows.

Returning to the case of rapid rotation, and assuming a time-factor $e^{i\sigma t}$ in (4), we find

$$\left.\begin{aligned} \left(\sigma^2-p^2-\frac{Cn}{I}\omega\cos\lambda\right)\theta+i\sigma\left(\frac{Cn}{I}-\omega\cos\lambda\right)\phi&=0,\\ -i\sigma\left(\frac{Cn}{J}-\omega\cos\lambda\right)\theta+\left(\sigma^2-\frac{Cn}{J}\omega\cos\lambda\right)\phi&=0,\end{aligned}\right\} \quad \text{......(8)}$$

whence

$$\sigma^4-\sigma^2\left(\frac{C^2n^2}{IJ}+p^2+\omega^2\cos^2\lambda\right)+\frac{Cn\omega\cos\lambda}{J}\left(p^2+\frac{Cn}{I}\omega\cos\lambda\right)=0. \quad \text{...(9)}$$

In practice n is enormous compared with ω and p, whilst p^2I is large compared with $Cn\omega$*, so that we may write

$$\sigma^4-\frac{C^2n^2}{IJ}\sigma^2+\frac{Cn}{J}p^2\omega\cos\lambda=0, \quad \text{......(10)}$$

to a high degree of approximation. The roots of this are

$$\sigma_1^2=\frac{C^2n^2}{IJ}, \quad \sigma_2^2=\frac{I\omega\cos\lambda}{Cn}p^2, \quad \text{......(11)†}$$

very nearly. The former root belongs to a very rapid vibration in which

$$\phi/\theta=-i\sqrt{(I/J)}. \quad \text{......(12)}$$

This would be quickly damped. The second and more important root gives a slow vibration in which

$$\frac{\theta}{\phi}=\frac{i\omega\cos\lambda}{\sigma_2}=i\sqrt{\left(\frac{Cn\omega\cos\lambda}{p^2I}\right)}, \quad \text{......(13)}$$

approximately. Since we have already assumed p^2I to be large compared with $Cn\omega$, this ratio is very small. In practice the period $2\pi/\sigma_2$ is comparable with an hour.

* In the Anschütz form $C=136000$, $n/2\pi=20000/60$, $p^2B=15\,.\,10^6$, in C.G.S. units. Thus $p^2B/Cn\omega=725$.

† With the preceding numerical data we find (at the equator) $\sigma_2/\omega=27$, about.

In the actual apparatus provision is made for damping; this introduces some new features in the theory*.

61. Gyrostatic Control of the Rolling of a Ship.

There are many other practical applications of gyroscopic principles, as for instance the steering of a torpedo, the turn-indicator of an aeroplane, the Brennan monorail, and so on. The only one which we shall here notice is the contrivance for steadying the rolling of a ship, invented by Schlick (1904). This can be readily understood without going into technical details.

A flywheel maintained in rapid rotation is carried by a frame which can swing about an axis at right angles to the medial plane of the vessel. The axis of the flywheel itself moves in this plane, and its standard position is upright, this being the position of stable equilibrium when the ship is at rest, the frame being weighted with this object. The swinging of the frame about the transverse axis is resisted by frictional brakes. Briefly, the principle of the contrivance is that the rolling of the ship produces a deviation of the axis of the flywheel in the medial plane, and a consequent absorption of energy by friction, which is so much lost to the rolling vessel.

If the angular displacements are small, the equations of motion of the frame are obtained by a slight modification of the equations (1) of Art. 58 relating to the nearly vertical top. If x, y denote small rotations about transverse and longitudinal axes, respectively, we have

$$I\ddot{x} = -Kx - Cn\dot{y} - R\dot{x}, \quad J\ddot{y} = Cn\dot{x} + M, \quad \text{..................(1)}$$

where I, J are the moments of inertia of the flywheel and its frame about the transverse and longitudinal axes of the latter, Kx is the righting moment due to gravity, $R\dot{x}$ is the damping moment, and M is the couple exerted on the frame by the ship as it rolls. We write

$$Cn/I = \beta, \quad Cn/J = \gamma, \quad K/I = p^2, \quad R/I = k, \quad M/J = Y,$$

and therefore

$$\ddot{x} + p^2 x + \beta\dot{y} + k\dot{x} = 0, \quad \ddot{y} - \gamma\dot{x} = Y. \quad \text{....................(2)}$$

If we were to write down the equation of rotation of the vessel itself about its longitudinal axis, Y could be eliminated, and we might proceed in this way to the consideration of the free and forced oscillations. The discussion of the question in this form is complicated†, but the matter can be illustrated to a certain extent by examining the effect of a *prescribed* oscillation of the ship, and the consequent absorption of energy.

* Reference may be made to H. Crabtree, *Spinning Tops and Gyroscopic Motion*, London, 1914, and to Klein and Sommerfeld, *Theorie des Kreisels*, Leipzig, 1897–1910.

† See Klein and Sommerfeld, p. 794.

Assuming, then,

$$y = y_0 e^{i\sigma t}, \qquad \text{.........(3)}$$

we have

$$\left.\begin{aligned}(\sigma^2 - p^2 - ik\sigma)\,x - i\beta\sigma y &= 0,\\ -i\gamma\sigma x - \sigma^2 y &= Y,\end{aligned}\right\} \qquad \text{.........(4)}$$

whence

$$Y = -\sigma^2 y + \frac{\beta\gamma\sigma^2 y}{\sigma^2 - p^2 - ik\sigma}. \qquad \text{.........(5)}$$

In real form

$$Y = -\sigma^2 y_0 \cos\sigma t + \frac{\beta\gamma y_0}{\rho}\cos(\sigma t + \epsilon), \qquad \text{.........(6)}$$

if

$$\rho\cos\epsilon = 1 - p^2/\sigma^2, \quad \rho\sin\epsilon = k/\sigma. \qquad \text{.........(7)}$$

This gives the couple necessary to maintain the prescribed oscillation of the frame, viz.

$$y = y_0 \cos\sigma t. \qquad \text{.........(8)}$$

The rate of dissipation of energy is $JY\dot{y}$, the mean value of which is

$$\frac{1}{2}\frac{J\beta\gamma\sigma y_0^2 \sin\epsilon}{\rho} = \frac{1}{2}\frac{Jk\beta\gamma\sigma^4 y_0^2}{(\sigma^2 - p^2) + k^2\sigma^2} \qquad \text{.........(9)}$$

If K be the moment of inertia of the ship about a longitudinal axis through the mass-centre, its mean energy of rolling is $\frac{1}{2}K\sigma^2 y_0^2$. The ratio which the energy dissipated by the breaks in a period $2\pi/\sigma$ bears to this is

$$\frac{8\pi J}{K}\cdot\frac{k\sigma\beta\gamma}{(\sigma^2 - p^2)^2 + k^2\sigma^2} = \frac{8\pi C^2 n^2}{IK}\cdot\frac{k\sigma}{(\sigma^2 - p^2)^2 + k^2\sigma^2}. \qquad \text{.........(10)}$$

The absorption is small if k be very small or very great, the reason in the latter case being that the frame hardly swings at all. The value of k which makes the absorption a maximum is $|\sigma^2 - p^2|/\sigma$, and the absorption is then

$$\frac{4\pi C^2 n^2}{IK|\sigma^2 - p^2|}. \qquad \text{.........(11)}$$

62. Precession and Nutation of the Earth's Axis.

The attraction of the sun on the earth reduces to a force through the centre of mass, together with a couple in a plane containing the sun and the polar axis. An approximate expression for the couple has been found in Art. 19.

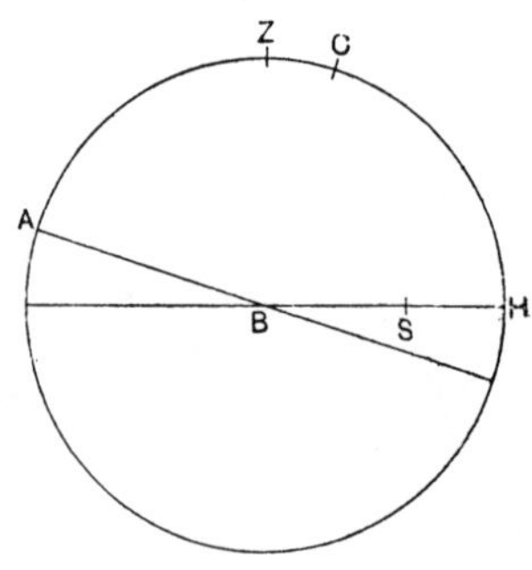

Fig. 53.

In the annexed figure, Z is the pole of the ecliptic, C that of the equator, so that the arc ZC measures the obliquity of the ecliptic, usually denoted by ω. Let B be the ascending node of the ecliptic, and let the point A be chosen on the equator so that OA, OB, OC shall be a right-handed system of axes. Let l be the longitude BS of the sun S, reckoned as usual from the node B. We have

$$\left.\begin{aligned}\cos SA &= -\cos\omega\sin l,\\ \cos SB &= \cos l,\\ \cos SC &= \sin\omega\sin l.\end{aligned}\right\} \qquad \text{.................(1)}$$

Hence if r be the sun's distance from the earth, its coordinates relative to OA, OB, OC will be

$$x = -r\cos\omega\sin l, \quad y = r\cos l, \quad z = r\sin\omega\sin l. \text{......(2)}$$

Hence if we write

$$\kappa = \frac{3}{2}\frac{\gamma S(C-A)}{r^3}, \text{........................(3)}$$

where S is the sun's mass, the expressions (23) of Art. 19 for the components of couple become, on putting $B = A$,

$$L = \kappa\sin\omega\sin 2l, \quad M = \kappa\sin\omega\cos\omega\,(1-\cos 2l), \quad N = 0. \text{...(4)}$$

Owing to the revolution of the sun in the ecliptic, the periodic functions $\sin 2l$ and $\cos 2l$ go through their phases twice in a year, during which time the change of direction of the earth's axis is very slight. The *average* effect is therefore due to a couple

$$M = \kappa\sin\omega\cos\omega \text{..........................(5)}$$

about OB, tending to move C towards Z. Inserting this in the formula (3) of Art. 54, and changing the sign to allow for the altered tendency, we get a precession

$$\dot{\psi} = -\frac{M}{Cn\sin\omega} = -\frac{\kappa}{Cn}\cos\omega, \text{..................(6)}$$

the square of $\dot{\psi}$ being neglected. The negative sign indicates that the motion of the earth's axis is retrograde. The average effect may be illustrated by a gyroscope weighted so that its centre of gravity is below the fixed point.

The neglected functions of $2l$ in (4) produce a semi-annual oscillation of the earth's axis about its mean position, but the amplitude is found on computation to be so small as to be almost insignificant.

If n' be the angular velocity of the earth in its orbit (assumed for the present purpose to be circular), we have $\gamma S/r^2 = n'^2 r$, so that (6) becomes

$$\dot{\psi} = -\frac{3}{2}\frac{n'^2}{n}\frac{C-A}{C}\cos\omega. \qquad \text{..................(7)}$$

The total precession in a year $(2\pi/n')$ due to solar action is accordingly

$$3\pi \frac{n'}{n} \frac{C-A}{C} \cos \omega. \quad \text{.....................(8)}$$

Putting

$$\frac{n'}{n} = \frac{1}{366{\cdot}25}, \quad \frac{C-A}{C} = {\cdot}00327, \quad \omega = 23^\circ\ 27', \quad \text{......(9)}$$

the result is $15{\cdot}9''$.

There is a similar and much greater influence of the moon, but the effect is complicated by the fact that the plane of the lunar orbit is inclined to the ecliptic, and that its line of intersection with the latter is continually regressing. If the orbit were fixed the lunar precession would be about its pole, but owing to solar perturbation of the moon's motion the pole in question describes on the celestial sphere a circle of about $5^\circ\ 9'$ radius about the pole of the ecliptic, backwards, in the comparatively short period of 18·6 years. The result is a mean precession about the pole of the ecliptic, with a periodic inequality going through its phases in the period mentioned. This inequality is known as the 'lunar nutation.'

It is not difficult to calculate, very approximately, the mean value of the lunar precession. Let Z be the pole of the ecliptic C that of the earth, M that of the lunar orbit. The mean velocity of C, due to the moon, is

$$-\frac{\kappa_1}{Cn} \sin MC \cos MC \quad \text{...........(10)}$$

at right angles to MC, where

$$\kappa_1 = \frac{3}{2} \frac{\gamma M (C-A)}{r_1{}^3} = \frac{3}{2} \frac{M}{E} n''^2 (C-A). \quad \text{...(11)}$$

Here M denotes the moon's mass, E that of the earth, r_1 the moon's distance from the earth, n'' the angular velocity in her orbit.

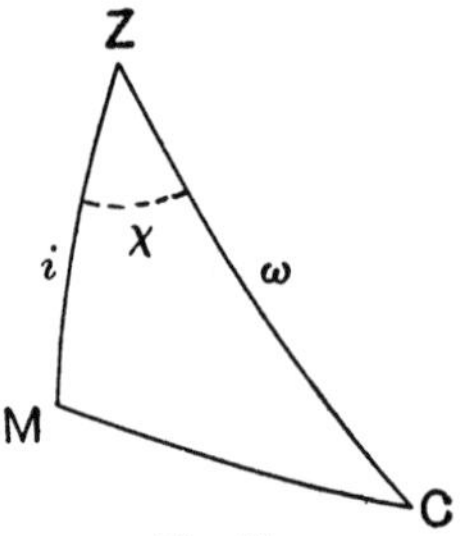

Fig. 54.

Resolving (10) at right angles to ZC, and dividing by $\sin \omega$, we find the rate of precession about Z to be

$$-\frac{\kappa_1}{Cn} \cdot \frac{\sin MC \cos MC \cos ZCM}{\sin \omega}. \quad \text{.....................(12)}$$

Denoting the inclination ZM of the lunar orbit to the ecliptic by i, and putting $MZC = \chi$, we have

$$\cos i = \cos MC \cos \omega + \sin MC \sin \omega \cos ZCM, \quad \text{...........(13)}$$

$$\cos MC = \cos i \cos \omega + \sin i \sin \omega \cos \chi, \quad \text{.................(14)}$$

whence

$$\sin MC \cos ZCM = \cos i \sin \omega - \sin i \cos \omega \cos \chi. \quad \text{...........(15)}$$

Thus (12) becomes

$$\frac{\kappa_1}{Cn} \cdot \frac{(\cos i \sin \omega - \sin i \cos \omega \cos \chi)(\cos i \cos \omega + \sin i \sin \omega \cos \chi)}{\sin \omega}. \quad \text{......(16)}$$

The angle χ goes through its phases in about 19 years, whilst i and ω are approximately constant. Hence, omitting periodic terms, the mean value of the above expression is

$$-\frac{\kappa_1}{Cn} \cdot \frac{3\cos^2 i - 1}{2} \cos\omega. \quad \text{.........................(17)}$$

Referring to (6), we find that the ratio of this to the solar precession is

$$\frac{M}{E}\left(\frac{n''}{n'}\right)^2 \frac{3\cos^2 i - 1}{2}. \quad \text{.........................(18)}$$

Putting $$\frac{M}{E} = \frac{1}{81 \cdot 4}, \quad \frac{n''}{n'} = 13 \cdot 4, \quad i = 5^\circ\, 9', \quad \text{.....................(19)}$$

the ratio is 2·18. With the numerical value already obtained for the solar precession, this gives a total precession of $50''\cdot 5$ per annum, which is very close to the true amount.

The calculation rests on the assumed value of $(C - A)/C$ in (9), and may be regarded as a verification of this. Actually, the value of this important constant is determined by the inverse process, viz. by comparison of the theoretical formula for the precession with the observed amount.

This Chapter has been restricted to an account of some of the simpler and more important illustrations of gyroscopic theory. Some additional examples will be found in Chap. X. For a more extensive treatment, with analytical developments, we may refer to the treatises cited below*.

EXAMPLES. XI.

1. The weight of the rotary part of an aeroplane engine is 300 lbs., and its radius of gyration is 1 ft. The weight of the propeller is 35 lbs., and its radius of gyration is 2·5 ft. The engine makes 20 revolutions per second. If the aeroplane is describing a horizontal circle at the rate of a complete turn in 10 secs., find the pitching moment due to the turning. [1270 ft.-lbs.]

2. A circular disk is spinning, with angular velocity n about its axis, on a smooth table. Prove that in steady motion the two rates of precession are given by

$$\tfrac{1}{2}\kappa^2 \sin\alpha\cos\alpha\,\dot\psi^2 - \kappa^2 n \sin\alpha\,\dot\psi + ga\cos\alpha = 0,$$

where a is the radius of the disk, κ the radius of gyration about the axis, and α the inclination of the axis to the vertical.

3. If the axis of the flywheel of a frictionless gyroscope is clamped at a given inclination to the vertical, the slightest force is sufficient to turn the vertical ring about its axis; but if the axis is *free* a considerable force is at first necessary. Explain this.

* Klein and Sommerfeld, and Greenhill, *ll.c. ante*; A. Gray, *Gyrostatics and Rotational Motion*, London, 1919.

4. The ring which carries the flywheel of a gyroscope is clamped at a given inclination a to the vertical. The whole being initially at rest, the vertical ring is turned through four right angles. Find the angle through which the flywheel would turn relative to its frame, if there were no friction at the pivots. [$-2\pi \cos a$.]

5. A top spins on a *smooth* horizontal plane. Prove that the inclination of the axis to the vertical oscillates between two fixed limits.

Prove that the condition for steady motion has the same form as in Art. 54 (5), provided A, A, C now denote the principal moments of inertia at the mass-centre.

6. Prove that *cusps* can occur in the path of the pole of a top, or a gyroscopic pendulum, only at the upper of the two limiting circles.

7. A slight impulse whose angular momentum is λ is given about a horizontal axis to a spinning top whose axis is vertical. Prove that in the subsequent motion the angle which the axis makes with the vertical is given by

$$\theta = \frac{\lambda}{\sigma A} \sin \sigma t,$$

and that the plane of this angle revolves with the angular velocity ρ, where ρ, σ have the values given in Art. 58.

8. If the axis of a top passes through the vertical position, prove that

$$\dot{\theta}^2 = \sigma^2 + \frac{4Mgh}{A} \sin^2 \tfrac{1}{2}\theta - \frac{\nu^2}{A^2} \tan^2 \tfrac{1}{2}\theta,$$

where σ is the value of $\dot{\theta}$ when $\theta = 0$.

Verify that there is one and only one maximum value of θ.

9. In the preceding example, prove that if σ be small the solution valid for small values of θ has, in general, one or other of the forms

$$\theta = \frac{\sigma}{\beta} \cos(\beta t + \epsilon), \quad \theta = \frac{\sigma}{\beta} \sinh(\beta t + \epsilon);$$

and interpret the two cases.

Prove that, under a special condition, the path of the pole approaches the highest point asymptotically; and examine the nature of the path in this case.

10. A top is spinning very rapidly about its axis, which is inclined at an angle a to the vertical, and momentarily at rest. Prove that the axis will descend through an angle

$$\frac{2AMgh}{C^2 n^2} \sin a,$$

approximately, before again ascending.

Work out numerically, assuming

$M = 1000$, $C = \tfrac{2}{5}A = 12500$, $h = 5$, $a = 30°$, $n/2\pi = 50$ (C.G.S.). [34′.]

11. Prove that the axis of a top whose principal moments are A, A, C, and spin n, will keep step with that of a top whose moments are each equal to A, and whose spin is Cn/A, provided the relation $M'h' = Mh$ is satisfied, and that the latter top is suitably started. Also that the Eulerian angles ϕ, ϕ' in the two cases are connected by the relation

$$\phi' = \phi + \frac{C - A}{A} nt. \qquad \text{(Darboux.)}$$

12. Prove that the result (13) of Art. 57 may be put in the forms

$$p^2 = \frac{A^2\omega^4 - 2AMgh\omega^2 \cos a + M^2g^2h^2}{A^2\omega^2},$$

$$p^2 = \frac{(\mu - \nu \cos a)^2 - 2(\mu - \nu \cos a)(\nu - \mu \cos a)\cos a + (\nu - \mu \cos a)^2}{A^2 \sin^4 a}.$$

(Ferrers; Klein and Sommerfeld.)

13. Prove that the kinetic energy of a gyroscope, when the inertia of the movable rings (assumed symmetrical) is taken into account, is given by

$$2T = (A + B_1)\dot{\theta}^2 + \{(A + A_1)\sin^2\theta + B_1\cos^2\theta + I\}\dot{\psi}^2 + C(\dot{\phi} + \cos\theta\dot{\psi})^2,$$

where A_1, B_1, I are certain constants relating to the rings.

14. The direction-cosines of the axis of a flywheel relative to fixed co-ordinate axes through its centre are l, m, n; prove that the components of angular momentum are

$$\kappa l + A(m\dot{n} - \dot{m}n), \text{ etc.,}$$

where κ is the angular momentum about the axis of symmetry.

15. If in Gilbert's barygyroscope (Art. 59) the horizontal line of pivots makes an angle β with the meridian, prove that the equilibrium deviation from the vertical is given by

$$\tan\chi = \frac{Cn\omega\cos\lambda\sin\beta}{Mgh + Cn\omega\sin\lambda}.$$

16. Prove that if a ship is in motion with a speed whose northerly component is v, the gyrostatic compass will (when in equilibrium) point to the W. of true N. by an amount

$$\frac{v}{\omega a \cos\lambda},$$

where a is the earth's radius.

Calculate the deviation for a northerly speed of 30 knots in latitude 60°.

[3° 50′.]

CHAPTER IX

MOVING AXES

63. Fundamental Equations.

The use of a moving system of coordinate axes has already been illustrated in various cases, and in particular in the deduction of Euler's equations (Art. 49). In order to secure the constancy of the moments of inertia the axes were there taken to be fixed in the moving body. We now proceed to ascertain the form which various fundamental equations assume when the axes are moving in a perfectly general manner.

Let us suppose in the first instance that the origin O is fixed in space. We denote the component angular velocities of the axes about their instantaneous positions by p', q', r', the symbols p, q, r being reserved to denote as before the angular velocities of a rigid body. To calculate the changes which are taking place in the projections x, y, z of a vector OP which is, actually, fixed in space, we remark that if the point P were moving with the axes its component displacements in the time δt would be

$$q'\delta t \,.\, z - r'\delta t \,.\, y, \quad r'\delta t \,.\, x - p'\delta t \,.\, z, \quad p'\delta t \,.\, y - q'\delta t \,.\, x, \quad \ldots(1)$$

by Art. 9 (3). Hence to restore P to its original position in space we must give it the *apparent* displacements

$$\delta x = (r'y - q'z)\,\delta t, \quad \delta y = (p'z - r'x)\,\delta t, \quad \delta z = (q'x - p'y)\,\delta t. \quad \ldots(2)$$

We have therefore the relations

$$\frac{dx}{dt} = r'y - q'z, \quad \frac{dy}{dt} = p'z - r'x, \quad \frac{dz}{dt} = q'x - p'y, \quad \ldots\ldots(3)$$

as the expression of the fact that the vector (x, y, z) is really fixed, and that its projections on the coordinate axes are only changing in consequence of the motion of the axes themselves.

As a particular case these equations hold if we replace x, y, z by the direction-cosines of a line (such as the vertical) which is fixed in direction.

In the case of a variable vector we have to add to the second members of (2) the terms $u\,\delta t$, $v\,\delta t$, $w\,\delta t$ (say) which represent the actual displacements of P parallel to the momentary directions of the axes. Thus

$$\frac{dx}{dt}=u+r'y-q'z, \quad \frac{dy}{dt}=v+p'z-r'x, \quad \frac{dz}{dt}=w+q'x-p'y. \quad \ldots(4)$$

We can at once write down a number of important formulæ. Thus if OP be the position vector of a point P relative to the (fixed) origin, the velocities of P parallel to the coordinate axes are

$$u=\frac{dx}{dt}-r'y+q'z, \quad v=\frac{dy}{dt}-p'z+r'x, \quad w=\frac{dz}{dt}-q'x+p'y. \quad \ldots(5)$$

When the origin is in motion these formulæ give the velocities relative to O. To find the absolute velocities we must add to the second members the terms u', v', w', respectively, which represent the velocity of O.

Again, if OV be drawn to represent the (absolute) velocity (u, v, w) of a moving point, the component velocities of V, i.e. the component accelerations of the moving point, will be

$$\alpha=\frac{du}{dt}-r'v+q'w, \quad \beta=\frac{dv}{dt}-p'w+r'u, \quad \gamma=\frac{dw}{dt}-q'u+p'v. \quad \ldots(6)$$

Again, the momenta of the various particles of any dynamical system are equivalent to a localized vector (ξ, η, ζ) representing the linear momentum, passing through a fixed origin O, and a free vector (λ, μ, ν) representing the angular momentum. The change in the linear momentum is equal to the impulse of the externa forces, whence

$$\frac{d\xi}{dt}-r'\eta+q'\zeta=X, \quad \frac{d\eta}{dt}-p'\zeta+r'\xi=Y, \quad \frac{d\zeta}{dt}-q'\xi+p'\eta=Z, \quad \ldots(7)$$

where (X, Y, Z) is the force-resultant of the total external force on the system.

The change in the angular momentum is given in like manner by the formulæ

$$\frac{d\lambda}{dt}-r'\mu+q'\nu=L, \quad \frac{d\mu}{dt}-p'\nu+r'\lambda=M, \quad \frac{d\nu}{dt}-q'\lambda+p'\mu=N, \quad \ldots(8)$$

where L, M, N are the moments of the external forces about the coordinate axes.

Since the linear momentum is the same whatever the position of O, the formulæ (7) will hold even if the origin be in motion.

The equations (8), on the other hand, will in general be affected. We have seen, however, in Chap. VI that we may take moments about the mass-centre as if it were at rest. The equations will therefore stand when the moving origin coincides with the mass-centre, provided (λ, μ, ν) denote the angular momentum relative to the mass-centre, and (L, M, N) the moment of the external forces relative to this point.

The modification required in (8) when the origin has any given velocity (u', v', w') is as follows. At the time $t+\delta t$ the linear momentum $(\xi+\delta\xi, \eta+\delta\eta, \zeta+\delta\zeta)$ is a localized vector through the new position of the origin. Its moment about the original position of the axis of x is therefore

$$\zeta \,.\, v'\delta t - \eta \,.\, w'\delta t,$$

terms of the second order being omitted. This expression is to be added to the value of $\delta\lambda$. In this way we find

$$\left.\begin{aligned} \frac{d\lambda}{dt} - r'\mu + q'\nu - w'\eta + v'\zeta &= L, \\ \frac{d\mu}{dt} - p'\nu + r'\lambda - u'\zeta + w'\xi &= M, \\ \frac{d\nu}{dt} - q'\lambda + p'\mu - v'\xi + u'\eta &= N. \end{aligned}\right\} \quad \text{............(9)}$$

If the moving origin coincides always with the centre of mass we have

$$\xi = M_0 u', \quad \eta = M_0 v', \quad \zeta = M_0 w', \quad \text{............(10)}$$

where M_0 is the total mass of the system. The equations (9) then reduce, as they ought, to the forms (8).

Ex. 1. In the particular case where the origin and the axis of z are fixed we have $p'=0$, $q'=0$, $r'=\omega$ (say). The equations (5) then reduce to

$$u = \dot{x} - \omega y, \quad v = \dot{y} + \omega x, \quad w = \dot{z}. \quad \text{....................(11)}$$

If ω be constant the accelerations are

$$\left.\begin{aligned} \alpha &= \dot{u} - \omega v = \ddot{x} - 2\omega\dot{y} - \omega^2 x, \\ \beta &= \dot{v} + \omega u = \ddot{y} + 2\omega\dot{x} - \omega^2 y, \\ w &= \ddot{z}. \end{aligned}\right\} \quad \text{.......................(12)}$$

These are known results [D. 33].

Ex. 2. In the case of axes fixed in a moving solid, and coincident with the principal axes of inertia at the origin,

$$p' = p, \quad q' = q, \quad r' = r, \quad \text{..........................(13)}$$

$$\lambda = Ap, \quad \mu = Bq, \quad \nu = Cr. \quad \text{.......................(14)}$$

Substituting in (8) we obtain Euler's equations. When there are no external forces, these express that the vector (Ap, Bq, Cr) which represents the angular momentum is fixed in space. (Arts. 46, 49.)

The application of the notation of Vector Analysis to this case may be noticed. Let OH $(=\mathbf{H})$ be a vector drawn from O to represent the angular momentum, and let ω be the angular velocity. If the point H were moving with the axes its velocity would be $[\omega\mathbf{H}]$, by Art. 23. Hence when H is fixed in space

$$\frac{d\mathbf{H}}{dt} = -[\omega\mathbf{H}], \quad \text{.................................(15)}$$

which is equivalent to Euler's three equations. Writing

$$\mathbf{H} = Ap\mathbf{i} + Bq\mathbf{j} + Cr\mathbf{k}, \quad \text{...........................(16)}$$

$$\omega = p\mathbf{i} + q\mathbf{j} + r\mathbf{k}, \quad \text{.................................(17)}$$

and reducing by means of the formulæ (10) of Art. 23, we obtain the usual Cartesian forms.

Ex. 3. To find the possible modes of steady rotation about a fixed point O of a body subject to gravity.

Since the instantaneous axis is supposed fixed in the body it will be fixed in space. It is easy, then, to see, apart from the calculation, that it must coincide with the vertical.

Let the coordinates of the centre of gravity (G) relative to the principal axes at O be (a, b, c); and let (l, m, n) be the direction of gravity relative to the same axes. Then if A, B, C be the principal moments at O, we have, since by hypothesis $\dot{p}, \dot{q}, \dot{r} = 0$,

$$\left.\begin{aligned} (B-C)\,qr &= Mg\,(nb - mc), \\ (C-A)\,rp &= Mg\,(lc - na), \\ (A-B)\,pq &= Mg\,(ma - lb). \end{aligned}\right\} \quad \text{........................(18)}$$

Also, since the direction OG is fixed in space,

$$\dot{l} = rm - qn, \quad \dot{m} = pn - rl, \quad \dot{n} = ql - pm, \quad \text{..................(19)}$$

by Art. 63 (3). From the first of equations (18), we have $\dot{n}b - \dot{m}c = 0$, whence substituting the values of $\dot{m}, \dot{n}$ from (19),

$$p\,(la + mb + nc) = l\,(pa + qb + rc).$$

Hence

$$\frac{p}{l} = \frac{q}{m} = \frac{r}{n}, = \omega \quad \text{..................................(20)}$$

where ω is the resultant angular velocity. The axis of rotation must therefore coincide with the vertical, as already anticipated. From (18) we now derive

$$(B-C)\,amn + (C-A)\,bnl + (A-B)\,clm = 0, \quad \text{..............(21)}$$

shewing that the axis of rotation must be a generator of a certain quadric cone having O as vertex. This cone contains the principal axes at O, the line OG, and the line whose equations are

$$x : y : z = \frac{a}{A} : \frac{b}{B} : \frac{c}{C}, \quad \text{..........................(22)}$$

and is accordingly determined by these five generators.

If the body is first turned so that any assigned generator is vertical, and set in rotation about this line with the appropriate angular velocity, it will continue so to rotate.

The angular velocity is given by any one of the equations

$$\omega^2 = \frac{Mg(nb-mc)}{(B-C)mn} = \frac{Mg(lc-na)}{(C-A)nl} = \frac{Mg(ma-lb)}{(A-B)lm}, \quad \text{.........(23)}$$

which are equivalent, by (18). If these expressions are negative, they become real on reversing the signs of l, m, n, which means a reversal of the body with respect to the upward and downward directions.

The above theory is due to Staude (1894), and the cone in question is known by his name. It is the locus of the lines through O each of which is a principal axis at some point on it.

In the case of symmetry about an axis ($A=B$, $a=b=0$) the cone is indeterminate. *Any* line through O can then be made an axis of steady rotation.

64. Motion relative to the Earth.

An important application of the theory of moving axes is to the case of motions relative to the rotating earth. The origin being taken at a point near the earth's surface, in latitude λ, let us suppose that the axis of x is directed horizontally to the east, that of y northwards, and that of z vertically upwards. The terms 'horizontal' and 'vertical' have of course reference to the direction of *apparent* gravity. If ω be the earth's angular velocity of rotation, the component angular velocities of the axes about their instantaneous positions will be

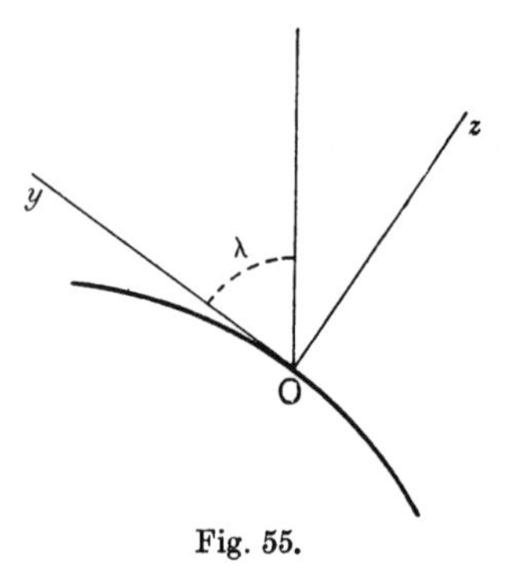

Fig. 55.

$$p' = 0, \quad q' = \omega \cos\lambda, \quad r' = \omega \sin\lambda. \quad \text{...............(1)}$$

The component velocities of a moving point are therefore, by Art. 63 (5),

$$\left.\begin{aligned} u &= \frac{dx}{dt} - \omega y \sin\lambda + \omega z \cos\lambda, \\ v &= \frac{dy}{dt} + \omega x \sin\lambda, \\ w &= \frac{dz}{dt} - \omega x \cos\lambda. \end{aligned}\right\} \quad \text{...............(2)}$$

The component accelerations are

$$\left.\begin{aligned} \alpha &= \frac{du}{dt} - \omega v \sin\lambda + \omega w \cos\lambda, \\ \beta &= \frac{dv}{dt} + \omega u \sin\lambda, \\ \gamma &= \frac{dw}{dt} - \omega u \cos\lambda. \end{aligned}\right\} \quad \ldots\ldots\ldots\ldots\ldots(3)$$

The equations of motion of a particle m, subject to forces X, Y, Z, as well as to apparent gravity, are then

$$m\alpha = X, \quad m\beta = Y, \quad m\gamma = Z - mg. \quad \ldots\ldots\ldots\ldots(4)$$

A remark is called for here. We have tacitly treated the origin O as fixed, whereas it is really in motion in a diurnal circle, and has consequently a velocity $\omega a \cos\lambda$ in this circle. This might be rectified by the addition of suitable terms in (2) and (3), which as they stand express velocities and accelerations relative to O; but the changes would finally disappear from the equations (4) when written out at length, owing to the fact that the axis of z is assumed to coincide with the direction of *apparent* gravity (g), as affected by 'centrifugal force,' and that λ is accordingly taken to denote the 'geographical' as distinguished from the 'geocentric' latitude.

Substituting from (2) and (3) in (4), we have

$$\left.\begin{aligned} &\frac{d^2x}{dt^2} - 2\omega\frac{dy}{dt}\sin\lambda + 2\omega\frac{dz}{dt}\cos\lambda - \omega^2 x = \frac{X}{m}, \\ &\frac{d^2y}{dt^2} + 2\omega\frac{dx}{dt}\sin\lambda - \omega^2 y \sin^2\lambda + \omega^2 z \sin\lambda\cos\lambda = \frac{Y}{m}, \\ &\frac{d^2z}{dt^2} - 2\omega\frac{dx}{dt}\cos\lambda + \omega^2 y \sin\lambda\cos\lambda - \omega^2 z \cos^2\lambda = \frac{Z}{m} - g. \end{aligned}\right\} \ldots(5)$$

The terms in these equations* which contain $\omega^2 x$, $\omega^2 y$, $\omega^2 z$ are however utterly negligible in any practical case. The values of x, y, z are always insignificant in comparison with the earth's radius (a), and $\omega^2 a$ is itself small compared with g. Indeed if we were to retain the terms in question, we ought for consistency to include the local variation of g.

* Due to Poisson (1838).

65. Foucault's Pendulum.

To apply the equations to the (small) motion of a pendulum, we put

$$X = -Tx/l, \quad Y = -Ty/l, \quad Z = -Tz/l, \quad \text{.........(1)}$$

where l is the length, T is the tension of the string and x, y are the horizontal components of the displacement of the bob from its mean position. The third equation of Art. 64 (5) shews that $T = mg$, approximately. Hence, neglecting terms of the second order,

$$\left.\begin{aligned} \frac{d^2x}{dt^2} - 2\omega \frac{dy}{dt} \sin \lambda &= -n^2 x, \\ \frac{d^2y}{dt^2} + 2\omega \frac{dx}{dt} \sin \lambda &= -n^2 y, \end{aligned}\right\} \quad \text{..............(2)}$$

where $$n^2 = g/l.$$

If we put $\zeta = x + iy$, these combine into the equation

$$\frac{d^2\zeta}{dt^2} + 2i\omega \frac{d\zeta}{dt} \sin \lambda + n^2 \zeta = 0. \quad \text{.................(3)}$$

Hence $$\zeta e^{i\omega t \sin \lambda} = He^{i\sigma t} + Ke^{-i\sigma t}, \quad \text{................(4)}$$

where $$\sigma = \sqrt{(n^2 + \omega^2 \sin^2 \lambda)}, \quad \text{....................(5)}$$

and the (complex) constants H, K are arbitrary. This represents motion in an ellipse which turns round the vertical with the constant angular velocity $\omega \sin \lambda$ in the negative direction, the period in the ellipse being $2\pi/\sigma$. In any practical case the distinction between σ and n is of course negligible.

The simple view of the matter is that the earth is rotating about the vertical at the rate $\omega \sin \lambda$, in the positive direction, beneath the pendulum.

66. Relative Motion of a Particle under Gravity.

In the case of motion under apparent gravity alone, the equations, with the omissions referred to in Art. 64, become

$$\left.\begin{aligned} \frac{d^2x}{dt^2} - 2\omega \frac{dy}{dt} \sin \lambda + 2\omega \frac{dz}{dt} \cos \lambda &= 0, \\ \frac{d^2y}{dt^2} + 2\omega \frac{dx}{dt} \sin \lambda &= 0, \\ \frac{d^2z}{dt^2} - 2\omega \frac{dx}{dt} \cos \lambda &= -g. \end{aligned}\right\} \quad \text{........(1)}$$

The second and third of these give

$$\left.\begin{aligned}\frac{dy}{dt} &= v_0 - 2\omega x \sin\lambda,\\ \frac{dz}{dt} &= w_0 - gt + 2\omega x \cos\lambda,\end{aligned}\right\} \qquad \ldots\ldots\ldots\ldots\ldots\ldots(2)$$

where v_0, w_0 are constants of integration. Substituting in the first equation we have

$$\frac{d^2x}{dt^2} + 4\omega^2 x = 2\omega\,(v_0 \sin\lambda - w_0 \cos\lambda) + 2\omega g t \cos\lambda, \quad \ldots(3)$$

the integral of which is

$$x = A\cos 2\omega t + B\sin 2\omega t + \frac{v_0 \sin\lambda - w_0\cos\lambda}{2\omega} + \frac{gt\cos\lambda}{2\omega}. \quad \ldots(4)$$

The values of y and z are then to be found by substituting this value of x in (2), and integrating.

In the case of free fall from (relative) rest at the origin the initial values of x, y, z, dx/dt, dy/dt, dz/dt are zero. Hence

$$v_0 = 0, \quad w_0 = 0, \quad A = 0, \quad B = -\frac{g\cos\lambda}{4\omega^2}. \quad \ldots\ldots\ldots\ldots\ldots\ldots(5)$$

Since the angle ωt through which the earth turns during the descent is very small, we expand $\sin 2\omega t$ in (4), and retain only the more important part of the result. The easterly deviation from the vertical through the starting-point is accordingly

$$x = \tfrac{1}{3} g\omega t^3 \cos\lambda. \quad \ldots\ldots\ldots\ldots\ldots\ldots(6)$$

Substituting in (2), we find for the northerly deviation

$$y = -\tfrac{1}{6} g\omega^2 t^4 \sin\lambda\cos\lambda, \quad \ldots\ldots\ldots\ldots\ldots\ldots(7)$$

but the ratio of this to (6) is of the order ωt, and the effect is therefore altogether negligible. Again, we have

$$z = -\tfrac{1}{2} g t^2, \quad \ldots\ldots\ldots\ldots\ldots\ldots(8)$$

approximately, the omitted term being of the same order as (7). The only effect of the earth's rotation which could be expected to be at all appreciable is the easterly deviation given by (6). In a fall of 100 metres this would amount to about 2·2 cos λ centimetres. The deviation is so slight, and it is so easily affected by accidental disturbances, that it is difficult to verify it with certainty by experiment. The attempt has however been made, with some success*.

* For references see Routh, *Dynamics of a Particle*, Art. 627.

In the application to the unresisted flight of projectiles the initial conditions may be taken to be

$$\left.\begin{array}{lll} x=0, & y=0, & z=0, \\ dx/dt=u_0, & dy/dt=v_0, & dz/dt=w_0. \end{array}\right\} \quad \text{.................(9)}$$

Hence in (4)

$$A=-\frac{v_0 \sin\lambda - w_0 \cos\lambda}{2\omega}, \quad B=\frac{u_0}{2\omega}-\frac{g\cos\lambda}{4\omega^2}. \quad \text{...........(10)}$$

Thus

$$x=\frac{v_0 \sin\lambda - w_0 \cos\lambda}{2\omega}(1-\cos 2\omega t)+\frac{g\cos\lambda}{4\omega^2}(2\omega t-\sin 2\omega t)+\frac{u_0}{2\omega}\sin 2\omega t. \quad \text{...(11)}$$

If we substitute in (2), and adjust the constants of integration so as to make y and $\dot{z}$ vanish for $t=0$, we find

$$y=v_0 t-\frac{\sin\lambda\,(v_0\sin\lambda-w_0\cos\lambda)}{2\omega}(2\omega t-\sin 2\omega t)$$
$$-\frac{g\sin\lambda\cos\lambda}{4\omega^2}(2\omega^2t^2+\cos 2\omega t-1)-\frac{u_0\sin\lambda}{2\omega}(1-\cos 2\omega t), \quad \text{...(12)}$$

$$z=w_0 t-\tfrac{1}{2}gt^2+\frac{\cos\lambda\,(v_0\sin\lambda-w_0\cos\lambda)}{2\omega}(2\omega t-\sin 2\omega t)$$
$$+\frac{g\cos^2\lambda}{4\omega^2}(2\omega^2t^2+\cos 2\omega t-1)+\frac{u_0\cos\lambda}{2\omega}(1-\cos 2\omega t). \quad \text{...(13)}$$

These formulæ are exact, so far as variations in the value of g can be neglected; but a sufficient approximation is obtained if we expand the circular functions, and retain only the more important terms. In this way we find

$$x=u_0 t+(v_0\sin\lambda-w_0\cos\lambda)\,\omega t^2+\tfrac{1}{3}g\cos\lambda\,.\,\omega t^3, \quad \text{...........(14)}$$
$$y=v_0 t-u_0\sin\lambda\,.\,\omega t^2, \quad \text{..(15)}$$
$$z=w_0 t-\tfrac{1}{2}gt^2+u_0\cos\lambda\,.\,\omega t^2. \quad \text{..................................(16)}$$

The terms here omitted are of the order ω^2t^2 as compared with those retained.

If the vertical plane of projection make an angle β with the plane zx, we have

$$u_0=u_0'\cos\beta, \quad v_0=u_0'\sin\beta, \quad \text{.......................(17)}$$

where u_0' is the initial horizontal (relative) velocity. The horizontal distance travelled in the plane of projection is

$$x'=x\cos\beta+y\sin\beta=u_0't+(\tfrac{1}{3}gt-w_0)\cos\beta\cos\lambda\,.\,\omega t^2, \quad \text{.........(18)}$$

and the deviation from this plane is

$$y'=y\cos\beta-x\sin\beta=\{(w_0-\tfrac{1}{3}gt)\sin\beta\cos\lambda-u_0'\sin\lambda\}\,\omega t^2. \quad \text{......(19)}$$

This is positive when the deviation is to the left.

67. Rolling of a Solid on a Fixed Surface.

The method of rotating axes is also convenient in problems relating to the rolling or spinning of solids on fixed surfaces. Considerable attention has at one time or another been bestowed on such questions by mathematicians. It is not that the phenomena, though familiar and often interesting, were held to be specially important, but it was regarded rather as a point of honour to shew how the mathematical formulation could be effected, even if the solution should prove to be impracticable, or difficult of interpretation.

The problems of this kind which are most tractable are, naturally, cases of steady motion, or of small oscillations about such a state. We proceed to the discussion of a few of the simpler examples.

We take first the steady motion of a sphere rolling on a surface of revolution whose axis is vertical.

We adopt a right-handed system of axes through the centre G, of which Gz is normal to the fixed surface, Gx is in the vertical plane through Gz, and Gy is horizontal. Let a be the radius of the sphere, c that of the horizontal circle which is assumed to be described by G, $\dot{\psi}$ the angular velocity in this circle, θ the constant inclination of Gz to the upward vertical.

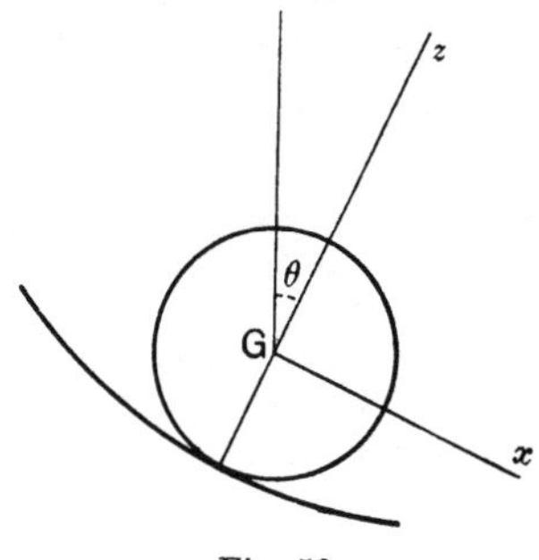

Fig. 56.

Let (X, Y, Z) be the reaction at the point of contact. Since the acceleration of G is $c\dot{\psi}^2$, towards the centre of its horizontal circle,

$$X = -Mg\sin\theta + Mc\dot{\psi}^2\cos\theta, \quad Y=0, \quad Z = Mg\cos\theta + Mc\dot{\psi}^2\sin\theta. \quad \text{...(1)}$$

Since our axes are rotating about the vertical with the angular velocity $\dot{\psi}$ we have

$$p' = -\dot{\psi}\sin\theta, \quad q'=0, \quad r'=\dot{\psi}\cos\theta. \quad \text{..................(2)}$$

If (p, q, r) be the angular velocity of the sphere itself we have the kinematical relations

$$ap = c\dot{\psi}, \quad q=0. \quad \text{.................................(3)}$$

The components of angular momentum are

$$\lambda = Ip, \quad \mu = 0, \quad \nu = Ir, \quad \text{.........................(4)}$$

where I is the moment of inertia about a diameter. The equations (8) of Art. 63 now reduce to

$$\frac{d\lambda}{dt} = Ya = 0, \quad -p'\nu + r'\lambda = -Xa, \quad \frac{d\nu}{dt} = 0. \quad \text{..............(5)}$$

The first of these is satisfied already, since we have assumed that $\dot{\psi}$ is constant. The third shews that $r =$ const., $= n$, say. The remaining equation leads to

$$(I + Ma^2)\frac{c}{a}\dot{\psi}^2 \cos\theta + In\,\dot{\psi}\sin\theta - Mga\sin\theta = 0, \quad \text{...........(6)}$$

which is the required condition for steady motion.

For given values of c, θ, and n there are two possible values of $\dot{\psi}$ of different magnitudes, the motions not being reversible unless we reverse also the spin.

The above investigation applies, as the figure indicates, to the case of a sphere rolling on the inside of a concave surface, and the two values of $\dot{\psi}$, when real, have then opposite signs. To make it applicable to the rolling of a sphere on a convex surface we must reverse the signs of c and $\dot{\psi}$.

68. Rolling of a Sphere on a Spherical Surface.

The general equations of motion of a sphere rolling on a fixed *spherical* surface may be obtained as follows.

We take as the standard case that indicated by Fig. 56. With the same arrangement of the moving axes as in the preceding Art., let (u, v, w) be the velocity of G. Then

$$u = -b\dot{\theta}, \qquad v = -b\sin\theta\dot{\psi}, \quad w = 0, \quad \text{..................(1)}$$

$$p' = -\sin\theta\dot{\psi}, \quad q' = \dot{\theta}, \qquad r' = \cos\theta\dot{\psi}, \quad \text{............(2)}$$

where b is the distance of G from the centre O of the fixed surface.

The remaining kinematical relations are

$$ap = b\sin\theta\dot{\psi}, \quad aq = -b\dot{\theta}. \quad \text{..........................(3)}$$

The components of momentum are

$$\xi = Mu, \quad \eta = Mv, \quad \zeta = 0, \quad \text{..........................(4)}$$

$$\lambda = Ip, \quad \mu = Iq, \quad \nu = Ir. \quad \text{..........................(5)}$$

The equations of linear momentum (Art. 63 (7)) become

$$\left.\begin{aligned} M\left(\frac{du}{dt} - r'v\right) &= X + Mg\sin\theta, \\ M\left(\frac{dv}{dt} + r'u\right) &= Y, \end{aligned}\right\} \quad \text{.....................(6)}$$

together with an equation which serves merely to determine the normal reaction Z.

The third equation of angular momentum makes $dr/dt=0$, whence $r=n$, a constant. The remaining equations are

$$\left.\begin{aligned}\frac{d\lambda}{dt}-r'\mu+q'\nu&=Ya,\\ \frac{d\mu}{dt}-p'\nu+r'\lambda&=-Xa.\end{aligned}\right\}\quad\text{.........................(7)}$$

If we eliminate X and Y between (6) and (7), and substitute the values of u, v, p, q, p', q', r' from (1), (2), and (3), we are led to the following equations:

$$\left.\begin{aligned}\left(1+\frac{\kappa^2}{a^2}\right)(\ddot{\theta}-\sin\theta\cos\theta\dot{\psi}^2)-\frac{\kappa^2}{ab}n\dot{\psi}\sin\theta&=-\frac{g}{b}\sin\theta,\\ \left(1+\frac{\kappa^2}{a^2}\right)\frac{d}{dt}(\sin^2\theta\dot{\psi})+\frac{\kappa^2}{ab}n\dot{\theta}\sin\theta&=0,\end{aligned}\right\}\quad\text{.........(8)}$$

where $\kappa^2=I/M$.

If we put $\kappa=0$ we get the equations of the spherical pendulum [D. 103].

The condition for steady motion is obtained by making $\theta=\text{const.}$, $\dot{\psi}=\text{const.}$ The result agrees with Art. 67 (6) if we put $c=b\sin\theta$.

If we multiply the equations (8) by $\dot{\theta}$ and $\dot{\psi}$, respectively, and add, we find on integration

$$\frac{1}{2}\left(1+\frac{\kappa^2}{a^2}\right)(\dot{\theta}^2+\sin^2\theta\dot{\psi}^2)=\frac{g}{b}\cos\theta+\text{const.},\quad\text{...............(9)}$$

which is easily recognized as the equation of energy.

The second of equations (8) is immediately integrable, but it should be noticed that it does *not* imply the constancy of angular momentum about the vertical through O. Since Y does not vanish, this angular momentum is in fact variable.

The case of a sphere rolling in a *circular cylinder* is simpler. We take the axis of z parallel to the axis of the cylinder, and the axes of x, y as in the figure. We have then

$$p'=0,\quad q'=0,\quad r'=\dot{\theta}.\quad\text{..........................(10)}$$

The kinematical relations are

$$u=-ar=b\dot{\theta},\quad v=0,\quad w=ap,\quad\text{.....................(11)}$$

where b is the distance of G from the axis of the cylinder.

Suppose in the first instance that there are no external forces other than the reaction (X, Y, Z) at the point of contact. The dynamical equations then reduce to

$$M\dot{u}=X,\quad Mu\dot{\theta}=Y,\quad M\dot{w}=Z,\quad\text{.....................(12)}$$

$$I(\dot{p}-q\dot{\theta})=-Za,\quad I(\dot{q}+p\dot{\theta})=0,\quad I\dot{r}=Xa.\text{...............(13)}$$

Eliminating X and r, we find $\dot{u}=0$, $\dot{r}=0$, whence $\dot{\theta}=\omega$, a constant. Again, eliminating Z and p,

$$(I+Ma^2)\,\dot{w}-Ia\omega q=0. \qquad (14)$$

Eliminating q between this and the second equation of (13), and having regard to (11), we find

$$(I+Ma^2)\,\ddot{w}+I\omega^2 w=0. \qquad (15)$$

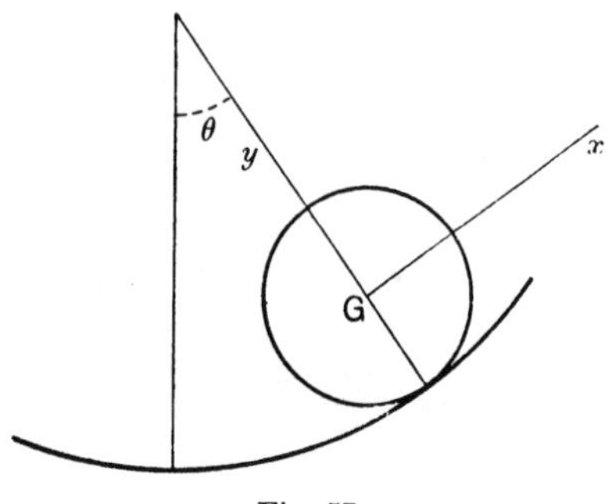

Fig. 57.

The longitudinal velocity of G is therefore periodic, viz.,

$$w=F\cos(\sigma t+\epsilon), \qquad (16)$$

where

$$\sigma=\sqrt{\left(\frac{I}{I+Ma^2}\right)}\cdot\omega. \qquad (17)$$

The solution is completed by the relations

$$p=\frac{w}{a},\quad q=\text{const.}-\frac{F\omega}{a\sigma}\sin(\sigma t+\epsilon),\quad r=-\frac{b}{a}\omega. \qquad (18)$$

When gravity acts we must write, in (12), $X-Mg\sin\theta$ in place of X, the axis of the cylinder being supposed horizontal. Eliminating X, we now have

$$\ddot{\theta}=-\frac{Ma^2}{I+Ma^2}\frac{g}{b}\sin\theta, \qquad (19)$$

exactly as if the motion were two-dimensional [D. 63].

The equation (14) is replaced by

$$(I+Ma^2)\,\dot{w}-Ia\dot{\theta}q=0. \qquad (20)$$

In the case of a small disturbance of the state of steady motion in which the sphere rolls along the lowest generator of the cylinder, θ is small. If in the steady motion the sphere has no spin about the vertical, $\dot{w}$ is small, of the second order, and the longitudinal velocity of G is therefore very approximately constant. If, on the other hand, there is a finite spin n about the vertical, we have

$$w=\text{const.}+\frac{Ian}{I+Ma^2}\theta, \qquad (21)$$

approximately. The longitudinal motion of the centre is therefore affected by a small periodic term.

69. Rolling Wheel.

We take next the case of a wheel rolling on a horizontal plane.

Let Gz be the normal to the plane of the wheel, making an angle θ with the vertical, whilst Gx passes through the point of contact. If ψ be the azimuth of the plane zx, we have

$$p' = -\sin\theta\dot{\psi},\quad q' = \dot{\theta},\quad r' = \cos\theta\dot{\psi}. \quad\text{.....................(1)}$$

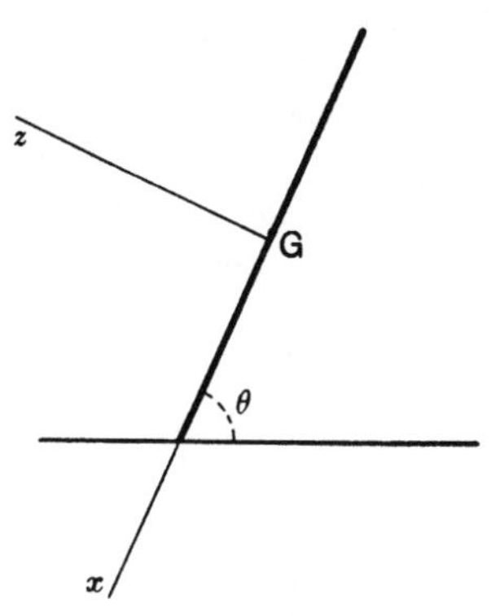

Fig. 58.

Also, considering the motion of a point on the axis of the wheel, fixed relatively to it, we have

$$p = p',\quad q = q'. \quad\text{..................................(2)}$$

Denoting by (u, v, w) the velocity of the centre, we have the kinematical relations

$$u = 0,\quad v = -ar,\quad w = aq, \quad\text{.........................(3)}$$

where a is the radius.

The equations of linear momentum, Art. 63 (7), reduce to

$$\left.\begin{aligned} M(arr' + aq^2) &= X + Mg\sin\theta, \\ M(-a\dot{r} - apq) &= Y, \\ M(a\dot{q} - arp) &= Z - Mg\cos\theta, \end{aligned}\right\} \quad\text{.........................(4)}$$

where X, Y, Z are the components of the reaction of the plane.

The principal moments at G being A, A, C, the equations of angular momentum become

$$\left.\begin{aligned} &A\dot{p} - Aqr' + Cqr = 0, \\ &A\dot{q} - Crp + Apr' = -Za, \\ &C\dot{r} \qquad\qquad\quad = Ya. \end{aligned}\right\} \quad\text{...........................(5)}$$

Eliminating Y, Z, and writing, for shortness,

$$A' = A + Ma^2,\quad C' = C + Ma^2, \quad\text{.........................(6)}$$

we have

$$\left.\begin{aligned} A'\dot{q} - C'rp + Apr' &= -Mga\cos\theta, \\ C'\dot{r} + Ma^2pq &= 0. \end{aligned}\right\} \quad\text{.....................(7)}$$

Substituting from (1) in these, and in the first of (5), we find

$$\left.\begin{aligned} A'\ddot{\theta} - A\sin\theta\cos\theta\dot{\psi}^2 + C'r\sin\theta\dot{\psi} = -Mga\cos\theta, \\ \frac{d}{dt}(A\sin^2\theta\dot{\psi}) - Cr\dot{\theta}\sin\theta = 0, \\ C'\dot{r} - Ma^2\sin\theta\dot{\theta}\dot{\psi} = 0. \end{aligned}\right\} \quad \text{..........(8)}$$

The equations are complicated but one or two problems of slightly disturbed steady motion may be noticed.

If the wheel be rolling on a nearly straight course, with its plane nearly vertical, we put $\theta = \frac{1}{2}\pi + \chi$, and assume χ and ψ to be small. Neglecting terms of the second order, we have $\dot{r} = 0$, or $r = n$ (say), and

$$\left.\begin{aligned} A'\ddot{\chi} + C'n\dot{\psi} - Mga\chi = 0, \\ A\ddot{\psi} - Cn\dot{\chi} = 0. \end{aligned}\right\} \quad \text{..........(9)}$$

The latter equation makes

$$A\dot{\psi} - Cn\chi = \text{const.} \quad \text{..........(10)}$$

Hence, omitting the constant,

$$A'\ddot{\chi} + \left(\frac{CC'n^2}{A} - Mga\right)\chi = 0. \quad \text{..........(11)}$$

The variation of χ is simple harmonic, and the steady motion is therefore stable, provided

$$n^2 > \frac{AMga}{CC'}. \quad \text{..........(12)}$$

Thus in the case of a hoop, the condition is

$$n^2 > \frac{1}{4}\frac{g}{a}. \quad \text{..........(13)}$$

If we retain the constant in (10), but assume it to be small, the solution will represent an oscillation about a state of steady motion in which the centre describes a horizontal circle of very large radius.

If the wheel be set spinning about a vertical diameter, with angular velocity ω, and slightly disturbed, χ and r are to be treated as small. The equations (8) make

$$\left.\begin{aligned} A'\ddot{\chi} + A\omega^2\chi + C'\omega r - Mga\chi = 0, \\ C'\dot{r} - Ma^2\omega\dot{\chi} = 0, \end{aligned}\right\} \quad \text{..........(14)}$$

approximately. From the latter equation we have

$$C'r = Ma^2\omega\chi, \quad \text{..........(15)}$$

an additive constant being omitted. Hence

$$A'\ddot{\chi} + (A'\omega^2 - Mga)\chi = 0. \quad \text{..........(16)}$$

For stability we must have

$$\omega^2 > Mga/A'. \quad \text{..........(17)}$$

In the case of a hoop this becomes

$$\omega^2 > \frac{2}{3}\frac{g}{a}. \quad \text{..........(18)}$$

The retention of a small arbitrary constant in (15) would merely involve small constant additions to the values of χ and r.

70. General scheme of Equations for the Rolling of a Solid on a Horizontal Plane.

In the preceding examples the treatment has been influenced to some extent by the special circumstances of each case. It may be worth while to give the general scheme of equations applicable to any case of a body moving, under gravity, in contact with a horizontal plane.

Taking the principal axes at the mass-centre G as axes of coordinates we have, in the first place, Euler's equations

$$\left.\begin{aligned} A\dot{p} - (B - C)\, qr &= yZ - zY, \\ B\dot{q} - (C - A)\, rp &= zX - xZ, \\ C\dot{r} - (A - B)\, pq &= xY - yX, \end{aligned}\right\} \quad \ldots\ldots\ldots\ldots\ldots(1)$$

where x, y, z are the coordinates of the point of contact, and X, Y, Z the components of the reaction there. Again, denoting by (u, v, w) the velocity of G, we have

$$\left.\begin{aligned} M(\dot{u} - rv + qw) &= X + Mgl, \\ M(\dot{v} - pw + ru) &= Y + Mgm, \\ M(\dot{w} - qu + pv) &= Z + Mgn, \end{aligned}\right\} \quad \ldots\ldots\ldots\ldots\ldots(2)$$

where l, m, n are the direction-cosines of the downward vertical. Since this is also the direction of the normal to the surface of the solid at the point of contact, l, m, n are known functions of x, y, z, which (again) are connected by the equation of the surface.

Since (l, m, n) is a fixed direction in space, we have, by Art. 63,

$$\left.\begin{aligned} \dot{l} - rm + qn &= 0, \\ \dot{m} - pn + rl &= 0, \\ \dot{n} - ql + pm &= 0. \end{aligned}\right\} \quad \ldots\ldots\ldots\ldots\ldots\ldots\ldots\ldots\ldots\ldots(3)$$

If the solid rolls without slipping, the velocity of that point which is in contact with the plane must vanish; thus

$$\left.\begin{aligned} u - ry + qz &= 0, \\ v - pz + rx &= 0, \\ w - qx + py &= 0. \end{aligned}\right\} \quad \ldots\ldots\ldots\ldots\ldots\ldots\ldots\ldots\ldots\ldots(4)$$

If on the other hand the plane is smooth, we have in place of (4)

$$X = lR, \quad Y = mR, \quad Z = nR, \quad \ldots\ldots\ldots\ldots\ldots(5)$$

where R is the normal pressure, together with the kinematical relation

$$l(u - ry + qz) + m(v - pz + rx) + n(w - qx + py) = 0. \ldots(6)$$

It is easy to satisfy ourselves that the problem, as thus formulated, is in each of the above cases mathematically determinate.

A general discussion of the equations is out of the question, but problems of disturbed steady motion are more or less tractable. For example, if the body was originally rotating about a principal axis (Gz) which was vertical and coincident with the normal at the point of contact, then in a slight disturbance from this state the quantities

$$p, q, x, y, X, Y, u, v, l, m$$

will be small. Neglecting terms of the second order, the third of equations (1) shews that $\dot{r}=0$, whence $r=\text{const.}$ We may also put

$$w=0, \quad n=1, \quad Z=-Mg, \quad z=\text{const.}=h, \text{ say.} \qquad \text{.........(7)}$$

If the lines of curvature of the surface of the solid at the original point of contact are parallel to Gx, Gy, the equation of the surface in this neighbourhood will be of the form

$$z=h-\tfrac{1}{2}(\alpha x^2+\beta y^2), \qquad \text{.........(8)}$$

whence

$$l=\alpha x, \quad m=\beta y, \qquad \text{.........(9)}$$

approximately.

The resulting equations are now linear. The results for one or two cases are given in the examples at the end of this Chapter.

71. The Stability of Aircraft.

The method of moving axes is frequently employed in investigations on the stability of an airship or aeroplane. The following is an outline of the procedure for the case of straight horizontal flight.

We adopt a right-handed system of axes, fixed in the body. The axis of x is drawn forwards from the centre of gravity, parallel to the axis of the screw, and therefore in the plane of symmetry; that of y is drawn normal to this plane, towards the right, or starboard, side. Accordingly when these axes are horizontal that of z is directed vertically downwards.

We denote by A, B, C, F, G, H the moments and products of inertia with respect to these axes, but it is obvious that on account of the symmetry with respect to the plane zx the products F and H vanish. The components of linear and angular momentum are accordingly, in our previous notation,

$$\left.\begin{aligned} \xi &= M_0 u, & \eta &= M_0 v, & \zeta &= M_0 w, \\ \lambda &= Ap - Gr, & \mu &= Bq, & \nu &= Cr - Gp. \end{aligned}\right\} \qquad \text{.........(1)}$$

The equations of motion of an aeroplane* are therefore

$$\left.\begin{aligned} M_0\left(\frac{du}{dt} - rv + qw\right) &= M_0 gl + P - X, \\ M_0\left(\frac{dv}{dt} - pw + ru\right) &= M_0 gm - Y, \\ M_0\left(\frac{dw}{dt} - qu + pv\right) &= M_0 gn - Z, \end{aligned}\right\} \quad \ldots\ldots\ldots\ldots(2)$$

$$\left.\begin{aligned} A\frac{dp}{dt} - G\frac{dr}{dt} - (B - C)\,qr - Gpq &= -L, \\ B\frac{dq}{dt} - (C - A)\,rp - G\,(p^2 - r^2) &= -M, \\ C\frac{dr}{dt} - G\frac{dp}{dt} - (A - B)\,pq + Gqr &= -N, \end{aligned}\right\} \quad \ldots\ldots(3)$$

where M_0 is the total mass†, l, m, n are the direction-cosines of the downward vertical, P is the thrust of the screw, and $-X$, $-Y$, $-Z$, $-L$, $-M$, $-N$ are the force- and couple-components of air-resistance‡.

To these we must add the kinematical equations

$$\left.\begin{aligned} \frac{dl}{dt} - rm + qn &= 0, \\ \frac{dm}{dt} - pn + rl &= 0, \\ \frac{dn}{dt} - ql + pm &= 0, \end{aligned}\right\} \quad \ldots\ldots\ldots\ldots\ldots\ldots\ldots\ldots(4)$$

which express the fixity of the vertical.

Assuming that in steady flight the axis of x is horizontal, the equations must be satisfied by $u = U$, say, whilst v, w, p, q, r, l, m all vanish, and $n = 1$. Hence

$$X_1 = P, \quad Y_1 = 0, \quad Z_1 = M_0 g, \quad L_1 = 0, \quad M_1 = 0, \quad N_1 = 0, \ldots(5)$$

where the unit suffix relates to the steady state. To adapt the

* In the case of an airship the terms involving g do not occur in equations (2), gravity being compensated by the buoyancy. On the other hand the equations (3) must be modified by the introduction of the couples due to gravity and buoyancy combined, acting through the respective centres.

† In the case of an airship the 'virtual mass,' which includes a correction for the inertia of the air, may be important.

‡ A further modification is required in the case of an airship, owing to the fact that the thrust of the screws is in a line considerably below the centre of mass.

equations to the case of a slight disturbance we write $U+u$ for u and omit terms of the second order in u, v, w, p, q, r, l, m, putting therefore $n=1$. Thus

$$\left.\begin{aligned} M_0\frac{du}{dt} - M_0 gl + X - X_1 = 0, \\ M_0\left(\frac{dv}{dt} + Ur\right) - M_0 gm + Y - Y_1 = 0, \\ M_0\left(\frac{dw}{dt} - Uq\right) + Z - Z_1 = 0, \end{aligned}\right\} \quad \text{.........(6)}$$

$$\left.\begin{aligned} A\frac{dp}{dt} - G\frac{dr}{dt} + L - L_1 = 0, \\ B\frac{dq}{dt} \qquad\quad + M - M_1 = 0, \\ C\frac{dr}{dt} - G\frac{dp}{dt} + N - N_1 = 0, \end{aligned}\right\} \quad \text{.............(7)}$$

with
$$\frac{dl}{dt} + q = 0, \quad \frac{dm}{dt} - p = 0. \quad \text{..................(8)}$$

Owing to the altered velocity of the air relative to the various parts of the machine, the quantities $X - X_1, \ldots, \ldots, L - L_1, \ldots, \ldots$ will involve the values of u, v, w, p, q, r. It is assumed that for small values of these quantities the functions are linear; thus we write

$$\left.\begin{aligned} X - X_1 = X_u u + X_v v + X_w w + X_p p + X_q q + X_r r, \\ \ldots\ldots\ldots\ldots\ldots\ldots\ldots\ldots\ldots\ldots\ldots\ldots \\ L - L_1 = L_u u + L_v v + L_w w + L_p p + L_q q + L_r r, \\ \ldots\ldots\ldots\ldots\ldots\ldots\ldots\ldots\ldots\ldots\ldots\ldots \end{aligned}\right\} \quad \text{...(9)}$$

where the thirty-six coefficients are constants depending on the structure of the machine and on the value of the undisturbed velocity U. They were called 'resistance-derivatives' by Bryan*, by whom they were first introduced. In a symmetrical machine, such as we are considering, their number reduces to eighteen. For instance, it is obvious that the value of $X - X_1$ must be unaffected by a reversal of the sign of v or p or r. Hence $X_v, X_p, X_r = 0$. By considerations of this kind we learn that X, M, Z will not occur with the suffixes v, p, r and that Y, L, N will not occur with the suffixes u, w, q.

* *Stability in Aviation*, London, 1911. For further investigations reference may be made to Bairstow, *Applied Aerodynamics*, London, 1920.

The values of the derivatives for a particular machine are inferred partly from theoretical considerations, and partly from experiments on models in a wind channel.

Our equations thus fall into two independent groups. In the first group we have

$$\left.\begin{aligned} M_0\frac{du}{dt} - M_0 gl + X_u u + X_w w + X_q q &= 0,\\ M_0\left(\frac{dw}{dt} - Uq\right) + Z_u u + Z_w w + Z_q q &= 0,\\ B\frac{dq}{dt} \qquad + M_u u + M_w w + M_q q &= 0,\\ \frac{dl}{dt} + q &= 0. \end{aligned}\right\} \quad \ldots\ldots(10)$$

These involve only u, w, q, l.

In the second group

$$\left.\begin{aligned} M_0\left(\frac{dv}{dt} + Ur\right) - M_0 gm + Y_v v + Y_p p + Y_r r &= 0,\\ A\frac{dp}{dt} - G\frac{dr}{dt} \qquad + L_v v + L_p p + L_r r &= 0,\\ C\frac{dr}{dt} - G\frac{dp}{dt} \qquad + N_v v + N_p p + N_r r &= 0,\\ \frac{dm}{dt} - p &= 0. \end{aligned}\right\} \ldots(11)$$

These involve only v, p, r, m.

Any small disturbance may therefore be resolved into two types. In one of these we have $v, p, r, m = 0$, whilst u, w, q, l are subject to the equations (10). The path of the mass-centre is confined to a vertical plane, and the only angular movement is that of 'pitching,' to which q refers.

In the second type we have $u, w, q, l = 0$, whilst v, p, r, m are subject to (11). The deviation from the steady state now consists in a 'side-slip' (v) accompanied by 'rolling' (p) and 'yawing' (r).

To investigate the stability of the straight flight when affected by disturbances of either type we assume as usual that the dependent variables are proportional to $e^{\lambda t}$, and proceed to ascertain the admissible values of λ. In the case of (10) we are led to the determinant

$$\begin{vmatrix} M_0\lambda + X_u, & X_w, & X_q + \dfrac{M_0 g}{\lambda} \\ Z_u, & M_0\lambda + Z_w, & -M_0 U + Z_q \\ M_u, & M_w, & B\lambda + M_q \end{vmatrix} = 0. \ldots (12)$$

In like manner the equations (11) give

$$\begin{vmatrix} M_0\lambda + Y_v, & Y_p - \dfrac{M_0 g}{\lambda}, & M_0 U + Y_r \\ L_v, & A\lambda + L_p, & -G\lambda + L_r \\ N_v, & -G\lambda + N_p, & C\lambda + N_r \end{vmatrix} = 0. \ldots (13)$$

In each case the result, when developed, is a biquadratic in λ. A real positive root of this would involve in the expressions for the velocities terms of the form $ae^{\alpha t}$, which are capable of indefinite increase. A pair of conjugate imaginary roots with real positive parts would introduce terms of the types $ae^{\alpha t}\cos\beta t$, $be^{\alpha t}\sin\beta t$, indicating an oscillation of continually increasing amplitude. It is therefore essential for stability that the roots should either be real and negative, or have their real parts negative. It is true that roots which are pure imaginaries might be held to indicate stability, as in the case of the small oscillations of a conservative system (Chap. XI), but in the present case we cannot assume that a slight alteration in the conditions might not replace them by complex roots, the real parts of which might be positive.

Let us suppose the equation (12) written out in the form

$$\lambda^4 + p_1\lambda^3 + p_2\lambda^2 + p_3\lambda + p_4 = 0, \quad \ldots\ldots\ldots\ldots\ldots (14)$$

and let us denote by H the product of the six sums of pairs of roots, viz.,

$$H = p_1 p_2 p_3 - p_1^2 p_4 - p_3^2. \quad \ldots\ldots\ldots\ldots\ldots\ldots (15)^*$$

It is easily found on examination of the various possible cases that if the roots of (14) are to be of the required character the quantities

* If α, β, γ, δ be the roots we find

$$(\alpha+\delta)(\beta+\delta)(\gamma+\delta) = -(p_1\delta^2 + p_3),$$

$$(\beta+\gamma)(\gamma+\alpha)(\alpha+\beta) = -(p_1\delta^2 + p_1^2\delta + p_1 p_2 - p_3).$$

Multiplying these expressions, and taking account of the fact that δ is a root of (14), we obtain the formula (15).

p_1, p_2, p_3, p_4, and H must all be positive. These conditions are moreover sufficient as well as necessary. For real positive roots of (14) are excluded by the supposition that the coefficients are positive. Again, if there are two pairs of imaginary roots, say $\alpha \pm i\beta$ and $\alpha' \pm i\beta'$, we have

$$H = 4\alpha\alpha' \{(\alpha + \alpha')^2 + (\beta + \beta')^2\} \{(\alpha + \alpha')^2 + (\beta - \beta')^2\}. \quad \ldots(16)$$

If H is positive, α and α' must have the same sign, and since $p_1 = -2(\alpha + \alpha')$ this sign must be negative if p_1 is positive. If the roots are two imaginary and two real, say $\alpha \pm i\beta$, γ, δ, we have

$$H = 2\alpha(\gamma + \delta) \{(\alpha + \gamma)^2 + \beta^2\} \{(\alpha + \delta)^2 + \beta^2\}. \quad \ldots\ldots(17)$$

Hence α and $\gamma + \delta$ must have the same sign. Since $p_1 = -(2\alpha + \gamma + \delta)$ this sign must be negative, and since $p_4 = (\alpha^2 + \beta^2)\gamma\delta$, γ and δ must both be negative*.

In this way the 'longitudinal stability,' as it is called, of the machine is determined. A similar treatment applied to the equation (13) gives the conditions for 'lateral stability.'

72. Equations of Motion of a Deformable Body.

It is sometimes convenient to refer the motion of an imperfectly rigid body, or of an assemblage of discrete particles, to a system of moving axes, especially when the changes of the *relative* configuration are slight. We have to decide, in the first place, on the principle which shall guide our choice of axes, and define their motion. The origin will naturally be taken at the mass-centre, and we need only consider motions relative to this point.

One method which suggests itself is to make the mean square of the deviations of the particles in a short time δt, from the positions which they would have occupied if they had been fixed relatively to the axes, a minimum. Let (p, q, r) be the angular velocity of our moving axes of x, y, z, and u, v, w the component velocities of a particle m. The deviation of this particle, in the above sense, has for components

$$(u - qz + ry)\,\delta t, \quad (v - rx + pz)\,\delta t, \quad (w - py + qx)\,\delta t. \quad \ldots(1)$$

* The criteria, and the argument, are due to Routh.

Hence, omitting the factor δt^2, the quantity which is to be made a minimum by variation of p, q, r is

$$\Sigma m \{(u - qz + ry)^2 + (v - rx + pz)^2 + (w - py + qx)^2\}. \quad \ldots(2)$$

The kinetic energy is

$$T = \tfrac{1}{2} \Sigma m (u^2 + v^2 + w^2), \quad \ldots\ldots(3)$$

and the components of angular momentum are

$$\lambda = \Sigma m (yw - zv), \quad \mu = \Sigma m (zu - xw), \quad \nu = \Sigma m (xv - yu). \quad \ldots(4)$$

The expression (2) is therefore equivalent to

$$2T + Ap^2 + Bq^2 + Cr^2 - 2Fqr - 2Grp - 2Hpq - 2(\lambda p + \mu q + \nu r), \quad \ldots(5)$$

with the usual notation for coefficients of inertia.

The conditions for a minimum as regards variations of p, q, r are therefore

$$\left.\begin{aligned} Ap - Hq - Gr &= \lambda, \\ -Hp + Bq - Fr &= \mu, \\ -Gp - Fq + Cr &= \nu. \end{aligned}\right\} \quad \ldots\ldots(6)$$

These equations determine the instantaneous values of p, q, r if those of λ, μ, ν, and of the inertia coefficients, are known. It will be noted that they have exactly the same form as for a rigid body (Art. 31).

Axes chosen so as to fulfil the above condition may be called 'mean axes.' Their determination is of course by no means unique. We may start at a given instant with *any* rectangular system, provided its subsequent motion be governed by the formulæ (6).

The equations of angular momentum have the forms

$$\left.\begin{aligned} \frac{d\lambda}{dt} - r\mu + q\nu &= L, \\ \frac{d\mu}{dt} - p\nu + r\lambda &= M, \\ \frac{d\nu}{dt} - q\lambda + p\mu &= N; \end{aligned}\right\} \quad \ldots\ldots(7)$$

but it must be remembered, in substituting the values of λ, μ, ν from (6), that A, B, C, F, G, H are now variable.

The preceding equations lend themselves to an investigation of the free Eulerian nutation of a slightly deformable body which in its normal state is kinetically symmetrical about an axis. We will suppose that the body is slightly disturbed from a state of steady rotation with angular velocity n about the axis of kinetic symmetry.

In the disturbed state p, q, F, G, H will be small quantities, and the changes in the principal moments and in the value of r will also be small. Hence, neglecting small quantities of the second order, we find, from (6) and (7),

$$\left.\begin{aligned} A\frac{dp}{dt}+(C-A)\,nq-\frac{dG}{dt}n+Fn^2=0,\\ A\frac{dq}{dt}-(C-A)\,np-\frac{dF}{dt}n-Gn^2=0.\end{aligned}\right\} \quad \text{..................(8)}$$

We have in view principally the case of the earth, where the ratio $(C-A)/A$ is small, and where the period of the nutation was consequently found, on the hypothesis of absolute rigidity, to be large compared with $2\pi/n$ (Art. 47). The products of inertia, F and G, in (8), may then be taken to have the statical values due to the instantaneous centrifugal force, as modified by the gravitational attraction of the deformed body, which tends to restore the normal state.

Now the 'centrifugal' potential per unit mass of a body rotating with angular velocity ω about an axis (l', m', n') is

$$\tfrac{1}{2}\omega^2\{x^2+y^2+z^2-(l'x+m'y+n'z)^2\},$$

and if the axis in question is nearly coincident with the axis of z this becomes

$$\tfrac{1}{2}\omega^2(x^2+y^2)-n\,(px+qy)\,z,$$

small quantities of the second order being neglected.

The first term in this expression is already operative in the undeformed state, and its effect may be supposed to be allowed for in the values of A and C. We may therefore write, for the disturbing potential,

$$V=-n\,(px+qy)\,z. \quad \text{.............................(9)}$$

If we trace the equipotential surfaces due to p alone, it is easy to see that if n be positive the tendency of the component p of the rotation is to diminish the value of G, and similarly that the influence of q is to diminish that of F. We assume therefore, as at all events a rough representation of the conditions,

$$F=-\beta nq, \quad G=-\beta np, \quad \text{.........................(10)*}$$

where β is a positive constant involving the elasticity of the material, the distribution of density, and the constant of gravitation, as well as the dimensions of the body.

On substituting in (8), we find that the equations combine into

$$(A+\beta n^2)\frac{d\zeta}{dt}-i\,(C-A-\beta n^2)\,n\zeta=0, \quad \text{..................(11)}$$

where

$$\zeta=p+iq. \quad \text{....................................(12)}$$

Hence, assuming that ζ varies as $e^{i\sigma t}$, we find

$$\frac{\sigma}{n}=\frac{C-A-\beta n^2}{A+\beta n^2}, \quad \text{..............................(13)}$$

shewing that the period is *lengthened* by the deformation.

* This is confirmed by more detailed examination. See Love, *Proc. R.S.* vol. 82 (1909).

If external forces are operative whose moments about the coordinate axes are L, M, 0, we have in place of (11),

$$(A+\beta n^2)\frac{d\zeta}{dt}-i(C-A-\beta n^2)\,n\zeta=L+iM. \quad\text{..............(14)}$$

Hence, if L, M vary as $e^{i\sigma t}$, the forced oscillations are given by

$$\zeta=\frac{i(L+iM)}{Cn-(A+\beta n^2)(\sigma+n)}\cdot \quad\text{........................(15)}$$

We may apply this to the precession of the earth's axis. Referring to Fig. 32, p. 83, where B' is that pole of the great circle ZC which lies on the side of A, the average solar effect is a couple $-\kappa\sin\omega\cos\omega$ about OB' where ω is the obliquity of the ecliptic and κ is given by Art. 62 (3). Since $\cos B'A=\sin\phi$, $\cos B'B=\cos\phi$, we have

$$L+iM=-\kappa\sin\omega\cos\omega\,(\sin\phi+i\cos\phi)=-i\kappa\sin\omega\cos\omega\, e^{-int}, \quad\text{...(16)}$$

very approximately. Hence, putting $\sigma=-n$ in (15),

$$\zeta=\frac{\kappa}{Cn}\sin\omega\cos\omega\, e^{-int}, \quad\text{..........................(17)}$$

or

$$p=\frac{\kappa}{Cn}\sin\omega\cos\omega\cos nt, \quad q=-\frac{\kappa}{Cn}\sin\omega\cos\omega\sin nt. \quad\text{.........(18)}$$

Since these results do not involve β, the motion of the earth's instantaneous axis is approximately the same as if the earth were rigid.

The precession is found from (18) by the formulæ of Art. 33. Thus

$$\dot\psi=\frac{q\sin\phi-p\cos\phi}{\sin\omega}=-\frac{\kappa}{Cn}\cos\omega, \quad\text{..................(19)}$$

in agreement with Art. 62 (6).

EXAMPLES. XII.

1. A particle is constrained by a smooth straight tube which rotates with constant angular velocity ω about an axis not in the same plane with it. Prove that the motion of the particle relative to the tube is the same as if the tube were at rest and the particle were repelled from the position of relative equilibrium by a force varying as the distance.

2. A tidal current is running northwards along a channel of breadth b, in latitude λ. Prove that the height of the tide on the E. coast exceeds that on the W. coast by $2bv\omega\sin\lambda/g$, where v is the velocity of the water, and ω the earth's angular velocity.

Work out the result for the case of the southern Irish Channel, taking $b=48$ sea-miles, $v=3{\cdot}2$ knots, $\lambda=52°$. [5·7 ft.]

3. A shot is fired eastwards in latitude 60°, with a velocity of 1000 ft./sec. and elevation 20°. Find the deviation due to the earth's rotation, neglecting the resistance of the air. [27 ft. to S.]

4. A particle is projected vertically upwards *in vacuo* with velocity v. Prove that when it returns to the ground there is a deviation

$$(\tfrac{4}{3}\,\omega v^3\cos\lambda)/g^2$$

to the W. (Laplace.)

5. A particle is projected with velocity V on a smooth horizontal plane. Prove that owing to the rotation of the earth it will describe an arc of a circle of radius $V/(2\omega \sin \lambda)$ with this constant velocity.

6. Deduce the formulæ for the component accelerations in spherical polar coordinates (Art. 34) from the theory of moving axes.

7. Apply the theory of moving axes to deduce the following formulæ, where l_1, m_1, n_1 are the direction-cosines of the tangent line to a twisted curve relative to fixed axes, l_2, m_2, n_2 those of the principal normal, l_3, m_3, n_3 those of the binormal, and ρ and σ are the radii of curvature and torsion, respectively:

$$\frac{dl_1}{ds} = \frac{l_2}{\rho}, \quad \frac{dl_2}{ds} = \frac{l_3}{\sigma} - \frac{l_1}{\rho}, \quad \frac{dl_3}{ds} = -\frac{l_2}{\sigma},$$

with similar formulæ for dm_1/ds, dn_1/ds, etc.

(Take as moving axes the tangent line, the principal normal, and the binormal.)

8. A particle is constrained by a smooth rigid tube in the form of a plane curve, which rotates in its own plane about the origin O, with angular velocity ω. If v be the velocity relative to the tube prove that the tangential and normal accelerations of the particle are

$$v\frac{dv}{ds} - \omega^2 r\frac{dr}{ds} + \dot{\omega}p, \quad \frac{v^2}{\rho} + 2\omega v + \omega^2 p + \dot{\omega}r\frac{dr}{ds},$$

respectively, where r is the radius vector, and p the perpendicular from the origin to the tangent.

9. A circular disk is set spinning with angular velocity ω about a vertical diameter, on a *smooth* horizontal plane; prove that the condition for stability is

$$\omega^2 > Mga/A,$$

where A is the moment of inertia about a diameter.

10. A sphere whose centre of mass coincides with its centre of figure, but whose principal moments at that point are A, A, C, is set rolling along a horizontal plane with angular velocity ω, in a direction at right angles to its axis of unequal moment, which is horizontal. Prove that if this state of steady motion be slightly disturbed, the period of a small oscillation about it is

$$\sqrt{\left\{\frac{A\,(A+Ma^2)}{C\,(C+Ma^2)}\right\}} \cdot \frac{2\pi}{\omega},$$

where M is the mass, and a the radius.

11. A solid of revolution has an equatorial plane of symmetry, and is rolling in steady motion on a horizontal plane, with angular velocity ω about its axis which is horizontal. If the motion be slightly disturbed, the period of a small oscillation is $2\pi/\sigma$, where

$$\sigma^2 = \frac{CC'\omega^2 - Mg\,(a-\rho)\,A}{AA'},$$

where a is the radius of the equatorial circle, ρ the radius of curvature of a meridian at the equator, and $A' = A + Ma^2$, $C' = C + Ma^2$.

12. A solid of revolution is placed with its axis vertical on a rough hori zontal plane, the height h of the centre of gravity being greater than the radius of curvature ρ at the point of contact, so that the position is unstable in the absence of rotation. Prove that if it be set spinning about the vertical axis with angular velocity ω, the condition for stability is

$$\omega^2 > \frac{4A' Mg(h-\rho)}{(C+Mh\rho)^2},$$

if A', A', C are the principal moments of inertia at the point of contact.

13. If in Ex. 12 the plane be smooth, the condition is

$$\omega^2 > \frac{4AMg(h-\rho)}{C^2},$$

if A, A, C are the principal moments at the mass-centre.

14. Investigate the small oscillations of a rolling wheel about a state of steady motion, in the two cases of Art. 69, by the intrinsic equations of Art. 55

15. A rigid body contains rotating flywheels, whose axes are fixed relatively to it. Prove that the modified form of Euler's equations is

$$A\frac{dp}{dt} - (B-C)qr = L + \beta r - \gamma q,$$

$$B\frac{dq}{dt} - (C-A)rp = M + \gamma p - \alpha r,$$

$$C\frac{dr}{dt} - (A-B)pq = N + \alpha q - \beta p.$$

Give the precise definitions of A, B, C, α, β, γ. (Volterra.)

16. A system of particles is moving in any manner, and axes Gx, Gy, Gz move about the mass-centre so as always to coincide with the principal axes of inertia at that point. Prove that the requisite angular velocities are

$$\frac{\Sigma m(yw+zv)}{C-B}, \quad \frac{\Sigma m(zu+xw)}{A-C}, \quad \frac{\Sigma m(xv+yu)}{B-A},$$

where (u, v, w) is the velocity of a particle m at (x, y, z), and A, B, C are the instantaneous values of the principal moments.

Prove that $\dfrac{dA}{dt} = 2\,\Sigma m(yv+zw)$, etc., etc.

17. Prove that in the case of Fig. 50, Art. 59, the acceleration of the pole C of the gyroscope along the meridian is

$$\ddot{\theta} - \omega^2 \sin\theta\cos\theta.$$

Also that in the case of Fig. 51 the acceleration of C in the horizontal plane is

$$\ddot{\phi} - \omega^2\cos^2\lambda\sin\phi\cos\phi + \omega^2\sin\lambda\cos\lambda\cos^2\phi.$$

CHAPTER X

GENERALIZED EQUATIONS OF MOTION

73. Generalized Coordinates and Velocities.

Suppose we have a dynamical system composed of material particles or rigid bodies, whether movable independently or connected in any way, subject to mutual forces, and also to the action of any given 'extraneous' forces, i.e. forces exerted from without on the system. Any given configuration* can be completely specified by the values assumed by a certain finite number n of independent quantities, called the 'generalized coordinates' of the system. These coordinates may be chosen in an endless variety of ways, but their number is determinate, and expresses the number of degrees of freedom of the system. We denote the coordinates by $q_1, q_2, \ldots, q_n$.

It is implied in the above description that the Cartesian coordinates x, y, z of any given particle m are definite functions of the q's, these functions varying of course from one particle to another. Hence

$$\left.\begin{aligned} \dot{x} &= \frac{\partial x}{\partial q_1}\dot{q}_1 + \frac{\partial x}{\partial q_2}\dot{q}_2 + \ldots + \frac{\partial x}{\partial q_n}\dot{q}_n, \\ \dot{y} &= \frac{\partial y}{\partial q_1}\dot{q}_1 + \frac{\partial y}{\partial q_2}\dot{q}_2 + \ldots + \frac{\partial y}{\partial q_n}\dot{q}_n, \\ \dot{z} &= \frac{\partial z}{\partial q_1}\dot{q}_1 + \frac{\partial z}{\partial q_2}\dot{q}_2 + \ldots + \frac{\partial z}{\partial q_n}\dot{q}_n. \end{aligned}\right\} \quad \ldots\ldots\ldots\ldots\ldots(1)$$

The velocity of every particle of the system at a given instant is thus completely determined when we know the configuration, and the time-rates of variation, viz. $\dot{q}_1, \dot{q}_2, \ldots, \dot{q}_n$, of the coordinates. These rates are called 'generalized components of velocity.'

If we employ the sign Σ to denote a summation extending over

* This term is meant to include position in space, as well as the relative positions of the parts of the system.

all the particles of the system, the kinetic energy T is given by the formula

$$2T = \Sigma m(\dot{x}^2 + \dot{y}^2 + \dot{z}^2) = a_{11}\dot{q}_1^2 + a_{22}\dot{q}_2^2 + \ldots + 2a_{12}\dot{q}_1\dot{q}_2 + \ldots, \quad \ldots(2)$$

where

$$\left.\begin{aligned} a_{rr} &= \Sigma m \left\{ \left(\frac{\partial x}{\partial q_r}\right)^2 + \left(\frac{\partial y}{\partial q_r}\right)^2 + \left(\frac{\partial z}{\partial q_r}\right)^2 \right\}, \\ a_{rs} &= \Sigma m \left\{ \frac{\partial x}{\partial q_r}\frac{\partial x}{\partial q_s} + \frac{\partial y}{\partial q_r}\frac{\partial y}{\partial q_s} + \frac{\partial z}{\partial q_r}\frac{\partial z}{\partial q_s} \right\} = a_{sr}. \end{aligned}\right\} \quad \ldots\ldots(3)$$

Thus T is expressed as a homogeneous quadratic function of the generalized velocities.

The coefficients a_{rr}, a_{rs} are called the 'coefficients of inertia' of the system; they are in general not constants, but functions of the coordinates q_r, and accordingly vary with the configuration. They are subject to certain algebraic limitations, which are necessary to ensure that the expression (2) shall be > 0 for all values of the velocities other than zero. If we write

$$\Delta_n = \begin{vmatrix} a_{11}, & a_{12}, & \ldots, & a_{1n} \\ a_{21}, & a_{22}, & \ldots, & a_{2n} \\ \ldots & \ldots & \ldots & \ldots \\ a_{n1}, & a_{n2}, & \ldots, & a_{nn} \end{vmatrix}, \quad \ldots\ldots\ldots\ldots\ldots\ldots(4)$$

and denote by Δ_{n-1} the determinant obtained by omitting the last row and column, by Δ_{n-2} that obtained by omitting the last two rows and columns, and so on, the necessary and sufficient conditions are that the determinants

$$\Delta_1, \Delta_2, \ldots, \Delta_{n-1}, \Delta_n$$

should all be > 0.

The simplest proof is as follows*. It is known that in any linear transformation of a quadratic form

$$\phi(x_1, x_2, \ldots, x_n) = a_{11}x_1^2 + a_{22}x_2^2 + \ldots + 2a_{12}x_1x_2 + \ldots, \quad \ldots\ldots\ldots(5)$$

the discriminant is merely altered in value by the square of the determinant of transformation as a factor. Now the terms in ϕ which contain x_1 are all included in the expression

$$\frac{1}{a_{11}}(a_{11}x_1 + a_{12}x_2 + \ldots + a_{1n}x_n)^2,$$

* Burnside and Panton, *Theory of Equations*.

and if this be subtracted from (5) the remainder will be a homogeneous quadratic function of the remaining variables $x_2, x_3, \ldots, x_n$. The terms of this remainder which contains x_2 can be separated in like manner, and in this way ϕ can be expressed as a sum of squares, thus

$$\phi = p_1 (x_1 + \ldots)^2 + p_2 (x_2 + \ldots)^2 + \ldots + p_n x_n^2, \quad \ldots\ldots\ldots\ldots\ldots(6)$$

or, say,

$$\phi = p_1 X_1^2 + p_2 X_2^2 + \ldots + p_n X_n^2, \quad \ldots\ldots\ldots\ldots\ldots\ldots\ldots(7)$$

and we note that $p_1 = a_{11} = \Delta_1$. The discriminant of the form (7) is

$$p_1 p_2 \ldots p_n,$$

and the determinant of transformation is obviously unity. Hence

$$\Delta_n = p_1 p_2 \ldots p_n. \quad \ldots\ldots\ldots\ldots\ldots\ldots\ldots\ldots\ldots\ldots(8)$$

The same argument applies to the function $\phi(x_1, x_2, \ldots, x_m, 0, 0, \ldots)$ where all the variables after x_m are made to vanish. Thus

$$\Delta_1 = p_1, \quad \Delta_2 = p_1 p_2, \quad \Delta_3 = p_1 p_2 p_3, \ldots \quad \ldots\ldots\ldots\ldots\ldots(9)$$

Hence the necessary and sufficient conditions that $p_1, p_2, \ldots, p_n$ should all be positive are as stated above.

It is to be observed that in our application no determinant of the series

$$\Delta_1, \Delta_2, \ldots, \Delta_n$$

can vanish. For if Δ_m, for instance, were zero, the equation

$$a_{r1} \dot{q}_1 + a_{r2} \dot{q}_2 + \ldots + a_{rm} \dot{q}_m = 0$$

could be satisfied for $r = 1, 2, \ldots, m$ by values of $\dot{q}_1, \dot{q}_2, \ldots, \dot{q}_m$ other than zero. Substituting these in the expression for T in Art. 74 (7), and making

$$\dot{q}_{m+1} = \dot{q}_{m+2} = \ldots = \dot{q}_n = 0,$$

we should have a zero value of the energy, with velocities other than zero.

74. Components of Momentum and Impulse.

Any given state of motion of the system through a given configuration may be imagined to be generated instantaneously from rest by the application of suitable impulsive forces. If (X', Y', Z') be the requisite impulse on a particle m we have

$$m\dot{x} = X', \quad m\dot{y} = Y', \quad m\dot{z} = Z'. \quad \ldots\ldots\ldots\ldots\ldots(1)$$

If we multiply these equations by $\partial x/\partial q_r, \partial y/\partial q_r, \partial z/\partial q_r$, respectively, and add, and then perform the summation Σ we have

$$p_r = P_r', \quad \ldots\ldots\ldots\ldots\ldots\ldots\ldots\ldots\ldots\ldots(2)$$

where

$$p_r = \Sigma m \left(\dot{x} \frac{\partial x}{\partial q_r} + \dot{y} \frac{\partial y}{\partial q_r} + \dot{z} \frac{\partial z}{\partial q_r} \right), \quad \ldots\ldots\ldots\ldots\ldots(3)$$

and

$$P_r' = \Sigma \left(X' \frac{\partial x}{\partial q_r} + Y' \frac{\partial y}{\partial q_r} + Z' \frac{\partial z}{\partial q_r} \right). \quad \ldots\ldots\ldots\ldots(4)$$

The quantity p_r is called a 'generalized component of momentum,' and P_r' a 'generalized component of impulse.' It is evident that in calculating P_r' we may omit all forces X', Y', Z' which are such that finite forces proportional to them would do no work in any infinitesimal displacement consistent with the connections of the system. Cf. Art. 77.

If in (3) we substitute the values of $\dot{x}, \dot{y}, \dot{z}$ from Art. 73 (1) we find

$$p_r = a_{1r}\dot{q}_1 + a_{2r}\dot{q}_2 + \ldots + a_{nr}\dot{q}_n, \qquad \text{.............(5)}$$

but the momenta in any given case are most easily calculated from the formula

$$p_r = \frac{\partial T}{\partial \dot{q}_r}, \qquad \text{.........................(6)}$$

which is obviously equivalent to (5).

By Euler's theorem of homogeneous functions we have

$$T = \frac{1}{2}\left(\dot{q}_1 \frac{\partial T}{\partial \dot{q}_1} + \dot{q}_2 \frac{\partial T}{\partial \dot{q}_2} + \ldots + \dot{q}_n \frac{\partial T}{\partial \dot{q}_n}\right)$$
$$= \tfrac{1}{2}(p_1\dot{q}_1 + p_2\dot{q}_2 + \ldots + p_n\dot{q}_n). \qquad \text{.................(7)}$$

Ex. Take the case of a single particle referred to spherical polar coordinates.

In the notation of Art. 34 the kinetic energy is given by

$$T = \tfrac{1}{2}m(\dot{r}^2 + r^2\dot{\theta}^2 + r^2\sin^2\theta\dot{\psi}^2). \qquad \text{.......................(8)}$$

The generalized components of momentum are accordingly

$$\frac{\partial T}{\partial \dot{r}} = m\dot{r}, \quad \frac{\partial T}{\partial \dot{\theta}} = mr^2\dot{\theta}, \quad \frac{\partial T}{\partial \dot{\psi}} = mr^2\sin^2\theta\dot{\psi}. \qquad \text{..................(9)}$$

The first of these expressions is recognized as the momentum of the particle in the direction of r, the second as the angular momentum about an axis through O perpendicular to the plane of θ, and the third as the angular momentum about the line OZ from which θ is measured. They might have been obtained, though less simply, from the formula (3), writing

$$x = r\sin\theta\cos\psi, \quad y = r\sin\theta\sin\psi, \quad z = r\cos\theta. \qquad \text{...........(10)}$$

75. Reciprocal Relations.

The formula (5) of Art. 74 leads to an important reciprocal relation between any two states of motion through the same configuration $(q_1, q_2, \ldots, q_n)$. Let the velocities and the momenta in one of these states be denoted as above, whilst those corresponding to the other state are distinguished by accents. The relation in question is

$$p_1\dot{q}_1' + p_2\dot{q}_2' + \ldots + p_n\dot{q}_n' = p_1'\dot{q}_1 + p_2'\dot{q}_2 + \ldots + p_n'\dot{q}_n. \quad \ldots(1)$$

Each of these expressions is in fact equal to

$$a_{11}\dot{q}_1\dot{q}_1' + a_{22}\dot{q}_2\dot{q}_2' + \ldots + a_{12}(\dot{q}_1\dot{q}_2' + \dot{q}_1'\dot{q}_2) + \ldots. \quad \ldots\ldots(2)$$

If we assume that all the components of momentum vanish except those corresponding to the coordinates q_r, q_s, we have

$$p_r\dot{q}_r' + p_s\dot{q}_s' = p_r'\dot{q}_r + p_s'\dot{q}_s. \quad \ldots\ldots\ldots\ldots(3)$$

If, further, we make $p_s = 0$, $p_r' = 0$, we have

$$\frac{\dot{q}_s}{p_r} = \frac{\dot{q}_r'}{p_s'}. \quad \ldots\ldots\ldots\ldots\ldots\ldots(4)$$

The interpretation of this result is simplest when the coordinates q_r, q_s are both of the same kind, e.g. both lines, or both angles. The theorem then asserts that the velocity of type s generated by an impulse of type r is equal to the velocity of type r generated by an equal impulse of type s.

Ex. 1. Suppose we have any number of bars AB, BC, CD, ..., freely jointed at the points B, C, D, ..., any one of which may be supposed fixed. For simplicity we suppose the bars to be in a straight line. The two coordinates q_r, q_s may be taken to be the displacements, at right angles to the lengths, at any two points P, Q of the system. The theorem then asserts that the velocity of Q due to an impulse at P is equal to the velocity of P due to an equal impulse at Q. Again, if we use angular coordinates, the angular velocity of any bar HK due to an impulsive couple applied to any other bar BC is equal to the angular velocity of BC due to an equal couple applied to HK. Finally, as an instance where the coordinates are of different kinds, we infer that if an impulse ξ applied at any point P of BC generates an angular velocity ω in HK, an impulsive couple ξa applied to HK would produce the velocity ωa at P [D. 108].

Ex. 2. If an impulsive couple H, applied to a free rigid body about an axis (l, m, n), generates an angular velocity whose component about an axis (l', m', n') is ω, the same couple about (l', m', n') will produce an equal angular velocity ω about (l, m, n).

If the principal axes at the mass-centre be the axes of coordinates, we find

$$\omega = H\left(\frac{ll'}{A} + \frac{mm'}{B} + \frac{nn'}{C}\right). \quad \ldots\ldots\ldots\ldots\ldots\ldots(5)$$

The symmetry of this relation verifies the theorem.

76. Theorems of Delaunay and Kelvin.

If T and T' denote the kinetic energies of the two states of motion we have, as an algebraical identity,

$$T' - T = \tfrac{1}{2}\Sigma_r (p_r'\dot{q}_r' - p_r\dot{q}_r)$$
$$= \tfrac{1}{4}\Sigma_r (p_r' + p_r)(\dot{q}_r' - \dot{q}_r) + \tfrac{1}{4}\Sigma_r (p_r' - p_r)(\dot{q}_r' + \dot{q}_r), \quad \ldots(1)$$

where the symbol Σ_r denotes a summation embracing all the suffixes*. The two parts of the latter expression are equal, by the reciprocal theorem, since the $p_r' + p_r$ are the momenta corresponding to the velocities $\dot{q}_r' + \dot{q}_r$, whilst the $p_r' - p_r$ are the momenta corresponding to the velocities $\dot{q}_r' - \dot{q}_r$. Hence we have the equivalent formulæ

$$T' - T = \tfrac{1}{2}\Sigma_r (p_r' + p_r)(\dot{q}_r' - \dot{q}_r), \quad \ldots\ldots\ldots\ldots(2)$$
$$T' - T = \tfrac{1}{2}\Sigma_r (p_r' - p_r)(\dot{q}_r' + \dot{q}_r). \quad \ldots\ldots\ldots\ldots(3)$$

From these we deduce two important theorems in which the kinetic energy of the system when set in motion by given impulses, or (on the other hand) with prescribed velocities, is compared with the energy acquired when the system is constrained in any way.

Let us suppose, in the first place, that the system is started by given *impulses* of certain types, but is otherwise free. We may suppose that by a linear transformation, if necessary, the coordinates are so adjusted that the constraint is expressed by the vanishing of certain coordinates of the remaining types. We have, from (3),

$$T - T' = \tfrac{1}{2}\Sigma_r (p_r - p_r')(\dot{q}_r + \dot{q}_r')$$
$$= \tfrac{1}{2}\Sigma_r (p_r - p_r')(\dot{q}_r - \dot{q}_r') + \Sigma_r (p_r - p_r')\dot{q}_r'. \quad \ldots\ldots\ldots(4)$$

If the accents be taken to refer to the constrained motion, we have $p_r' = p_r$ in those types for which the impulses are prescribed, whilst the velocities $\dot{q}_r'$ vanish in the remaining types. Thus

$$T - T' = \tfrac{1}{2}\Sigma_r (p_r - p_r')(\dot{q}_r - \dot{q}_r'). \quad \ldots\ldots\ldots\ldots(5)$$

The right-hand side is the energy of a motion in which the velocities are $\dot{q}_r - \dot{q}_r'$, and is therefore essentially positive. The energy generated by given *impulses* is therefore *greater* than if the system had been constrained, by an amount equal to the energy of the motion which is the difference between the free and the constrained motions. This theorem was given in its complete form by

* This will not be confused with the previous use of Σ (without a suffix) to denote a summation over the *particles* of the system.

Delaunay (1844). A good example is furnished by Euler's problem (Art. 44, Ex.).

Next suppose that the system is started with prescribed *velocities* of certain types, by means of suitable impulses of those types, but is otherwise free. From (2) we have

$$T' - T = \tfrac{1}{2}\Sigma_r (p_r' + p_r)(\dot{q}_r' - \dot{q}_r)$$
$$= \tfrac{1}{2}\Sigma_r (p_r' - p_r)(\dot{q}_r' - \dot{q}_r) + \Sigma_r p_r (\dot{q}_r' - \dot{q}_r). \quad \text{.........(6)}$$

We have $\dot{q}_r' = \dot{q}_r$ in the types where the velocities are prescribed, whilst the momenta p_r vanish in the remaining types. Hence

$$T' - T = \tfrac{1}{2}\Sigma_r (p_r' - p_r)(\dot{q}_r' - \dot{q}_r). \quad \text{...............(7)}$$

The energy acquired when the system is started with prescribed *velocities* is therefore *less* than if the motion had been constrained, by an amount equal to the energy of the motion which is the difference between the free and the constrained motions. This theorem is due to Kelvin (1863).

Another proof of the above theorems is given in Art. 83. It will be noticed that apart from the present interpretation they are purely algebraical. They occur with other meanings in various branches of mathematical physics. The statical analogues are explained in Art. 88.

Ex. Take the case of a waggon containing a number of bodies which are free to roll about, or to swing like pendulums. A given horizontal impulse on it will generate more energy than if the relative motion of any of the contained bodies had been prevented. On the other hand, it requires a smaller expenditure of work to produce a given velocity of the waggon.

For instance, suppose that the waggon carries (transversally) a single cylindrical roller of mass m, radius a, and radius of gyration κ. It is easily found that the impulse necessary to produce a given velocity u of the waggon is

$$\xi = M'u, \quad \text{..................................(8)}$$

where

$$M' = M + \frac{m\kappa^2}{a^2 + \kappa^2}, \quad \text{..................................(9)}$$

if M denote the virtual mass of the waggon alone, allowance being made for the rotatory inertia of the wheels. The formula $\frac{1}{2}\xi u$ for the energy generated may be put in either of the forms

$$\frac{1}{2}\frac{\xi^2}{M'}, \text{ or } \frac{1}{2}M'u^2. \quad \text{.............................(10)}$$

Since $M' < M + m$, the theorems are verified.

This example illustrates Rayleigh's remark that the effect of a constraint is (in a general sense) to increase the *inertia* of a system.

77. Lagrange's Equations.

To form the general equations of continuous motion we start with the equations of motion of an individual particle, viz.

$$m\ddot{x} = X, \quad m\ddot{y} = Y, \quad m\ddot{z} = Z. \quad \dots\dots\dots\dots\dots\dots(1)$$

Multiplying these by $\partial x/\partial q_r$, $\partial y/\partial q_r$, $\partial z/\partial q_r$, respectively, and adding and summing the result for all the particles of the system,

$$\Sigma m \left(\ddot{x}\frac{\partial x}{\partial q_r} + \ddot{y}\frac{\partial y}{\partial q_r} + \ddot{z}\frac{\partial z}{\partial q_r}\right) = \Sigma\left(X\frac{\partial x}{\partial q_r} + Y\frac{\partial y}{\partial q_r} + Z\frac{\partial z}{\partial q_r}\right). \quad \dots(2)$$

Now, referring to Art. 73 (1), we have

$$\frac{d}{dt}\left(\frac{\partial x}{\partial q_r}\right) = \frac{\partial^2 x}{\partial q_1 \partial q_r}\dot{q}_1 + \frac{\partial^2 x}{\partial q_2 \partial q_r}\dot{q}_2 + \dots + \frac{\partial^2 x}{\partial q_n \partial q_r}\dot{q}_n$$

$$= \frac{\partial}{\partial q_r}\left(\frac{\partial x}{\partial q_1}\dot{q}_1 + \frac{\partial x}{\partial q_2}\dot{q}_2 + \dots + \frac{\partial x}{\partial q_n}\dot{q}_n\right) = \frac{\partial \dot{x}}{\partial q_r}, \quad \dots\dots(3)$$

the differentiation with respect to t being 'total.' Hence

$$\ddot{x}\frac{\partial x}{\partial q_r} = \frac{d}{dt}\left(\dot{x}\frac{\partial x}{\partial q_r}\right) - \dot{x}\frac{d}{dt}\left(\frac{\partial x}{\partial q_r}\right)$$

$$= \frac{d}{dt}\left(\dot{x}\frac{\partial x}{\partial q_r}\right) - \dot{x}\frac{\partial \dot{x}}{\partial q_r}, \quad \dots\dots\dots\dots\dots\dots\dots(4)$$

with similar relations. Thus

$$\Sigma m \left(\ddot{x}\frac{\partial x}{\partial q_r} + \ddot{y}\frac{\partial y}{\partial q_r} + \ddot{z}\frac{\partial z}{\partial q_r}\right)$$

$$= \frac{d}{dt}\Sigma m\left(\dot{x}\frac{\partial x}{\partial q_r} + \dot{y}\frac{\partial y}{\partial q_r} + \dot{z}\frac{\partial z}{\partial q_r}\right) - \Sigma m\left(\dot{x}\frac{\partial \dot{x}}{\partial q_r} + \dot{y}\frac{\partial \dot{y}}{\partial q_r} + \dot{z}\frac{\partial \dot{z}}{\partial q_r}\right)$$

$$= \frac{dp_r}{dt} - \frac{\partial T}{\partial q_r}, \quad \dots\dots\dots\dots\dots\dots\dots\dots\dots\dots\dots\dots(5)$$

by Art. 74 (3).

If we calculate the work done by the forces on the system in an infinitesimal change of configuration, we find

$$\Sigma(X\delta x + Y\delta y + Z\delta z) = P_1\delta q_1 + P_2\delta q_2 + \dots + P_n\delta q_n, \quad \dots(6)$$

where
$$P_r = \Sigma\left(X\frac{\partial x}{\partial q_r} + Y\frac{\partial y}{\partial q_r} + Z\frac{\partial z}{\partial q_r}\right). \quad \dots\dots\dots\dots(7)$$

These quantities P_r are called the 'generalized components of force' on the system. The symbols X, Y, Z in (1) are supposed to include all the forces, of whatever origin, which act on the particle m; but in calculating the value of any generalized component P_r

we may of course ignore all forces, such as the internal forces of a rigid body, or the reactions between smooth surfaces in contact, which on the whole do no work.

The equation (2) may now be written

$$\frac{dp_r}{dt} - \frac{\partial T}{\partial q_r} = P_r, \qquad \text{.....................(8)}$$

or

$$\frac{d}{dt}\left(\frac{\partial T}{\partial \dot{q}_r}\right) - \frac{\partial T}{\partial q_r} = P_r. \qquad \text{.....................(9)}$$

If in this we put $r = 1, 2, 3, \ldots, n$ in succession, we get the n independent equations of motion of the system, in the form given by Lagrange, with whom the first conception, as well as the formulation, of a general dynamical method applicable to all systems of finite freedom appears to have originated*.

To verify the relation between kinetic energy and work done by the forces, we have from Art. 74 (7)

$$2T = \Sigma_r p_r \dot{q}_r, \qquad \text{.......................(10)}$$

where Σ_r is used as before to denote a summation extending over all the coordinates. Hence

$$\begin{aligned} 2\frac{dT}{dt} &= \Sigma_r (\dot{p}_r \dot{q}_r + p_r \ddot{q}_r) \\ &= \Sigma_r \left(\frac{\partial T}{\partial q_r}\dot{q}_r + \frac{\partial T}{\partial \dot{q}_r}\ddot{q}_r + P_r \dot{q}_r\right) \\ &= \frac{dT}{dt} + \Sigma_r (P_r \dot{q}_r), \end{aligned}$$

or

$$\frac{dT}{dt} = P_1 \dot{q}_1 + P_2 \dot{q}_2 + \ldots + P_n \dot{q}_n. \qquad \text{...........(11)}$$

This expresses that the kinetic energy increases at a rate equal to that at which work is being done by the forces.

In any small change of a 'conservative' system which is free from extraneous force we have

$$\Sigma (X\delta x + Y\delta y + Z\delta z) = -\delta V, \qquad \text{..............(12)}$$

* *Mécanique analytique*, 1st ed., 1788. The original investigation of Lagrange is reproduced below in Art. 102. The proof in the text was given by Hamilton (*Phil. Trans.*, 1835, p. 96), and afterwards independently by Jacobi in his *Vorlesungen über Dynamik* (1842), by Bertrand in the notes to his edition of Lagrange (1853), and by Thomson and Tait (2nd ed., 1879).

where V is the potential energy. Hence in this case

$$P_r = -\frac{\partial V}{\partial q_r}, \quad \text{......................(13)}$$

and Lagrange's equations become

$$\frac{d}{dt}\left(\frac{\partial T}{\partial \dot{q}_r}\right) - \frac{\partial T}{\partial q_r} = -\frac{\partial V}{\partial q_r}, \quad \text{................(14)}$$

or

$$\frac{d}{dt}\left(\frac{\partial L}{\partial \dot{q}_r}\right) - \frac{\partial L}{\partial q_r} = 0, \quad \text{....................(15)}$$

if we write

$$L = T - V. \quad \text{..........................(16)}$$

This function L is called the 'Lagrangian Function.' The same function, but with the sign reversed, was called by Helmholtz the 'kinetic potential.'

If we substitute from (13) in (11) we get

$$\frac{dT}{dt} = -\frac{dV}{dt}, \text{ or } T + V = \text{const.}, \quad \text{..............(17)}$$

the equation of energy.

If we integrate the equation (8) from $t = 0$ to $t = \tau$, we have

$$[p_r] = \int_0^\tau \frac{\partial T}{\partial q_r} dt + \int_0^\tau P_r dt, \quad \text{..................(18)}$$

where the square brackets are meant to indicate that the excess of the final over the initial value of the enclosed quantity is to be taken. In the case of instantaneous generation of motion from rest the first integral disappears, since $\partial T/\partial q_r$ involves the coordinates and the velocities only, and is therefore essentially finite. Hence

$$p_r = P_r', \quad \text{.........................(19)}$$

where

$$P_r' = \int_0^\tau P_r dt = \Sigma\left(X' \frac{\partial x}{\partial q_r} + Y' \frac{\partial y}{\partial q_r} + Z' \frac{\partial z}{\partial q_r}\right), \quad \text{......(20)}$$

as in Art. 74.

78. Applications of Lagrange's Equations.

The use of Lagrange's equations may be exemplified by applying them to one or two problems which have already been treated in other ways.

1°. Thus, to form the equations of motion of a particle in terms of spherical polar coordinates, we have on reference to Art. 74 (8)

$$\frac{\partial T}{\partial r} = m\,(r\dot{\theta}^2 + r\sin^2\theta\dot{\psi}^2),\quad \frac{\partial T}{\partial \theta} = mr^2\sin\theta\cos\theta\dot{\psi}^2,\quad \frac{\partial T}{\partial \psi} = 0. \quad \ldots(1)$$

The equations of motion are therefore

$$\left.\begin{aligned} m\,(\ddot{r} - r\dot{\theta}^2 - r\sin^2\theta\dot{\psi}^2) &= R,\\ \frac{d}{dt}(mr^2\dot{\theta}) - mr^2\sin\theta\cos\theta\dot{\psi}^2 &= \Theta,\\ \frac{d}{dt}(mr^2\sin^2\theta\dot{\psi}) &= \Psi, \end{aligned}\right\} \quad \ldots\ldots\ldots\ldots(2)$$

where R, Θ, Ψ are the generalized components of force. The definition of these latter quantities is that the expression

$$R\delta r + \Theta\delta\theta + \Psi\delta\psi \ldots\ldots\ldots\ldots\ldots\ldots\ldots\ldots(3)$$

is to be equal to the work done in an arbitrary infinitesimal displacement. Hence R is the radial component of the force on the particle, Θ is the moment of the force about an axis through O normal to the plane of θ, and Ψ is the moment about OZ.

In the case of the spherical pendulum we have

$$r = l,\quad \Theta = -\,mgl\sin\theta,\quad \Psi = 0. \ldots\ldots\ldots\ldots\ldots(4)$$

The second and third of equations (2) then become

$$\ddot{\theta} - \sin\theta\cos\theta\dot{\psi}^2 = -\frac{g}{l}\sin\theta,\quad \sin^2\theta\dot{\psi} = h, \ldots\ldots\ldots(5)$$

which are the usual equations of the problem [D. 103]. The first of equations (2) determines the tension ($mg\cos\theta - R$) of the string.

2°. In the case of the top we have, by Art. 33 (6),

$$2T = A\,(\dot{\theta}^2 + \sin^2\theta\dot{\psi}^2) + C\,(\dot{\phi} + \cos\theta\dot{\psi})^2, \ldots\ldots\ldots(6)$$

$$V = Mgh\cos\theta. \quad \ldots\ldots\ldots\ldots\ldots\ldots\ldots\ldots\ldots\ldots\ldots(7)$$

The component momenta (λ, μ, ν, say) are

$$\lambda = \frac{\partial T}{\partial \dot{\theta}} = A\dot{\theta},\quad \mu = \frac{\partial T}{\partial \dot{\psi}} = A\sin^2\theta\dot{\psi} + C\cos\theta\,(\dot{\phi} + \cos\theta\dot{\psi}),$$

$$\nu = \frac{\partial T}{\partial \dot{\phi}} = C\,(\dot{\phi} + \cos\theta\dot{\psi}). \quad \ldots(8)$$

We recognize λ as the angular momentum about an axis through O normal to the plane of θ; μ is the angular momentum about the vertical through O; and ν that about the axis of symmetry. Again, we have

$$\left.\begin{aligned}\frac{\partial T}{\partial \theta} &= A \sin\theta \cos\theta \dot{\psi}^2 - C(\dot{\phi} + \dot{\psi}\cos\theta)\sin\theta\dot{\psi},\\ \frac{\partial T}{\partial \psi} &= 0, \quad \frac{\partial T}{\partial \phi} = 0.\end{aligned}\right\}\quad\text{......(9)}$$

The formula (14) of Art. 77 accordingly gives

$$\left.\begin{aligned}A\ddot{\theta} - A\sin\theta\cos\theta\dot{\psi}^2 + C(\dot{\phi} + \cos\theta\dot{\psi})\,\dot{\psi}\sin\theta &= Mgh\sin\theta,\\ \frac{d}{dt}\{A\sin^2\theta\dot{\psi} + C(\dot{\phi} + \cos\theta\dot{\psi})\cos\theta\} &= 0,\\ \frac{d}{dt}C(\dot{\phi} + \cos\theta\dot{\psi}) &= 0.\end{aligned}\right\}\quad(10)$$

The last two equations express the constancy of the components μ, ν of momentum. Hence

$$\left.\begin{aligned}A\ddot{\theta} - A\sin\theta\cos\theta\dot{\psi}^2 + \nu\sin\theta\dot{\psi} &= Mgh\sin\theta,\\ A\sin^2\theta\dot{\psi} + \nu\cos\theta &= \mu.\end{aligned}\right\}\quad\text{...(11)}$$

The theory of the top has been developed from these equations in Art. 57, but the following deduction for the case of the nearly upright top may be inserted for the sake of comparison with Art. 58.

We may now put

$$\sin\theta = \theta, \quad \cos\theta = 1 - \tfrac{1}{2}\theta^2,$$

approximately. The resulting equations may be written

$$\left.\begin{aligned}\ddot{\theta} - \theta\dot{\chi}^2 &= -\frac{\nu^2 - 4AMgh}{4A^2}\theta,\\ \theta^2\dot{\chi} &= \text{const.},\end{aligned}\right\}\quad\text{........................(12)}$$

where

$$\chi = \psi - \tfrac{1}{2}\nu t/A. \quad\text{..................................(13)}$$

We may regard θ, χ as the polar coordinates of the horizontal projection of a point on the axis of the top, relative to an initial line which revolves with the angular velocity $\frac{1}{2}\nu/A$. The relative motion of such a point is therefore elliptic-harmonic, with the period

$$\frac{4\pi A}{\sqrt{(\nu^2 - 4AMgh)}}, \quad\text{..............................(14)}$$

subject to the usual condition $\nu^2 > 4AMgh$.

Some further examples are appended.

Ex. 1. To deduce Euler's equations of motion of a rigid body about a fixed point.

In the notation of Art. 33 (3) and (4) we have

$$p=\dot{\theta}\sin\phi-\dot{\psi}\sin\theta\cos\phi,\quad q=\dot{\theta}\cos\phi+\dot{\psi}\sin\theta\sin\phi,\quad r=\dot{\phi}+\dot{\psi}\cos\theta. \quad \text{.........(15)}$$

Since
$$2T=Ap^2+Bq^2+Cr^2, \quad \text{.........(16)}$$

we have
$$\frac{\partial T}{\partial\dot{\phi}}=Cr\frac{\partial r}{\partial\dot{\phi}}=Cr, \quad \text{.........(17)}$$

$$\frac{\partial T}{\partial\phi}=Ap\frac{\partial p}{\partial\phi}+Bq\frac{\partial q}{\partial\phi}=(A-B)pq. \quad \text{.........(18)}$$

Also, if the work done in an infinitesimal displacement be denoted by

$$\Theta\delta\theta+\Psi\delta\psi+\Phi\delta\phi, \quad \text{.........(19)}$$

it is plain from Fig. 32, p. 83, that Φ is the moment of the external forces about the axis OC, usually denoted by N.

Lagrange's equation

$$\frac{d}{dt}\frac{\partial T}{\partial\dot{\phi}}-\frac{\partial T}{\partial\phi}=\Phi$$

accordingly becomes

$$C\frac{dr}{dt}-(A-B)pq=N. \quad \text{.........(20)}$$

Since it is indifferent which of the principal axes at the fixed point we denote by OC, the remaining equations of Euler's triad will also hold.

Ex. 2. In the steam-engine the driving power is usually controlled by some form of 'centrifugal governor.' The original type, introduced by Watt, is shewn in the annexed sketch. The spindle to which the arms carrying the two balls are hinged rotates at a rate proportional to the speed of the engine. When this rotation is uniform the balls, under the action of gravity and centrifugal force, take up a definite 'equilibrium' position depending on the speed. If the speed increases the balls diverge outwards, raising the collar c to which the lower arms are connected, and thus operating a system of levers which turn a valve so as to reduce the supply of steam. Conversely, when the speed diminishes, the collar descends, and the supply is reinforced.

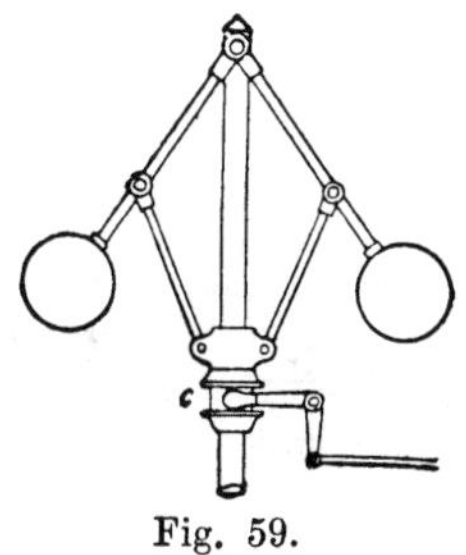

Fig. 59.

If θ denote the inclination of the upper arms to the spindle, and $\dot{\psi}$ the angular velocity about the vertical, the expression for the kinetic energy has the form

$$2T=A\dot{\theta}^2+I\dot{\psi}^2, \quad \text{.........(21)}$$

where A and I are functions of θ. The coefficient I is supposed to include a term representing the inertia of the engine and of the train of machinery in connection with it. Lagrange's formula gives

$$\frac{d}{dt}(A\dot{\theta})-\frac{1}{2}\frac{\partial A}{\partial\theta}\dot{\theta}^2-\frac{1}{2}\frac{\partial I}{\partial\theta}\dot{\psi}^2=-\frac{\partial V}{\partial\theta}, \quad \text{.........(22)}$$

and
$$\frac{d}{dt}(I\dot{\psi})=\Psi, \quad \text{.........(23)}$$

where V is the potential energy of the governor, and Ψ represents the excess of driving power over resistance. If this excess vanishes when the valve has the position corresponding to $\theta=\alpha$ we may write, as an approximation,

$$\Psi=-\beta(\theta-\alpha). \qquad (24)$$

For steady motion we must have $\theta=\alpha$, $\dot{\psi}=\omega$, where ω is determined by

$$\frac{1}{2}\frac{\partial I}{\partial\theta}\omega^2=\frac{\partial V}{\partial\theta}. \qquad (25)$$

Since this involves the above value of θ, the speed ω will vary with any permanent change in the driving power. The contrivance does not therefore maintain a constant speed independent of variations in the driving power, and it was therefore suggested by Maxwell that it should properly be called a 'moderator' rather than a 'governor.'

To examine the effect of accidental disturbances of the steady motion, we write

$$\theta=\alpha+x, \quad \dot{\psi}=\omega+y, \qquad (26)$$

and treat x, y as small. If we cancel the terms which refer to the steady motion, the equations (22) and (23) become

$$A\ddot{x}-\tfrac{1}{2}I''\omega^2x-I'\omega y+V''x=0, \qquad (27)$$

$$I\dot{y}+I'\omega\dot{x}+\beta x=0, \qquad (28)$$

where accents indicate differentiations with respect to θ, and the coefficients are supposed to have the values corresponding to $\theta=\alpha$ and are therefore constants. Assuming that x and y vary as $e^{\lambda t}$, we find

$$AI\lambda^3+\{I(V''-\tfrac{1}{2}I'\omega^2)+I'^2\omega^2\}\lambda+I'\beta\omega=0. \qquad (29)$$

It is essential of course that the governor should be stable when the speed ω is maintained constant. The condition for this is*

$$V''-\tfrac{1}{2}I'\omega^2>0. \qquad (30)$$

This being satisfied, the coefficients in (29) are all positive. There is therefore one negative, and no positive root. Since the sum of the roots is zero, the remaining roots must be imaginary with positive real part. The complete solution of (27) and (28) therefore consists of terms of the types $e^{-2\mu t}$, $e^{\mu t}\cos\nu t$, $e^{\mu t}\sin\nu t$. The latter pair indicate an oscillation of continually increasing amplitude.

This instability is checked to some extent by the inevitable friction between various parts of the mechanism, but in order definitely to eliminate it a viscous resistance is sometimes expressly introduced, opposing variations of θ. This may be represented by inserting a term $-\gamma d\theta/dt$ on the right-hand side of (22), and therefore a term $\gamma\dot{x}$ in (27). The resulting equation in λ is

$$AI\lambda^3+\gamma I\lambda^2+\{I(V''-\tfrac{1}{2}I'\omega^2)+I'^2\omega^2\}\lambda+I'\beta\omega=0. \qquad (31)$$

* In the problem of Art. 79, Ex. 1, we have

$$V=-mga\cos\theta, \quad I=ma^2\sin^2\theta,$$

and therefore

$$V''-\tfrac{1}{2}I'\omega^2=m\omega^2a\sin^2\theta,$$

in the position of relative equilibrium, where $\cos\theta=g/\omega^2a$. A closer analogy to the circumstances of Watt's governor is furnished by Ex. 4 of Art. 79.

There is obviously one negative root, as before. The condition that the remaining roots should be negative, or imaginary with negative real part, is*

$$\gamma I\{I(V'' - \tfrac{1}{2}I'\omega^2) + I'^2\omega^2\} > AII'\beta\omega, \quad \text{.................(32)}$$

which is satisfied if the frictional coefficient γ is sufficiently great.

79. Varying Relations.

It was assumed in the proof of Lagrange's equations that the geometrical relations which determine the absolute position of any given particle of the system in terms of the generalized coordinates are of invariable form. It was shewn however by Vieille (1849) that the equations retain their form when the relations in question vary continuously with the time, being now of the type

$$x = f(t, q_1, q_2, \ldots, q_n), \quad y = g(t, q_1, q_2, \ldots, q_n), \quad z = h(t, q_1, q_2, \ldots, q_n). \quad \text{.........(1)}$$

The functions are of course different for the various particles.

We have, in place of Art. 73 (1),

$$\left.\begin{aligned} \dot{x} &= \frac{\partial x}{\partial t} + \frac{\partial x}{\partial q_1}\dot{q}_1 + \frac{\partial x}{\partial q_2}\dot{q}_2 + \ldots + \frac{\partial x}{\partial q_n}\dot{q}_n, \\ \dot{y} &= \frac{\partial y}{\partial t} + \frac{\partial y}{\partial q_1}\dot{q}_1 + \frac{\partial y}{\partial q_2}\dot{q}_2 + \ldots + \frac{\partial y}{\partial q_n}\dot{q}_n, \\ \dot{z} &= \frac{\partial z}{\partial t} + \frac{\partial z}{\partial q_1}\dot{q}_1 + \frac{\partial z}{\partial q_2}\dot{q}_2 + \ldots + \frac{\partial z}{\partial q_n}\dot{q}_n. \end{aligned}\right\} \quad \text{...........(2)}$$

Hence

$$\begin{aligned} 2T &= \Sigma m(\dot{x}^2 + \dot{y}^2 + \dot{z}^2) \\ &= 2T_0 + 2(\alpha_1\dot{q}_1 + \alpha_2\dot{q}_2 + \ldots + \alpha_n\dot{q}_n) \\ &\qquad + a_{11}\dot{q}_1^2 + a_{22}\dot{q}_2^2 + \ldots + 2a_{12}\dot{q}_1\dot{q}_2 + \ldots, \end{aligned} \quad \text{......(3)}$$

where

$$2T_0 = \Sigma m\left\{\left(\frac{\partial x}{\partial t}\right)^2 + \left(\frac{\partial y}{\partial t}\right)^2 + \left(\frac{\partial z}{\partial t}\right)^2\right\}, \quad \text{...............(4)}$$

$$\alpha_r = \Sigma m\left(\frac{\partial x}{\partial t}\frac{\partial x}{\partial q_r} + \frac{\partial y}{\partial t}\frac{\partial y}{\partial q_r} + \frac{\partial z}{\partial t}\frac{\partial z}{\partial q_r}\right), \quad \text{...........(5)}$$

* If α, β, γ be the roots of the cubic

$$x^3 + p_1x^2 + p_2x + p_3 = 0,$$

we have

$$(\beta+\gamma)(\gamma+\alpha)(\alpha+\beta) = p_3 - p_1p_2.$$

Hence if all the roots be negative we have $p_1p_2 > p_3$. If they are $-\lambda$, $\mu \pm i\nu$, we have

$$2\mu\{(\mu-\lambda)^2 + \nu^2\} = p_3 - p_1p_2,$$

and the condition that μ should be negative is, again, $p_1p_2 > p_3$. Conversely, if $p_1p_2 > p_3$, we cannot have imaginary roots with real part positive, whilst positive real roots are excluded if all the coefficients of the cubic are positive.

and the coefficients a_{rr}, a_{rs} have the same algebraical forms as in Art. 73 (3), but now involve the time explicitly, as well as the coordinates.

We write also

$$p_r = \Sigma m \left(\dot{x} \frac{\partial x}{\partial q_r} + \dot{y} \frac{\partial y}{\partial q_r} + \dot{z} \frac{\partial z}{\partial q_r} \right)$$
$$= \alpha_r + a_{1r}\dot{q}_1 + a_{2r}\dot{q}_2 + \ldots + a_{nr}\dot{q}_n, \quad \ldots\ldots\ldots\ldots(6)$$

whence

$$p_r = \frac{\partial T}{\partial \dot{q}_r}. \quad \ldots\ldots\ldots\ldots\ldots\ldots(7)$$

Again, we have

$$\frac{d}{dt}\left(\frac{\partial x}{\partial q_r}\right) = \frac{\partial^2 x}{\partial t \partial q_r} + \frac{\partial^2 x}{\partial q_1 \partial q_r}\dot{q}_1 + \frac{\partial^2 x}{\partial q_2 \partial q_r}\dot{q}_2 + \ldots + \frac{\partial^2 x}{\partial q_n \partial q_r}\dot{q}_n, \ldots(8)$$

or

$$\frac{d}{dt}\left(\frac{\partial x}{\partial q_r}\right) = \frac{\partial \dot{x}}{\partial q_r}, \quad \ldots\ldots\ldots\ldots\ldots\ldots(9)$$

by (2). The investigation of Art. 77 after equation (3) then applies without further modification to shew that

$$\frac{d}{dt}\frac{\partial T}{\partial \dot{q}_r} - \frac{\partial T}{\partial q_r} = P_r, \quad \ldots\ldots\ldots\ldots\ldots\ldots(10)$$

as before.

It is to be noticed, however, that the formulæ (11) and (17) of Art. 77 no longer hold. In particular the energy of a conservative system is no longer constant.

It was pointed out by Hayward that the present case can be brought under the type contemplated by Lagrange by the introduction of a new coordinate ϕ in place of t, so far as t appears *explicitly* in (1) and in the formulæ derived from them. Putting

$$\left.\begin{aligned} \alpha_0 &= \Sigma m \left\{ \left(\frac{\partial x}{\partial \phi}\right)^2 + \left(\frac{\partial y}{\partial \phi}\right)^2 + \left(\frac{\partial z}{\partial \phi}\right)^2 \right\}, \\ \alpha_r &= \Sigma m \left(\frac{\partial x}{\partial \phi}\frac{\partial x}{\partial q_r} + \frac{\partial y}{\partial \phi}\frac{\partial y}{\partial q_r} + \frac{\partial z}{\partial \phi}\frac{\partial z}{\partial q_r} \right), \end{aligned}\right\} \quad \ldots\ldots\ldots(11)$$

we have

$$2T = \alpha_0 \dot{\phi}^2 + 2(\alpha_1 \dot{q}_1 + \alpha_2 \dot{q}_2 + \ldots + \alpha_n \dot{q}_n)\dot{\phi}$$
$$+ a_{11}\dot{q}_1^2 + a_{22}\dot{q}_2^2 + \ldots + 2a_{12}\dot{q}_1\dot{q}_2 + \ldots. \quad \ldots(12)$$

Lagrange's equations (Art. 77) then lead to (10), with the additional equation

$$\frac{d}{dt}\frac{\partial T}{\partial \dot{\phi}} - \frac{\partial T}{\partial \phi} = \Phi, \quad \ldots\ldots\ldots\ldots\ldots\ldots(13)$$

where Φ is the generalized component of force corresponding to ϕ. We may suppose Φ to be so adjusted as to make $\dot{\phi} = 1$; and nothing will then be altered in the preceding equations if we write t for ϕ before, instead of after, the differentiations.

The reason why the ordinary equation of energy does not hold is now apparent. We should need to add to the formula (11) of Art. 77 a term $\Phi\dot{\phi}$. This represents the rate at which work is being done by the constraining forces which are required in order to keep $\dot{\phi}$ constant.

Ex. 1. A particle moves in a smooth circular tube which is free to rotate about a vertical diameter.

If θ be the angular distance of the particle from the lowest point, ψ the azimuth of the tube, we have

$$2T = ma^2(\dot{\theta}^2 + \sin^2\theta\dot{\psi}^2) + I\dot{\psi}^2, \quad \text{.....................(14)}$$

where a is the radius, and I the moment of inertia of the tube about the vertical diameter. The equations of motion are therefore

$$\left.\begin{aligned} ma^2(\ddot{\theta} - \sin\theta\cos\theta\dot{\psi}^2) &= \Theta, \\ \frac{d}{dt}(I + ma^2\sin^2\theta)\,\dot{\psi} &= \Psi. \end{aligned}\right\} \quad \text{.........................(15)}$$

Now suppose the tube constrained to rotate with the constant angular velocity ω. The last equation becomes

$$\Psi = 2ma^2\omega\sin\theta\cos\theta\dot{\theta}. \quad \text{..........................(16)}$$

This gives the varying couple which must be applied to the tube to maintain the angular velocity constant. The remaining equation

$$ma^2(\ddot{\theta} - \omega^2\sin\theta\cos\theta) = \Theta \quad \text{........................(17)}$$

is the same as if the tube were at rest, and a centrifugal force $m\omega^2 a\sin\theta$ were to act on the particle from the axis of rotation. It might have been obtained at once by applying Vieille's theorem to the expression for the kinetic energy *of the particle*, viz.

$$2T_1 = ma^2(\dot{\theta}^2 + \omega^2\sin^2\theta). \quad \text{........................(18)}$$

Since the potential energy is

$$V = -mga\cos\theta, \quad \text{.............................(19)}$$

we may verify that

$$\frac{d}{dt}(T_1 + V) = \Psi\omega, \quad \text{..............................(20}$$

in accordance with a remark made above.

Ex. 2. To form the general equations of motion of a particle referred to axes rotating about a fixed line.

If x, y, z be the relative coordinates, the axis of z being that of rotation, we have

$$2T = m\{(\dot{x} - \omega y)^2 + (\dot{y} + \omega x)^2 + \dot{z}^2\}, \quad \text{.....................(21)}$$

where ω is the angular velocity. Hence

$$\left.\begin{aligned} \frac{d}{dt} m(\dot{x} - \omega y) - \omega(\dot{y} + \omega x) &= X, \\ \frac{d}{dt} m(\dot{y} + \omega x) + \omega(\dot{x} - \omega y) &= Y, \\ \frac{d}{dt} m\dot{z} &= Z. \end{aligned}\right\} \quad \text{.....................(22)}$$

These reduce to well-known forms. In particular, if ω be constant, the first two equations become

$$\left.\begin{aligned} m(\ddot{x} - 2\omega\dot{y} - \omega^2 x) &= X, \\ m(\ddot{y} + 2\omega\dot{x} - \omega^2 y) &= Y. \end{aligned}\right\} \quad \text{...........................(23)}$$

Ex. 3. To calculate the motion of a simple pendulum whose length is varied in an arbitrary manner.

If r be the length of the string at any instant, and θ its inclination to the vertical, we have

$$2T = m(\dot{r}^2 + r^2\dot{\theta}^2), \quad V = -mgr\cos\theta. \quad \text{..................(24)}$$

Lagrange's rule gives

$$\frac{d}{dt}(r^2\dot{\theta}) + gr\sin\theta = 0, \quad \text{..........................(25)}$$

as is evident otherwise on taking moments about the point of suspension.

When the length varies at a constant rate, and the oscillations are of small amplitude, the investigation can be continued*. Writing $\dot{r} = c$, and changing to r as independent variable, we have

$$\frac{d^2}{dr^2}(r\theta) + k\theta = 0, \quad \text{................................(26)}$$

where $k = g/c^2$. This equation is reduced to a standard form by putting

$$x^2 = 4kr, \quad y = x\theta. \quad \text{..............................(27)}$$

We find

$$\frac{d^2y}{dx^2} + \frac{1}{x}\frac{dy}{dx} + \left(1 - \frac{1}{x^2}\right)y = 0, \quad \text{.......................(28)}$$

the solution of which is

$$y = AJ_1(x) + BY_1(x), \quad \text{..........................(29)}$$

in the notation of Bessel Functions.

When x is large, i.e. when r is large compared with c^2/g, the known asymptotic forms of these functions lead to the result

$$\theta = \frac{1}{r^{\frac{3}{4}}}\left\{A'\cos\frac{2\sqrt{(gr)}}{c} + B'\sin\frac{2\sqrt{(gr)}}{c}\right\}. \quad \text{..............(30)}$$

* Lecornu, *Dynamique appliquée*, Paris, 1925.

This shews that transits of the string across the vertical, in the same direction, recur when $\surd(gr)$ increases by πc. If τ be the interval between two such transits, and r_1, r_2 the values of r at the beginning and end of it, we have

$$\surd g\,(\surd r_2 - \surd r_1) = \pi c = \pi\,(r_2 - r_1)/\tau,$$

and therefore

$$\tau = \frac{\pi\,(\surd r_2 + \surd r_1)}{\surd g} = 2\pi\sqrt{\frac{l}{g}}, \quad \text{.......................(31)}$$

if

$$\surd l = \tfrac{1}{2}\,(\surd r_2 + \surd r_1). \quad \text{..............................(32)}$$

The angular amplitude varies as $1/r^{\frac{3}{4}}$. The amplitude of the *linear* oscillations, on the other hand, varies as $r^{\frac{1}{4}}$.

Ex. 4. A rigid body is free to swing about a horizontal axis COC', and is symmetrical with respect to a plane meeting this axis at right angles in O. This axis is constrained to rotate about the vertical through O, with the constant angular velocity ω. The line joining O to the mass-centre G is assumed to be a principal axis at O.

The angular velocities about the principal axes at O are found by putting $\theta = \frac{1}{2}\pi$, $\psi = \omega t$ in Art. 33 (3), (4). Hence

$$2T = A\omega^2 \sin^2\phi + B\omega^2\cos^2\phi + C\dot{\phi}^2, \quad \text{..................(33)}$$

where C is the moment of inertia about COC', B that about OG, A that about the normal at O to the plane GCC', and ϕ is the inclination of OG to the downward vertical. If $OG = h$, we have

$$V = -Mgh\cos\phi. \quad \text{..............................(34)}$$

Lagrange's formula then gives

$$C\ddot{\phi} - (A - B)\,\omega^2 \sin\phi\cos\phi = -Mgh\sin\phi. \quad \text{...............(35)}$$

The possible steady motions are given by $\phi = 0$, $\phi = \pi$, and

$$\cos\phi = \frac{Mgh}{(A - B)\,\omega^2}, \quad \text{...........................(36)}$$

but the last case can only occur if

$$|\,A - B\,|\,\omega^2 > |\,Mgh\,|. \quad \text{..........................(37)}$$

The reader will find it an interesting exercise to examine the stability of the various steady motions.

Ex. 5. A pendulum symmetrical about its axis hangs by a universal flexure joint from a vertical spindle, which is made to rotate with constant angular velocity ω.

In the notation of Art. 33, Ex. 3, we have

$$\left.\begin{aligned} 2T &= A\,(\dot{\theta}^2 + \sin^2\dot{\psi}\theta^2) + C\{\omega - (1 - \cos\theta)\dot{\psi}\}^2, \\ V &= -Mgh\cos\theta, \end{aligned}\right\} \quad \text{............(38)}$$

where h is the depth of the mass-centre below the suspension when the pendulum is vertical.

Lagrange's formula can now be applied, but we will consider only the case of small oscillations about the vertical. We write

$$x = \sin\theta\cos\psi, \quad y = \sin\theta\sin\psi, \quad \text{.....................(39)}$$

so that x, y are the horizontal coordinates of a point on the axis. Hence

$$\dot{x}^2 + \dot{y}^2 = \cos^2\theta\,\dot{\theta}^2 + \sin^2\theta\,\dot{\psi}^2, \quad x\dot{y} - y\dot{x} = \sin^2\theta\,\dot{\psi}. \quad \text{...........(40)}$$

We have, then, with sufficient approximation, since θ is assumed to be small,

$$\left.\begin{aligned} 2T &= A\,(\dot{x}^2 + \dot{y}^2) - C\omega\,(x\dot{y} - y\dot{x}) + \text{const.}, \\ 2V &= Mgh\,(x^2 + y^2) + \text{const.} \end{aligned}\right\} \quad \text{...............(41)}$$

The equations of small relative motion are therefore

$$\left.\begin{aligned} A\ddot{x} + C\omega\dot{y} + Mghx &= 0, \\ A\ddot{y} - C\omega\dot{x} + Mghy &= 0. \end{aligned}\right\} \quad \text{...........................(42)}$$

These are identical, as we should expect, with Art. 58 (1), if we allow for the different convention as to the sign of h.

80. Equations of a Rotating System.

The general form of the equations of motion of a dynamical system relative to axes which are rotating with a constant angular velocity is of interest. It has a bearing on such questions as the theory of the tides on a rotating planet.

Taking the axis of rotation as axis of z we write

$$\begin{aligned} 2T &= \Sigma m\,\{(\dot{x} - \omega y)^2 + (\dot{y} + \omega x)^2 + \dot{z}^2\} \\ &= 2T_0 + 2\omega\,\Sigma m\,(x\dot{y} - y\dot{x}) + 2\mathfrak{T}, \quad \text{...........(1)} \end{aligned}$$

where

$$2T_0 = \Sigma m\,(x^2 + y^2)\,\omega^2, \quad \text{....................(2)}$$

$$2\mathfrak{T} = \Sigma m\,(\dot{x}^2 + \dot{y}^2 + \dot{z}^2). \quad \text{....................(3)}$$

If the *relative* configuration be determined by the generalized coordinates $q_1, q_2, \ldots, q_n$, $\mathfrak{T}$ will be a homogeneous quadratic function of the corresponding velocities, viz.

$$2\mathfrak{T} = a_{11}\dot{q}_1^2 + a_{22}\dot{q}_2^2 + \ldots + 2a_{12}\dot{q}_1\dot{q}_2 + \ldots, \quad \text{........(4)}$$

where the coefficients a_{rr}, a_{rs} have the same forms as in Art. 73 (3). $\mathfrak{T}$ denotes, in fact, the kinetic energy which the system would possess in the absence of the rotation ω. T_0, on the other hand, denotes the energy of the system when rotating in relative rest, in a given configuration; it involves the relative coordinates only and not the corresponding velocities. It represents what we may call the centrifugal energy. As regards the remaining term in (1) we have

$$\omega \Sigma m (x\dot{y} - y\dot{x}) = \alpha_1 \dot{q}_1 + \alpha_2 \dot{q}_2 + \ldots + \alpha_n \dot{q}_n, \quad \text{.........(5)}$$

where

$$\alpha_r = \omega \Sigma m \left(x \frac{\partial y}{\partial q_r} - y \frac{\partial x}{\partial q_r} \right). \quad \text{..................(6)}$$

Hence

$$\frac{d}{dt}\frac{\partial T}{\partial \dot{q}_r} = \frac{d}{dt}\left(\frac{\partial \mathfrak{T}}{\partial \dot{q}_r} + \alpha_r\right)$$

$$= \frac{d}{dt}\frac{\partial \mathfrak{T}}{\partial \dot{q}_r} + \frac{\partial \alpha_r}{\partial q_1}\dot{q}_1 + \frac{\partial \alpha_r}{\partial q_2}\dot{q}_2 + \ldots + \frac{\partial \alpha_r}{\partial q_n}\dot{q}_n, \quad \text{.........(7)}$$

$$\frac{\partial T}{\partial q_r} = \frac{\partial \mathfrak{T}}{\partial q_r} + \frac{\partial \alpha_1}{\partial q_r}\dot{q}_1 + \frac{\partial \alpha_2}{\partial q_r}\dot{q}_2 + \ldots + \frac{\partial \alpha_n}{\partial q_r}\dot{q}_n + \frac{\partial T_0}{\partial q_r}. \quad \text{......(8)}$$

The typical equation of motion, Art. 79 (10), therefore assumes the form

$$\frac{d}{dt}\frac{\partial \mathfrak{T}}{\partial \dot{q}_r} - \frac{\partial \mathfrak{T}}{\partial q_r} + (r, 1)\dot{q}_1 + (r, 2)\dot{q}_2 + \ldots + (r, n)\dot{q}_n - \frac{\partial T_0}{\partial q_r} = P_r, \quad \text{...(9)*}$$

where

$$(r, s) = \frac{\partial \alpha_r}{\partial q_s} - \frac{\partial \alpha_s}{\partial q_r} = 2\omega \Sigma m \frac{\partial (x, y)}{\partial (q_s, q_r)}. \quad \text{............(10)}$$

It is particularly to be noticed that

$$(r, r) = 0, \quad (r, s) = -(s, r). \quad \text{..............(11)}$$

The formula may also be derived *ab initio* from Art. 79 (23) by the method of Art. 77.

Referring to the equations just cited, we have

$$\Sigma m (\dot{x}\ddot{x} + \dot{y}\ddot{y} + \dot{z}\ddot{z}) - \omega^2 \Sigma m (x\dot{x} + y\dot{y}) = \Sigma (X\dot{x} + Y\dot{y} + Z\dot{z}), \quad \text{.........(12)}$$

or, in generalized coordinates,

$$\frac{d}{dt}(\mathfrak{T} - T_0) = P_1 \dot{q}_1 + P_2 \dot{q}_2 + \ldots + P_n \dot{q}_n. \quad \text{.........(13)}$$

In the case of a conservative system free from external disturbance this leads to

$$\mathfrak{T} + V - T_0 = \text{const.}, \quad \text{....................(14)}$$

which takes the place of the equation of energy. The term $-T_0$ may be called the potential energy of centrifugal force.

The formula (13) may also be derived from the equations of type (9); cf. Art. 77. It follows also from a consideration of the rate at which work is done by the constraining forces; see Art. 79.

* Thomson and Tait, 2nd ed., Art. 319.

81. Velocities in terms of Momenta.

Returning to the case of Arts. 73–77 where the kinematical relations do not involve the time explicitly, we have seen that the momenta are given in terms of the velocities by the formula

$$p_r = a_{1r}\dot{q}_1 + a_{2r}\dot{q}_2 + \ldots + a_{nr}\dot{q}_n. \quad \text{...............(1)}$$

Since the determinant Δ_n of Art. 73 does not vanish, the n equations of this type can be used to determine the velocities in terms of the momenta. Thus

$$\dot{q}_r = A_{1r}p_1 + A_{2r}p_2 + \ldots + A_{nr}p_n, \quad \text{............(2)}$$

where $$A_{rs} = \alpha_{rs}/\Delta_n = A_{sr}, \quad \text{.....................(3)}$$

if α_{rs} denote the minor of the constituent a_{rs} in Δ_n.

The formulæ (2) may be regarded as giving the velocities due to given impulses applied to the system when at rest in a given configuration. The coefficients A_{rr}, A_{rs} may accordingly be called 'coefficients of mobility.'

The kinetic energy of the system can now be expressed as a homogeneous quadratic function of the *momenta*. When in this form we will denote it by T'. Thus

$$2T' = \sum_r p_r\dot{q}_r = A_{11}p_1^2 + A_{22}p_2^2 + \ldots + 2A_{12}p_1p_2 + \ldots, \quad \text{...(4)}$$

and we see that (2) is equivalent to

$$\dot{q}_r = \frac{\partial T'}{\partial p_r}. \quad \text{..........................(5)*}$$

It may be noted that this follows also from (3) of Art. 76, if we put $p_r' = p_r + \delta p_r$ and $p_s' = p_s$ except when $s = r$. Similarly our previous formula

$$p_r = \frac{\partial T}{\partial \dot{q}_r}$$

follows from Art. 76 (2).

If Δ_n' be the discriminant of $2T'$, as given by (4), it is easily seen that

$$\Delta_n \Delta_n' = 1. \quad \text{..........................(6)}$$

Ex. The expressions for the component momenta λ, μ, ν of a uniaxal body movable about a fixed point have been given in Art. 78. Hence we derive

$$\dot{\theta} = \frac{\lambda}{A}, \quad \dot{\psi} = \frac{\mu - \nu\cos\theta}{A\sin^2\theta}, \quad \dot{\phi} = \frac{\nu}{C} - \frac{(\mu - \nu\cos\theta)\cos\theta}{A\sin^2\theta}, \quad \text{.........(7)}$$

$$2T' = \frac{\lambda^2}{A} + \frac{(\mu - \nu\cos\theta)^2}{A\sin^2\theta} + \frac{\nu^2}{C}. \quad \text{.......................(8)}$$

* The formula is due to Sir W. R. Hamilton; see Art. 82.

82. Hamiltonian Equations of Motion.

We proceed to find the appropriate form of the equations of motion of a conservative system when we regard the state of motion at any instant as defined by the configuration and the *momenta*, instead of by the configuration and the velocities. The corresponding forms of the kinetic energy will be denoted as in the previous Art. by $T^{\backprime}$ and T.

Since these are merely different symbols for the same thing, we have the identity

$$T^{\backprime} + T = \sum_r p_r \dot{q}_r, \quad \text{.......................(1)}$$

by Art. 74. Hence, using the symbol δ to indicate a *complete* infinitesimal variation, affecting the coordinates as well as the velocities and momenta,

$$\delta T^{\backprime} = \sum_r (p_r \delta \dot{q}_r + \dot{q}_r \delta p_r) - \delta T$$

$$= \sum_r \left(p_r \delta \dot{q}_r + \dot{q}_r \delta p_r - \frac{\partial T}{\partial \dot{q}_r} \delta \dot{q}_r - \frac{\partial T}{\partial q_r} \delta q_r \right) \quad \text{.........(2)}$$

$$= \sum_r \left(\dot{q}_r \delta p_r - \frac{\partial T}{\partial q_r} \delta q_r \right). \quad \text{.........................(3)}$$

Since p_r, q_r are the independent variables in $T^{\backprime}$, it follows that

$$\dot{q}_r = \frac{\partial T^{\backprime}}{\partial p_r}, \quad \text{.........................(4)}$$

as already proved, and also that

$$\frac{\partial T^{\backprime}}{\partial q_r} = -\frac{\partial T}{\partial q_r}. \quad \text{.......................(5)}$$

The typical Lagrangian equation of motion of a conservative system free from extraneous force now transforms into

$$\dot{p}_r = -\frac{\partial}{\partial q_r}(T^{\backprime} + V). \quad \text{....................(6)}$$

Hence, if we write

$$H = T^{\backprime} + V, \quad \text{.........................(7)}$$

the equations (6) and (4) take the forms

$$\dot{p}_r = -\frac{\partial H}{\partial q_r}, \quad \dot{q}_r = \frac{\partial H}{\partial p_r}, \quad \text{....................(8)}$$

where H denotes the total energy of the system, expressed in terms of the coordinates and the momenta.

We have thus a complete system of $2n$ differential equations of the *first order*, to determine the n coordinates, and the n momenta, as functions of t. They have been termed the 'canonical form' of the equations of motion of a conservative system*.

The equation of energy is at once verified. Thus

$$\frac{dH}{dt} = \sum_r \left(\frac{\partial H}{\partial q_r} \dot{q}_r + \frac{\partial H}{\partial p_r} \dot{p}_r \right) = 0, \quad \ldots\ldots\ldots\ldots\ldots(9)$$

by (8).

Equations of the form (8) hold also in the case of *varying relations* (Art. 79), but the meaning of H is different. We write

$$H = \sum_r p_r \dot{q}_r - T + V, \quad \ldots\ldots\ldots\ldots\ldots(10)$$

and imagine this function to be expressed in terms of the momenta, the coordinates, and the time t. The *internal* forces of the system are supposed to be conservative, with a potential energy V which is a function of the coordinates only. Performing a variation δ on both sides of (10), we have

$$\delta H = \sum_r \left\{ \dot{q}_r \, \delta p_r + \left(p_r - \frac{\partial T}{\partial \dot{q}_r} \right) \delta \dot{q}_r - \frac{\partial}{\partial q_r} (T - V) \, \delta q_r \right\}$$

$$= \sum_r \{ \dot{q}_r \delta p_r - \dot{p}_r \, \delta q_r \}, \quad \ldots\ldots\ldots\ldots\ldots(11)$$

by Art. 79. Hence

$$\dot{q}_r = \frac{\partial H}{\partial p_r}, \quad \dot{p}_r = -\frac{\partial H}{\partial q_r}. \quad \ldots\ldots\ldots\ldots\ldots(12)$$

But instead of (9) we now have

$$\frac{dH}{dt} = \frac{\partial H}{\partial t} + \sum_r \left(\frac{\partial H}{\partial p_r} \dot{p}_r + \frac{\partial H}{\partial q_r} \dot{q}_r \right) = \frac{\partial H}{\partial t}. \quad \ldots\ldots\ldots\ldots\ldots(13)$$

83. The Routhian Function.

An ingenious combination of the methods of Lagrange and Hamilton, respectively, was made by Routh†.

Instead of expressing the kinetic energy in a given configuration in terms of the velocities alone, or the momenta alone, we may express it in terms of the velocities corresponding to a certain group of coordinates, say $q_1, q_2, \ldots, q_m$, and the momenta corresponding to the remaining $n - m$ coordinates, which for the sake of distinction we will denote by $\chi, \chi', \chi'', \ldots$. This plan is specially appropriate when we come (Art. 84) to the consideration of systems with 'latent motions.'

* Sir W. R. Hamilton, 'On a General Method in Dynamics,' *Phil. Trans.*, 1834.

† *Treatise on Stability of Motion*, London, 1877.

The momenta $\kappa, \kappa', \kappa'', \ldots$ corresponding to the coordinates $\chi, \chi', \chi'', \ldots$ are given primarily by the relations

$$\frac{\partial T}{\partial \dot{\chi}} = \kappa, \quad \frac{\partial T}{\partial \dot{\chi}'} = \kappa', \quad \frac{\partial T}{\partial \dot{\chi}''} = \kappa'', \; \ldots. \qquad \ldots\ldots\ldots\ldots(1)$$

Modifying the procedure of Hamilton (Art. 82), Routh introduces the function

$$R = T - \kappa\dot{\chi} - \kappa'\dot{\chi}' - \kappa''\dot{\chi}'' - \ldots, \qquad \ldots\ldots\ldots\ldots(2)$$

where T is supposed for the moment to be expressed in terms of the original coordinates $q_1, q_2, \ldots, q_m, \chi, \chi', \chi'', \ldots$, and the corresponding velocities. Hence, if δ denotes a complete infinitesimal variation, we have

$$\delta R = \Sigma\left(\frac{\partial T}{\partial \dot{q}}\delta\dot{q} + \frac{\partial T}{\partial q}\delta q + \frac{\partial T}{\partial \dot{\chi}}\delta\dot{\chi} + \frac{\partial T}{\partial \chi}\delta\chi - \kappa\delta\dot{\chi} - \dot{\chi}\delta\kappa\right), \ldots(3)$$

where the summation embraces all the degrees of freedom. Omitting the terms which cancel by (1), we have

$$\delta R = \Sigma\left(\frac{\partial T}{\partial \dot{q}}\delta\dot{q} + \frac{\partial T}{\partial q}\delta q + \frac{\partial T}{\partial \chi}\delta\chi - \dot{\chi}\delta\kappa\right). \qquad \ldots\ldots\ldots(4)$$

Now the equations (1), when written out in full, serve to determine $\dot{\chi}, \dot{\chi}', \dot{\chi}'', \ldots$ as linear functions of the velocities $\dot{q}_1, \dot{q}_2, \ldots, \dot{q}_m$ and the momenta $\kappa, \kappa', \kappa'', \ldots$; and if we make these substitutions in (2), R is obtained as a homogeneous quadratic function of these two sets of variables, with coefficients depending on the configuration. On this understanding the variations $\delta\dot{q}_r, \delta q_r, \delta\kappa, \delta\chi$ in (4) are independent. Hence we derive $2m$ relations of the types

$$\frac{\partial T}{\partial \dot{q}_r} = \frac{\partial R}{\partial \dot{q}_r}, \quad \frac{\partial T}{\partial q_r} = \frac{\partial R}{\partial q_r}, \qquad \ldots\ldots\ldots\ldots(5)$$

together with $2(n-m)$ relations of the forms

$$\dot{\chi} = -\frac{\partial R}{\partial \kappa}, \quad \frac{\partial T}{\partial \chi} = \frac{\partial R}{\partial \chi}. \qquad \ldots\ldots\ldots\ldots(6)$$

The formula (2) may now be written

$$T = R - \kappa\frac{\partial R}{\partial \kappa} - \kappa'\frac{\partial R}{\partial \kappa'} - \kappa''\frac{\partial R}{\partial \kappa''} - \ldots, \qquad \ldots\ldots\ldots(7)$$

giving T in terms of the new variables. It is evident that the terms in R which are bilinear in respect of the two sets of variables $\dot{q}_1, \dot{q}_2, \ldots, \dot{q}_m$ and $\kappa, \kappa', \kappa'', \ldots$ will cancel on the right-hand side of this formula. Hence, finally,

$$T = \mathfrak{T} + K, \qquad \ldots\ldots\ldots\ldots(8)$$

where $\mathfrak{T}$ is a homogeneous quadratic function of the velocities $\dot{q}_1, \dot{q}_2, \ldots, \dot{q}_m$ alone, whilst K is a similar function of the momenta $\kappa, \kappa', \kappa'', \ldots$.

This formula (8) leads to an immediate proof of the theorems of Delaunay and Kelvin discussed in Art. 76. Let us suppose in the first place that the system is started from rest by given impulses of certain types, but is in other respects free. The former of the two theorems asserts that the kinetic energy is greater than if by impulses of the remaining types the system were constrained to take any other course. We may imagine the coordinates to be so chosen that the constraint in question is expressed by the vanishing of the velocities $\dot{q}_1, \dot{q}_2, \ldots, \dot{q}_m$, whilst the given impulses are $\kappa, \kappa', \kappa'', \ldots$. Hence the energy of the actual motion is greater than in the constrained motion by $\mathfrak{T}$.

Again, suppose that the system is started with prescribed velocity-components $\dot{q}_1, \dot{q}_2, \ldots, \dot{q}_m$, by means of suitable impulses of the corresponding types, but is otherwise free, so that in the actual motion we have $\kappa = 0$, $\kappa' = 0$, $\kappa'' = 0, \ldots$, and therefore $K = 0$. The energy is therefore less than in any other motion consistent with the prescribed velocity-conditions, by the value which K assumes when $\kappa, \kappa', \kappa'', \ldots$ represent the impulses due to the constraints. This is Kelvin's theorem.

84. Cyclic Systems.

A 'cyclic' or 'gyrostatic' system is characterized by the following properties. In the first place there are certain coordinates, which we denote by $\chi, \chi', \chi'', \ldots$, whose absolute values do not enter into the expression for the kinetic energy, but only their rates of variation $\dot{\chi}, \dot{\chi}', \dot{\chi}'', \ldots$. Secondly, there are no forces of the types of these coordinates. This case arises, for instance, when the system includes frictionless gyrostats, the coordinates in question then being the angular coordinates of the flywheels relative to their frames.

Coordinates of this character are called 'cyclic.' The remaining coordinates of the system, say $q_1, q_2, \ldots, q_m$, may be distinguished as 'palpable'; in many practical questions they are the only coordinates which are directly in evidence.

The general theory of such systems has been developed by Thomson and Tait, and by Routh, the object being to obtain equations of motion in terms of the 'palpable' variables only. Since the cyclic coordinates as well as the corresponding velocities do not appear in these equations, they are sometimes described as 'ignored' coordinates, and the method to be explained is called that of 'ignoration of coordinates' (Thomson and Tait).

By hypothesis, the Lagrangian equations corresponding to the cyclic coordinates reduce to

$$\frac{d}{dt}\frac{\partial T}{\partial \dot{\chi}}=0, \quad \frac{d}{dt}\frac{\partial T}{\partial \dot{\chi}'}=0, \quad \frac{d}{dt}\frac{\partial T}{\partial \dot{\chi}''}=0, \ldots, \qquad \ldots\ldots\ldots(1)$$

whence

$$\frac{\partial T}{\partial \dot{\chi}}=\kappa, \quad \frac{\partial T}{\partial \dot{\chi}'}=\kappa', \quad \frac{\partial T}{\partial \dot{\chi}''}=\kappa'', \ldots, \qquad \ldots\ldots\ldots\ldots\ldots(2)$$

where $\kappa, \kappa', \kappa'', \ldots$ are constants, viz. the constant momenta of the types in question. a

Assuming

$$R=T-\kappa\dot{\chi}-\kappa'\dot{\chi}'-\kappa''\dot{\chi}''-\ldots, \qquad \ldots\ldots\ldots\ldots\ldots(3)$$

where R is supposed expressed in terms of the velocities $\dot{q}_1, \dot{q}_2, \ldots, \dot{q}_m$ corresponding to the palpable coordinates, and the momenta $\kappa, \kappa', \kappa'', \ldots$ corresponding to the ignored coordinates, we have, as in Art. 83,

$$\frac{\partial T}{\partial \dot{q}_r}=\frac{\partial R}{\partial \dot{q}_r}, \quad \frac{\partial T}{\partial q_r}=\frac{\partial R}{\partial q_r}, \qquad \ldots\ldots\ldots\ldots\ldots\ldots(4)$$

and

$$\dot{\chi}=-\frac{\partial R}{\partial \kappa}, \quad \dot{\chi}'=-\frac{\partial R}{\partial \kappa'}, \quad \dot{\chi}''=-\frac{\partial R}{\partial \kappa''}, \quad \ldots\ \ldots\ldots(5)$$

Substituting from (4) in Lagrange's formula, we have

$$\frac{d}{dt}\frac{\partial R}{\partial \dot{q}_r}-\frac{\partial R}{\partial q_r}=P_r \quad [r=1, 2, \ldots, m]. \qquad \ldots\ldots\ldots\ldots(6)$$

This form of the equations is due to Routh.

Ex. In the case of the top the cyclic coordinates are ψ and ϕ. With the notation of Art. 78 (8), we have

$$\begin{aligned} R &= T-\mu\dot{\psi}-\nu\dot{\phi} \\ &= \tfrac{1}{2}A\dot{\theta}^2-\tfrac{1}{2}\frac{(\mu-\nu\cos\theta)^2}{A\sin^2\theta}-\tfrac{1}{2}\frac{\nu^2}{C}. \qquad \ldots\ldots\ldots\ldots\ldots\ldots\ldots(7) \end{aligned}$$

The equation (6) then gives

$$A\ddot{\theta}+\frac{(\mu-\nu\cos\theta)(\nu-\mu\cos\theta)}{A\sin^3\theta}=Mgh\sin\theta. \qquad \ldots\ldots\ldots\ldots\ldots(8)$$

From this we could easily investigate the small oscillations about precessional motion.

Another form of the equations, shewing more explicitly the influence of the cyclic momenta, is due to Kelvin. The function R as above defined contains terms of three types, thus

$$R = R_{2,0} + R_{1,1} + R_{0,2}, \qquad \text{(9)}$$

where $R_{2,0}$ is a homogeneous quadratic function of the velocities $\dot{q}_1, \dot{q}_2, \ldots, \dot{q}_m$; $R_{0,2}$ is a similar function of the cyclic momenta $\kappa, \kappa', \kappa'', \ldots$; whilst $R_{1,1}$ is a bilinear function of these two sets of variables, say

$$R_{1,1} = \alpha_1 \dot{q}_1 + \alpha_2 \dot{q}_2 + \ldots + \alpha_m \dot{q}_m, \qquad \text{(10)}$$

where

$$\alpha_r = \beta_r \kappa + \beta_r' \kappa' + \beta_r'' \kappa'' + \ldots. \qquad \text{(11)}$$

The formula

$$T = R - \kappa \frac{\partial R}{\partial \kappa} - \kappa' \frac{\partial R}{\partial \kappa'} - \kappa'' \frac{\partial R}{\partial \kappa''} - \ldots, \qquad \text{(12)}$$

proved in Art. 83, gives, on substitution from (9),

$$T = R_{2,0} - R_{0,2}, \qquad \text{(13)}$$

a formula previously written as

$$T = \mathfrak{T} + K.$$

Hence, putting $R_{0,2} = -K$, and substituting from (9),

$$R = \mathfrak{T} - K + \alpha_1 \dot{q}_1 + \alpha_2 \dot{q}_2 + \ldots + \alpha_m \dot{q}_m. \qquad \text{(14)}$$

Now

$$\frac{d}{dt}\frac{\partial R}{\partial \dot{q}_1} = \frac{d}{dt}\left(\frac{\partial \mathfrak{T}}{\partial \dot{q}_r} + \alpha_r\right)$$

$$= \frac{d}{dt}\frac{\partial \mathfrak{T}}{\partial \dot{q}_r} + \frac{\partial \alpha_r}{\partial q_1}\dot{q}_1 + \frac{\partial \alpha_r}{\partial q_2}\dot{q}_2 + \ldots + \frac{\partial \alpha_r}{\partial q_m}\dot{q}_m, \qquad \text{(15)}$$

and

$$\frac{\partial R}{\partial q_r} = \frac{\partial \mathfrak{T}}{\partial q_r} - \frac{\partial K}{\partial q_r} + \frac{\partial \alpha_1}{\partial q_r}\dot{q}_1 + \frac{\partial \alpha_2}{\partial q_r}\dot{q}_2 + \ldots + \frac{\partial \alpha_m}{\partial q_r}\dot{q}_m. \qquad \text{(16)}$$

Hence, substituting in (5), we obtain the typical equation of motion of a gyrostatic system in the form

$$\frac{d}{dt}\frac{\partial \mathfrak{T}}{\partial \dot{q}_r} - \frac{\partial \mathfrak{T}}{\partial q_r} + (r, 1)\dot{q}_1 + (r, 2)\dot{q}_2 + \ldots + (r, m)\dot{q}_m + \frac{\partial K}{\partial q_r} = P_r, \qquad \text{(17)}$$

where

$$(r, s) = \frac{\partial \alpha_r}{\partial q_s} - \frac{\partial \alpha_s}{\partial q_r}. \qquad \text{(18)}$$

This form is due to Kelvin*.

It is to be noticed that

$$(r, r) = 0, \quad (r, s) = -(s, r). \qquad \text{(19)}$$

* Thomson and Tait, 2nd ed., Art. 319, Example (G).

Hence if in (17) we put $r = 1, 2, \ldots, m$ in succession, and multiply the resulting equations by $\dot{q}_1, \dot{q}_2, \ldots, \dot{q}_m$, respectively, we find, as in the proof of Art. 77 (11),

$$\frac{d}{dt}(\mathfrak{T} + K) = P_1\dot{q}_1 + P_2\dot{q}_2 + \ldots + P_m\dot{q}_m, \quad \ldots\ldots\ldots(20)$$

or, in the case of a conservative system free from extraneous force,

$$\mathfrak{T} + V + K = \text{const.}, \quad \ldots\ldots\ldots\ldots\ldots\ldots\ldots(21)$$

which is the equation of energy.

When $q_1, q_2, \ldots, q_m$ have been determined from (17) as functions of the time, the velocities corresponding to the cyclic coordinates can be found, if required, from the relations (6) of Art. 83, which give, on reference to (13) and (10) above,

$$\left.\begin{aligned} \dot{\chi} &= \frac{\partial K}{\partial \kappa} - \beta_1\dot{q}_1 - \beta_2\dot{q}_2 - \ldots - \beta_m\dot{q}_m, \\ \dot{\chi}' &= \frac{\partial K}{\partial \kappa'} - \beta_1'\dot{q}_1 - \beta_2'\dot{q}_2 - \ldots - \beta_m'\dot{q}_m, \\ &\ldots\ldots\ldots\ldots\ldots\ldots\ldots\ldots\ldots\ldots \end{aligned}\right\} \quad \ldots\ldots\ldots(22)$$

In the particular case where the cyclic momenta $\kappa, \kappa', \kappa'', \ldots$ are all zero, we have $\alpha_r = 0$, $K = 0$, and the typical equation (17) reduces to

$$\frac{d}{dt}\frac{\partial \mathfrak{T}}{\partial \dot{q}_r} - \frac{\partial \mathfrak{T}}{\partial q_r} = P_r. \quad \ldots\ldots\ldots\ldots\ldots\ldots(23)$$

The form is now the same as in Art. 77, and the system accordingly behaves, as regards the coordinates $q_1, q_2, \ldots, q_m$, exactly like the acyclic type there considered. These coordinates do not, however, now fix the position of every particle of the system. For example, if the system passes a second time through a given configuration, so far as this is defined by $q_1, q_2, \ldots, q_m$, the 'ignored' coordinates $\chi, \chi', \chi'', \ldots$ do not in general resume their original values.

It is to be noticed, again, that the terms in (17) which are linear in $\dot{q}_1, \dot{q}_2, \ldots, \dot{q}_m$ change sign with dt, whilst the rest do not. Hence the motion of a gyrostatic system is not reversible, unless indeed we reverse the cyclic motions as well as the velocities $\dot{q}_1, \dot{q}_2, \ldots, \dot{q}_m$ which relate to the palpable coordinates. For instance, as has been already remarked, the precessional motion of a top is not reversible unless we reverse the spin.

85. Kineto-Statics.

The conditions of 'equilibrium' of a system with latent cyclic motions follow directly from the principle of energy. If the system be guided from (apparent) rest in the configuration $(q_1, q_2, \ldots, q_m)$ to rest in an adjacent configuration under external forces $Q_1, Q_2, \ldots, Q_m$, the work done must be equal to the increment of the latent kinetic energy. Thus

$$\sum_r Q_r \delta q_r = \delta K, \quad \ldots\ldots\ldots\ldots\ldots\ldots(1)$$

or

$$Q_r = \frac{\partial K}{\partial q_r}. \quad \ldots\ldots\ldots\ldots\ldots\ldots(2)$$

This follows also from equations (17) of the preceding Art. if we put $\dot{q}_1 = \dot{q}_2 = \ldots = \dot{q}_m = 0$.

The formulæ (22) of Art. 84 now reduce to

$$\dot{\chi} = \frac{\partial K}{\partial \kappa}, \quad \dot{\chi}' = \frac{\partial K}{\partial \kappa'}, \quad \dot{\chi}'' = \frac{\partial K}{\partial \kappa''}, \ldots. \quad \ldots\ldots\ldots(3)$$

These equations determine $\kappa, \kappa', \kappa'', \ldots$ as linear functions of $\dot{\chi}, \dot{\chi}', \dot{\chi}'', \ldots$, and K can then be expressed otherwise as a homogeneous quadratic function of these latent *velocities*. Denoting this function by T_0, we have

$$\frac{\partial K}{\partial q_r} = -\frac{\partial T_0}{\partial q_r}, \quad \ldots\ldots\ldots\ldots\ldots\ldots(4)$$

by Art. 82 (5). The formula (2) then makes

$$Q_r = -\frac{\partial T_0}{\partial q_r}. \quad \ldots\ldots\ldots\ldots\ldots\ldots(5)$$

An example is furnished by the top. The cyclic coordinates being ψ, ϕ, we have (see Art. 84 (7))

$$\left.\begin{aligned} 2\mathfrak{T} &= A\dot{\theta}^2, \\ 2K &= \frac{(\mu - \nu\cos\theta)^2}{A\sin^2\theta} + \frac{\nu^2}{C}, \\ 2T_0 &= A\sin^2\theta\,\dot{\psi}^2 + C(\dot{\phi} + \cos\theta\,\dot{\psi})^2, \end{aligned}\right\} \quad \ldots\ldots\ldots(6)$$

whence the relation

$$\frac{\partial K}{\partial \theta} = -\frac{\partial T_0}{\partial \theta} \quad \ldots\ldots\ldots\ldots\ldots\ldots(7)$$

may be verified.

If V be the potential energy, the condition (1) of 'equilibrium' gives

$$\frac{\partial}{\partial \theta}(K+V)=0. \quad \text{.......................}(8)$$

This leads to the condition for steady precession (Art. 54).

EXAMPLES. XIII.

1. Assuming from Delaunay's Theorem (Art. 76) that the energy of a free rigid body set in motion by an impulsive couple (λ, μ, ν) is a maximum, prove that the component angular velocities are λ/A, μ/B, ν/C, the coordinate axes being the principal axes of inertia at the mass-centre.

2. In the case of two degrees of freedom, assuming

$$2T=a_{11}\dot{q}_1^2+2a_{12}\dot{q}_1\dot{q}_2+a_{22}\dot{q}_2^2,$$

prove that

$$2T'=\frac{a_{22}p_1^2-2a_{12}p_1p_2+a_{11}p_2^2}{a_{11}a_{22}-a_{12}^2};$$

and verify the relations

$$\dot{q}_1=\frac{\partial T'}{\partial p_1},\quad \dot{q}_2=\frac{\partial T'}{\partial p_2},\quad \frac{\partial T'}{\partial q_1}=-\frac{\partial T}{\partial q_1},\quad \frac{\partial T'}{\partial q_2}=-\frac{\partial T}{\partial q_2}.$$

3. Prove that in the notation of Art. 81

$$2T'=-\frac{1}{\Delta}\begin{vmatrix} a_{11}, & a_{12}, & \dots, & a_{1n}, & p_1 \\ a_{21}, & a_{22}, & \dots, & a_{2n}, & p_2 \\ \dots & \dots & \dots & \dots & \dots \\ a_{n1}, & a_{n2}, & \dots, & a_{nn}, & p_n \\ p_1, & p_2, & \dots, & p_n, & 0 \end{vmatrix},$$

where Δ is the discriminant of T.

4. A waggon of mass M carries a pendulum of mass m and length l which can swing in the direction of motion of the waggon. If $\dot{x}$ be the velocity of the waggon, θ the inclination of the pendulum to the vertical, prove that

$$2T=(M+m)\dot{x}^2+2ml\cos\theta\,\dot{x}\dot{\theta}+ml^2\dot{\theta}^2.$$

If ξ, κ be the components of momentum corresponding to x and θ, prove that

$$2T'=\frac{ml^2\xi^2-2ml\cos\theta\,\xi\kappa+(M+m)\kappa^2}{ml^2(M+m\sin^2\theta)}.$$

Also, express the kinetic energy in terms (i) of ξ and $\dot{\theta}$, and (ii) of $\dot{x}$ and κ; and verify Bertrand's and Kelvin's theorems.

5. A particle moves in the plane xy under an acceleration μr towards the origin, where r is the radius vector. Form Lagrange's equations in terms of the coordinates ξ, η defined by

$$x=c\cosh\xi\cos\eta,\quad y=c\sinh\xi\sin\eta;$$

and verify that they are satisfied by $\xi=\text{const.}$, $\dot{\eta}=\pm\sqrt{\mu}$.

Examine the case where the central force is repulsive.

6. A uniaxal solid whose principal moments are A, A, C is mounted so that it can swing about an equatorial axis, which is horizontal. This axis is carried by a vertical frame which is made to rotate with constant angular velocity ω about its vertical diameter, in which the mass-centre of the solid lies. Form the equation of motion; and prove that the vertical or horizontal position of the axis of symmetry is the stable one, according as $C \gtrless A$.

Prove that the period of a small oscillation about stable equilibrium is

$$\sqrt{\left\{\frac{A}{|C-A|}\right\}} \cdot \frac{2\pi}{\omega}.$$

Shew that the results are the same if the vertical frame be *free*.

7. Employ Lagrange's equations to obtain the equations of motion of a rigid body about a fixed point in the forms

$$\frac{d}{dt}\frac{\partial T}{\partial p} - r\frac{\partial T}{\partial q} + q\frac{\partial T}{\partial r} = L, \text{ etc., etc.,}$$

the coordinate axes being supposed fixed in the body, but not coincident with the principal axes of inertia at the point.

8. A right circular cone of semi-angle α is placed on its side, at rest, on a rough horizontal plane which then begins to rotate about the vertical through the apex O. Prove that when the angular velocity of the plane is ω the axis of the cone revolves about the vertical with the angular velocity $C\omega/I$, and that its motion relative to the plane is at the rate $(C-A)/I \cdot \omega \sin^2 \alpha$, where A, A, C are the principal moments at O, and I is the moment of inertia about a generator.

9. A pendulum symmetrical about a longitudinal axis OA hangs by a Hooke's joint from a vertical spindle which is made to rotate with a constant angular velocity ω. Let θ be the inclination to the vertical of that arm (COC') of the cross to which the pendulum is attached, and ϕ the angle which the plane AOC makes with the vertical plane through OC. Prove that

$$2T = A(\dot{\theta}\sin\phi - \omega\sin\theta\cos\phi)^2 + B(\dot{\theta}\cos\phi + \omega\sin\theta\sin\phi)^2 + B(\dot{\phi} + \omega\cos\theta)^2,$$
$$V = -Mgh\sin\theta\cos\phi,$$

h being the distance of the mass-centre from O.

10. In the previous question, putting $\theta = \frac{1}{2}\pi - \xi$, $\phi = \eta$, and treating ξ, η as small, prove that

$$\left.\begin{aligned} B\ddot{\xi} - (2B-A)\omega\dot{\eta} + \{(A-B)\omega^2 + Mgh\}\xi = 0, \\ B\ddot{\eta} + (2B-A)\omega\dot{\xi} + \{(A-B)\omega^2 + Mgh\}\eta = 0. \end{aligned}\right\}$$

Solve these equations, and interpret the solution.

11. A body whose principal moments A, B, C are unequal is spinning about the axis OC with very great angular velocity n, which is maintained constant. If this axis be made to revolve in one plane about O, with angular velocity v, the requisite constraining couples are, *on the average*, $\frac{1}{2}(A+B)\dot{v}$ in that plane, and Cnv at right angles to it.

12. A chain of equal bars loosely jointed at their extremities is laid out in a straight line. The mass of each bar is M, its length is $2a$, and its radius of

gyration about the centre (which is the centre of mass) is κ. If the chain be set in motion by impulses at right angles to the length, and $\dot{y}_n$ be the initial velocity of the nth joint, prove that the terms in the initial kinetic energy which involve $\dot{y}_n$ are

$$\tfrac{1}{4}M\left(1+\frac{\kappa^2}{a^2}\right)\dot{y}_n{}^2+\tfrac{1}{4}M\left(1-\frac{\kappa^2}{a^2}\right)\dot{y}_n\,(\dot{y}_{n+1}+\dot{y}_{n-1}),$$

except in the case of the end joints of the chain.

If the chain extend to infinity in one direction, and if the finite end ($n=0$) be free, prove that the value of $\dot{y}_n$ due to a lateral impulse ξ_0 on this end is

$$(-)^n\frac{2\xi_0 a}{M\kappa}\left(\frac{a-\kappa}{a+\kappa}\right)^n.$$

13. Prove that Lagrange's equations (Art. 77) when written out in full have the form

$$\sum_s a_{rs}\,\ddot{q}_s+\sum_r\sum_s\begin{bmatrix} sk \\ r\end{bmatrix}\dot{q}_s\,\dot{q}_k=P_r,$$

where

$$\begin{bmatrix} sk \\ r\end{bmatrix}=\frac{1}{2}\left(\frac{\partial a_{rs}}{\partial q_k}+\frac{\partial a_{rk}}{\partial q_s}-\frac{\partial a_{sk}}{\partial q_r}\right).$$

14. A string attached to a fixed point passes through a ring vertically beneath, and carries a particle m at its lower end. The lower portion of the string (of length r) can swing like a pendulum. Prove that the equation of angular motion when the ring is moved vertically in an arbitrary manner is

$$r\ddot{\theta}+2\dot{r}\dot{\theta}=-(g+\ddot{r})\sin\theta.$$

15. A particle is in a smooth circular tube which rotates with angular velocity ω about a vertical diameter. If it starts from relative rest at an angular distance α from the lowest point, prove that in the subsequent motion

$$\omega t=F_1(k)-F(k,\phi),$$

where

$$k=\cos\alpha,\quad \sin\phi=\cos\theta/\cos\alpha.$$

If the particle makes complete revolutions, prove that

$$\surd(\omega^2+n^2)\,t=F(k,\theta),$$

where t is the time reckoned from the lowest point, n is the value of $\dot{\theta}$ at this point, and

$$k=\frac{\omega}{\surd(\omega^2+n^2)}.$$

16. Verify by the theorem of Art. 85 that the circular form of a whirling chain (Art. 99, Ex.) is stable.

[Prove that T_0 is a maximum.]

CHAPTER XI

THEORY OF VIBRATIONS

86. Conditions of Equilibrium. Stability.

In order that, in a conservative system free from extraneous force, the configuration $(q_1, q_2, \ldots, q_n)$ may be one of equilibrium, the equations of motion, viz.

$$\frac{d}{dt}\frac{\partial T}{\partial \dot{q}_r} - \frac{\partial T}{\partial q_r} = -\frac{\partial V}{\partial q_r}, \quad [r = 1, 2, \ldots, n] \quad \ldots\ldots\ldots(1)$$

must be satisfied by

$$\dot{q}_1, \dot{q}_2, \ldots, \dot{q}_r, \ldots, \dot{q}_n = 0, \quad \ldots\ldots\ldots\ldots\ldots\ldots\ldots(2)$$

for all values of t, whence

$$\frac{\partial V}{\partial q_r} = 0. \quad \ldots\ldots\ldots\ldots\ldots\ldots\ldots\ldots\ldots\ldots(3)$$

Hence the necessary and sufficient condition is that the potential energy should be 'stationary' for small variations of the coordinates.

If, further, V is a *minimum* in the configuration in question, the equilibrium is stable. For in the motion consequent on a slight disturbance the total energy $T + V$ is constant, and since T is essentially positive, it follows that V can never exceed its equilibrium value by more than a slight amount depending on the energy of the disturbance. This implies, on the present hypothesis, that there is an upper limit to the deviation of each coordinate from its equilibrium value; moreover, the limit diminishes indefinitely with the energy of the original disturbance*.

No such simple argument is available to shew without qualification that the above condition is *necessary*. If, however, we recognize the existence of dissipative forces which are called into play by any motion whatever of the system, the conclusion can be drawn as follows. However slight these forces may be, the total energy $T + V$ must continually diminish so long as the velocities $\dot{q}_1, \dot{q}_2, \ldots, \dot{q}_n$ differ from zero. Hence if the system be started from rest in a

* The argument is due to P. L. Dirichlet (1805–59).

configuration for which V is less than in the equilibrium configuration, the potential energy must still further decrease, since T cannot be negative. It follows either that the system will tend to come to rest in some new configuration of equilibrium, or that V will diminish indefinitely*.

87. Statical Relations between Forces and Displacements.

When extraneous forces act on the system, we may denote their generalized components by $Q_1, Q_2, \ldots, Q_n$. We must now add a term Q_r to the second member of Art. 86 (1).

The conditions of equilibrium are accordingly given by n equations of the type

$$Q_r = \frac{\partial V}{\partial q_r}. \qquad \text{.........................(1)}$$

These give directly the forces required to maintain specified displacements, and can of course be used conversely to find the static displacements due to given forces; see Art. 89.

When the displacements are small, it is convenient to adjust the coordinates $q_1, q_2, \ldots, q_n$ (e.g. by the addition of suitable constants) so that they shall vanish in the undisturbed configuration. The potential energy is then given with sufficient approximation for many purposes by an expression of the type

$$2V = c_{11} q_1^2 + c_{22} q_2^2 + \ldots + 2c_{12} q_1 q_2 + \ldots . \qquad \text{.........(2)}$$

A constant term in the expansion would be irrelevant, since it would disappear from (1), and terms of the first order are not admissible, since by hypothesis the equations (3) of the preceding Art. must be satisfied by zero values of the coordinates†.

We have then, from (1),

$$Q_r = c_{1r} q_1 + c_{2r} q_2 + \ldots + c_{nr} q_n. \qquad \text{...............(3)}$$

Any system in which the forces are assumed to be linear functions of the displacements from equilibrium, with constant coefficients,

* This argument is due to Lord Kelvin; see Thomson and Tait, *Natural Philosophy*, 2nd ed., Art. 345 ii.

† For stability it is necessary that the expression (2) should be essentially positive. The coefficients c_{rr}, c_{rs} are therefore subject to certain algebraical restrictions similar to those imposed on the coefficients a_{rr}, a_{rs} in Art. 73.

may be called by analogy an 'elastic' system, since this is the kind of relation which is ordinarily assumed as the basis of the Theory of Elasticity. The constants c_{rr}, c_{rs} may be called 'coefficients of elasticity,' or (for a reason which will appear) 'coefficients of stability.'

From (2) and (3) we derive

$$V = \tfrac{1}{2}(Q_1 q_1 + Q_2 q_2 + \ldots + Q_n q_n). \quad \ldots\ldots\ldots\ldots(4)$$

This has an interesting verification. Let us suppose that the forces are gradually increased from zero to their final values $Q_1, Q_2, \ldots, Q_n$, preserving throughout the same ratios to one another, so that their values at any stage of the process are $\theta Q_1, \theta Q_2, \ldots, \theta Q_n$, where θ is to range from 0 to 1. The corresponding values of the coordinates will be $\theta q_1, \theta q_2, \ldots, \theta q_n$. Hence, considering any one coordinate, the element of work will be

$$\theta Q_r . \delta(\theta q_r) = Q_r q_r . \theta\,\delta\theta. \quad \ldots\ldots\ldots\ldots\ldots\ldots(5)$$

The total work is therefore

$$(Q_1 q_1 + Q_2 q_2 + \ldots + Q_n q_n)\int_0^1 \theta\,d\theta. \quad \ldots\ldots\ldots\ldots(6)$$

Since the work done on the system has its equivalent in potential energy, the result (4) follows.

88. Reciprocal Relations.

The mathematical relations with which we are now concerned are the same as in the theory of initial motions and impulses (Arts. 75, 76), and the theorems there developed have accordingly their analogues in the present subject*. Thus we have the reciprocal relation

$$Q_1 q_1' + Q_2 q_2' + \ldots + Q_n q_n' = Q_1' q_1 + Q_2' q_2 + \ldots + Q_n' q_n \ \ldots(1)$$

between the forces and the displacements in any two cases, each of these expressions being equal to

$$c_{11} q_1 q_1' + c_{22} q_2 q_2' + \ldots + c_{12}(q_1 q_2' + q_1' q_2) + \ldots. \quad \ldots\ldots(2)$$

In particular, if all the forces vanish except Q_r, Q_s', we have

$$\frac{q_s}{Q_r} = \frac{q_r'}{Q_s'}. \ldots\ldots\ldots\ldots\ldots\ldots\ldots\ldots\ldots\ldots(3)$$

* Cf. Rayleigh, *Theory of Sound*, c. iv.

If the two coordinates here concerned are of the same character, e.g. both lines or both angles, the theorem asserts that the displacement of type s due to a force of type r is equal to the displacement of type r due to an equal force of type s.

The most interesting applications are to systems of infinite freedom, to which they are extended by analogy. The Theory of Elasticity furnishes a number of examples, some of which have been found useful in the Theory of Structures [S. 142].

Thus if a weight W suspended from a point A of a horizontal beam produces a deflection y at B, the same weight W suspended from B will produce an equal deflection y at A. Again, if a bending moment M applied at A produces a rotation θ at B, an equal couple M applied at B will produce a rotation θ at A. Further, if a couple M at A produces a deflection y at B, a force M/a at B will produce a rotation y/a at A.

Statements of this kind can of course be verified directly, but often only by rather elaborate calculation.

89. Potential Energy in terms of Disturbing Forces.

By the method of Art. 76 we derive the formulæ

$$V' - V = \tfrac{1}{2}\sum_r (Q_r' + Q_r)(q_r' - q_r), \quad \text{...........(1)}$$

$$V' - V = \tfrac{1}{2}\sum_r (Q_r' - Q_r)(q_r' + q_r), \quad \text{...........(2)}$$

which are of course equivalent, in virtue of the reciprocal theorem. The former of these may be obtained otherwise by calculating the work done as the forces vary from Q_r to Q_r', in such a way that at a certain stage in the process the forces are

$$Q_r + \theta(Q_r' - Q_r).$$

The corresponding displacements will then be

$$q_r + \theta(q_r' - q_r).$$

The element of work is accordingly

$$\{Q_r + \theta(Q_r' - Q_r)\}(q_r' - q_r)\,\delta\theta.$$

Integrating this from $\theta = 0$ to $\theta = 1$, and adding up the results for the forces of various types, we obtain the formula (1).

The theorems analogous to those of Delaunay and Kelvin (Art. 76) are included in the following statement:

The potential energy of the system when deformed by given *forces* is *greater*, whilst the work required to produce a given *deformation* is *less*, than if the freedom had been limited by the introduction of constraints. In other words the effect of the constraints is to increase the 'stiffness' of the system. It is unnecessary to give the proofs, which may be derived by mere literal substitutions from the investigation of Art. 76, writing Q_r for p_r and q_r for $\dot{q}_r$.

It will be noticed that if in (1) we make $q_r' - q_r$ infinitesimal, we reproduce the formula (1) of Art. 87.

Again, if we solve the n equations of the type (3) of Art. 87 for $q_1, q_2, \ldots, q_n$ in terms of $Q_1, Q_2, \ldots, Q_n$, and substitute in (4) of the same Art., we can express the potential energy as a function of the *forces* $Q_1, Q_2, \ldots, Q_n$. Denoting it (in this form) by V', we write

$$2V' = C_{11}Q_1^2 + C_{22}Q_2^2 + \ldots + 2C_{12}Q_1Q_2 + \ldots . \quad \ldots\ldots(3)$$

If in (2) we make $Q_r' - Q_r$ infinitesimal, we find

$$q_r = \frac{\partial V'}{\partial Q_r}, \quad \ldots\ldots(4)$$

or

$$q_r = C_{1r}Q_1 + C_{2r}Q_2 + \ldots + C_{nr}Q_n. \quad \ldots\ldots(5)$$

This gives the displacements due to prescribed forces. The coefficients C_{rr}, C_{rs} may be called 'coefficients of pliability.'

Finally, there is an analogy with the investigation of Art. 83. If we isolate a particular group of coordinates, say $\chi, \chi', \chi'', \ldots$, and denote by $\mathrm{X}, \mathrm{X}', \mathrm{X}'', \ldots$ the corresponding components of force, we have, as part of the conditions of equilibrium,

$$\mathrm{X} = \frac{\partial V}{\partial \chi}, \quad \mathrm{X}' = \frac{\partial V}{\partial \chi'}, \quad \mathrm{X}'' = \frac{\partial V}{\partial \chi''}, \ldots . \quad \ldots\ldots(6)$$

These relations enable us to express $\chi, \chi', \chi'', \ldots$ as linear functions of the remaining coordinates, which we denote by $q_1, q_2, \ldots, q_m$, and the *forces* $\mathrm{X}, \mathrm{X}', \mathrm{X}'', \ldots$. If these values be substituted in the function

$$W = V - \mathrm{X}\chi - \mathrm{X}'\chi' - \mathrm{X}''\chi'' - \ldots, \quad \ldots\ldots(7)$$

we find, exactly as in Art. 83,

$$Q_r = \frac{\partial W}{\partial q_r}, \quad \ldots\ldots(8)$$

and

$$\chi = -\frac{\partial W}{\partial \mathrm{X}}, \quad \chi' = -\frac{\partial W}{\partial \mathrm{X}'}, \quad \chi'' = -\frac{\partial W}{\partial \mathrm{X}''}, \ldots . \quad \ldots\ldots(9)$$

Since, from (7) and (9),

$$V = W - \left(\mathrm{X}\frac{\partial W}{\partial \mathrm{X}} + \mathrm{X}'\frac{\partial W}{\partial \mathrm{X}'} + \mathrm{X}''\frac{\partial W}{\partial \mathrm{X}''} + \ldots\right), \quad \ldots\ldots(10)$$

the potential energy, when similarly expressed, reduces to the sum of two homogeneous quadratic functions, viz. a function of $q_1, q_2, \ldots, q_m$, and a function of X, X′, X″, ..., since the terms which are bilinear in these quantities will cancel.

The expression W may be interpreted as follows. We may regard V as the internal elastic energy of the system, and the remaining terms in (7) as the energy in relation to a field of *constant* external forces X, X′, X″, ... The formula (7) then gives the total energy.

Let us first suppose that prescribed values are given to the coordinates $q_1, q_2, \ldots, q_m$ by suitable forces of the corresponding types, the forces X, X′, X″, ... being zero. This may be regarded as imposing a certain geometrical constraint on the system. The coordinates χ, χ', χ'', ... will in consequence assume certain values. Next let the forces X, X′, X″, ... be gradually applied, increasing always in the same proportions up to their final values, the displacements $q_1, q_2, \ldots, q_m$ being maintained unaltered. The *additional* changes in the coordinates χ, χ', χ'', ..., and consequently the work done by these forces in increasing the elastic energy of the system, will evidently be the same whatever the particular constrained configuration from which the start is made. We have thus a physical proof of the theorem, already proved analytically, that V can be resolved into the sum of a function of $q_1, q_2, \ldots, q_m$ and a function of X, X′, X″, ...

The matter becomes more interesting, as well as more obvious, if we consider a case of infinite freedom, for instance a horizontal beam, or a cantilever. First suppose that a prescribed deflection y is maintained at a point P by means of a force applied there. The consequent elastic energy has the form $\frac{1}{2}By^2$. Next let a load be applied at another point Q, gradually increasing from zero up to its final value w. If the position of P be maintained unchanged, the *additional* deflection (z) at Q due to the load there will be the same as if P had been fixed in its zero position, e.g. by means of a prop. Hence the work done by the load, and the consequent addition to the elastic energy of the beam, will be $\frac{1}{2}wz$, or $\frac{1}{2}w^2/C$, if $w = Cz$. Thus

$$V = \tfrac{1}{2}By^2 + \tfrac{1}{2}\frac{w^2}{C}. \qquad \text{.............................(11)}$$

90. Free Vibrations.

Proceeding now to the theory of small oscillations about equilibrium, we take first the case of free vibrations, extraneous forces being supposed absent.

The formula for the kinetic energy is of the same type as in Art. 73, viz.

$$2T = a_{11}\dot{q}_1^2 + a_{22}\dot{q}_2^2 + \ldots + 2a_{12}\dot{q}_1\dot{q}_2 + \ldots \qquad \text{.........(1)}$$

In the investigations which follow, the deviations from the equilibrium configuration are assumed to be small, so that the coefficients

a_{rr}, a_{rs} may be regarded as constants, and equal to the values which they have in the equilibrium configuration.

Again, if as in Art. 87 we suppose the coordinates to be adjusted so as to vanish in the equilibrium configuration, we may write

$$2V = c_{11}q_1^2 + c_{22}q_2^2 + \ldots + 2c_{12}q_1q_2 + \ldots. \quad \ldots\ldots\ldots(2)$$

The Lagrangian equations now take the form

$$\frac{d}{dt}\frac{\partial T}{\partial \dot{q}_r} + \frac{\partial V}{\partial q_r} = 0, \quad \ldots\ldots\ldots(3)$$

the term $\partial T/\partial q_r$ being omitted as of the second order in the velocities. Hence

$$a_{1r}\ddot{q}_1 + a_{2r}\ddot{q}_2 + \ldots + a_{nr}\ddot{q}_n + c_{1r}q_1 + c_{2r}q_2 + \ldots + c_{nr}q_n = 0. \ \ldots(4)$$

To solve this system of linear equations we assume

$$q_r = A_r e^{\lambda t}. \quad \ldots\ldots\ldots(5)$$

This gives n equations of the type

$$(a_{1r}\lambda^2 + c_{1r})A_1 + (a_{2r}\lambda^2 + c_{2r})A_2 + \ldots + (a_{nr}\lambda^2 + c_{nr})A_n = 0. \quad \ldots\ldots\ldots(6)$$

Eliminating the $n-1$ ratios

$$A_1 : A_2 : \ldots : A_n, \quad \ldots\ldots\ldots(7)$$

we obtain

$$\Delta(\lambda^2) = 0, \quad \ldots\ldots\ldots(8)$$

where

$$\Delta(\lambda^2) = \begin{vmatrix} a_{11}\lambda^2 + c_{11}, & a_{21}\lambda^2 + c_{21}, & \ldots, & a_{n1}\lambda^2 + c_{n1} \\ a_{12}\lambda^2 + c_{12}, & a_{22}\lambda^2 + c_{22}, & \ldots, & a_{n2}\lambda^2 + c_{n2} \\ \cdots & \cdots & \cdots & \cdots \\ a_{1n}\lambda^2 + c_{1n}, & a_{2n}\lambda^2 + c_{2n}, & \ldots, & a_{nn}\lambda^2 + c_{nn} \end{vmatrix}. \quad \ldots\ldots(9)$$

This is a symmetric determinant, of degree n in λ^2. It may be shewn that the n values of λ^2 are all real, and that if further the expression (2) for V is essentially positive they are all negative.

Consider the series formed by the determinant (9) and the determinants obtained by erasing the first row and column, the first two rows and columns, and so on, and denote them, with the addition of a positive constant Δ_0 at the end, by

$$\Delta_n, \ \Delta_{n-1}, \ \ldots, \ \Delta_1, \ \Delta_0, \quad \ldots\ldots\ldots(10)$$

respectively. When $\lambda^2 = +\infty$ these are all positive, in virtue of the essentially positive character of T (Art. 73). When $\lambda^2 = -\infty$, Δ_r is positive or negative according as r is even or odd; hence the signs in the series (10) will alternate. Thus as λ^2 diminishes from $+\infty$ to $-\infty$, n variations of sign are introduced in the series.

By a known theorem in Determinants, if $\Delta_r = 0$, where $r < n$, the signs of Δ_{r+1} and Δ_{r-1} (assumed not to vanish with Δ_r) will be opposed. Hence no variation of sign is introduced when in the above process λ^2 passes through a root of Δ_r

Variations can only arise through the vanishing of Δ_n, which must therefore occur for n real (and distinct) values of λ^2. Similarly, there are $n-1$ real roots of Δ_{n-1}, and so on.

Again the signs of the series are initially

$$+ + + + \ldots.$$

As λ^2 passes through the first root of Δ_n they become

$$- + + + \ldots,$$

but before the next root of Δ_n is reached the sign of Δ_{n-1} must have become $-$, for otherwise the signs would again become $+$ as the root of Δ_n is passed. This would leave n variations of sign still to be introduced. Hence the roots of Δ_{n-1} separate those of Δ_n. In the same way the roots of Δ_{n-2} separate those of Δ_{n-1}, and so on.

So far no condition has been imposed on the potential energy. But if V is essentially positive, the signs of the series (10) are all positive if $\lambda^2 = 0$. Hence the n variations are introduced as λ^2 diminishes from 0 to $-\infty$. The roots of (9) in λ^2 are therefore now not only real but all negative*.

The preceding argument assumes that adjacent members of the series (10) do not vanish simultaneously. This peculiarity, if it occurs, must be due to some special relation between the coefficients in (1) and (2), and can be made to disappear by a slight alteration in the constitution of the system, e.g. by a slight change in one of the coefficients of inertia. Regarding a double root of Δ_n as due in the limit, when the change referred to vanishes, to a coincidence of two previously distinct roots, we see that a double root of Δ_n will be a simple root of Δ_{n-1}. In like manner a triple root of Δ_n will be a double root of Δ_{n-1} and a simple root of Δ_{n-2}. And, generally, an r-fold root of Δ_n is an $(r-1)$-fold root of Δ_{n-1}, and so on.

It will be shewn later that a multiple root of Δ_n is characterized by the vanishing of *all* its first minors. We will suppose for the present that the roots are all distinct, and consequently that the solutions (12) below are determinate.

Hence if V be a minimum in the equilibrium configuration the values of λ will occur in pairs of the form

$$\lambda = \pm i\sigma. \qquad \text{.........................(11)}$$

The equations (6) corresponding to any value of λ are equivalent in virtue of (8) to $n-1$ relations determining the $n-1$ ratios (7), the absolute values of the constants being alone arbitrary. Thus

$$\frac{A_1}{a_1} = \frac{A_2}{a_2} = \ldots = \frac{A_n}{a_n}, = H, \text{ say}, \qquad \text{...........(12)}$$

* The proof is due to Routh. It is an extension of one given by Salmon for the 'equation of secular inequalities,' *Higher Algebra*, Art. 44. Another proof is indicated in Art. 94 below.

where $\alpha_1, \alpha_2, \ldots, \alpha_n$ are the minors of any one row of the determinant $\Delta(\lambda^2)$. Since these minors are functions of λ^2, and are therefore the same for values of λ which differ only in sign, the solution corresponding to a pair of roots of the form (11) will be

$$q_r = \alpha_r (He^{i\sigma t} + Ke^{-i\sigma t}), \quad \text{.................(13)}$$

or, in real form,

$$q_r = C\alpha_r \cos(\sigma t + \epsilon), \quad \text{.................(14)}$$

where the arbitrary constants H, K, or C, ϵ, are the same for all the coordinates*.

This solution, taken by itself, represents what is called a 'normal mode,' or 'fundamental mode,' of vibration. Each particle of the system executes a simple-harmonic vibration of period $2\pi/\sigma$. The direction of vibration of each particle is determinate, as also the relative amplitudes of the various particles, which keep step with one another, and so pass simultaneously through their equilibrium positions. The only arbitrary elements are the absolute amplitude, which depends upon C, and the phase-constant ϵ.

When all the values of λ^2 are negative (and distinct) there are n such normal modes, and the most general small motion of the system is obtained by superposition of these, with arbitrary amplitudes and phases; thus

$$q_r = C\alpha_r \cos(\sigma t + \epsilon) + C'\alpha_r' \cos(\sigma' t + \epsilon') + C''\alpha_r'' \cos(\sigma'' t + \epsilon'') + \ldots. \quad \text{.........(15)}$$

The $2n$ arbitrary constants $C, C', C'', \ldots, \epsilon, \epsilon', \epsilon'', \ldots$ enable us to adapt the solution to given initial (small) values of the n coordinates $q_1, q_2, \ldots, q_n$ and the n velocities $\dot{q}_1, \dot{q}_2, \ldots, \dot{q}_n$.

The periodic form of the solution (15) indicates that, since $C, C', C'', \ldots$ are assumed to be small, the configuration never departs far from the equilibrium configuration. A minimum value of V therefore indicates stability, in accordance with Dirichlet's argument (Art. 86).

When V is not an absolute minimum, so that the expression (2) may become negative, positive values of λ^2 will occur, leading to solutions of the type

$$q_r = \alpha_r (He^{\lambda t} + Ke^{-\lambda t}). \quad \text{.................(16)}$$

* The theory is due substantially to Lagrange, *Mécanique Analytique*, 2me partie, 6me sect.

Unless the initial conditions are specially adjusted, the value of q_r increases until the assumptions on which our approximate equations were based cease to be valid.

Unless otherwise stated we shall have in mind chiefly the case of stability.

91. Illustrations.

The following examples will serve to elucidate various points of the preceding theory.

Ex. 1. A particle oscillates about the lowest point of a smooth surface, to which it is confined. This is the case also of Blackburn's pendulum [D. 29].

Using rectangular coordinates, with the axis of z vertically upwards, the equation of the surface will have the form

$$2z = ax^2 + 2hxy + by^2 + \ldots.$$

Hence, with sufficient approximation,

$$2T = m(\dot{x}^2 + \dot{y}^2), \qquad \ldots\ldots(1)$$

$$2V = 2mgz = ax^2 + 2hxy + by^2. \qquad \ldots\ldots(2)$$

The equations of motion are therefore

$$m\ddot{x} + ax + hy = 0, \quad m\ddot{y} + hx + by = 0. \qquad \ldots\ldots(3)$$

Assuming

$$x = Fe^{i\sigma t}, \quad y = Ge^{i\sigma t}, \qquad \ldots\ldots(4)$$

we have

$$\left.\begin{array}{r}(m\sigma^2 - a)F - hG = 0, \\ -hF + (m\sigma^2 - b)G = 0,\end{array}\right\} \qquad \ldots\ldots(5)$$

whence

$$(m\sigma^2 - a)(m\sigma^2 - b) - h^2 = 0. \qquad \ldots\ldots(6)$$

In order that V may be a minimum for $x=0$, $y=0$, we must have

$$a>0, \quad b>0, \quad ab - h^2 > 0.$$

The values of σ^2 which satisfy (6) are then easily seen to be real and positive. For the expression on the left-hand side of (6) is positive for $\sigma^2 = +\infty$, negative for $\sigma^2 = a/m$ or b/m, and again positive for $\sigma^2 = -\infty$. Denoting the roots by σ^2, σ'^2, the complete solution is

$$\left.\begin{array}{l}x = C\cos(\sigma t + \epsilon) + C'\cos(\sigma' t + \epsilon'), \\ y = \dfrac{m\sigma^2 - a}{h} C\cos(\sigma t + \epsilon) + \dfrac{m\sigma'^2 - a}{h} C'\cos(\sigma' t + \epsilon').\end{array}\right\} \qquad \ldots\ldots(7)$$

If we eliminate σ^2 between the equations (5), we get

$$h(F^2 - G^2) = (a - b)FG. \qquad \ldots\ldots(8)$$

The directions of vibration in the two normal modes are therefore given by

$$h(x^2 - y^2) = (a - b)xy. \qquad \ldots\ldots(9)$$

This is the equation of the principal axes of the indicatrix of the surface, as was to be expected.

Ex. 2. Three equal particles are attached at equal intervals a to a tense string of length $4a$, which is fixed at the ends. It is assumed that the lateral deflections are so small that the tension (P) is not sensibly altered.

The statical forces required to maintain the deflections y_1, y_2, y_3 of the three particles are evidently

$$Y_1 = P\left(\frac{y_1}{a} + \frac{y_1 - y_2}{a}\right), \quad Y_2 = P\left(\frac{y_2 - y_1}{a} + \frac{y_2 - y_3}{a}\right), \quad Y_3 = P\left(\frac{y_3}{a} + \frac{y_3 - y_2}{a}\right). \quad (10)$$

The potential energy is therefore, by Art. 87 (4),

$$V = \tfrac{1}{2}(Y_1 y_1 + Y_2 y_2 + Y_3 y_3)$$

$$= \frac{P}{a}(y_1^2 + y_2^2 + y_3^2 - y_1 y_2 - y_2 y_3). \quad \text{.................(11)}$$

The equations of motion are accordingly

$$\left.\begin{aligned} m\ddot{y}_1 + \frac{P}{a}(2y_1 - y_2) &= 0, \\ m\ddot{y}_2 + \frac{P}{a}(2y_2 - y_1 - y_3) &= 0, \\ m\ddot{y}_3 + \frac{P}{a}(2y_3 - y_2) &= 0. \end{aligned}\right\} \quad \text{.......................(12)}$$

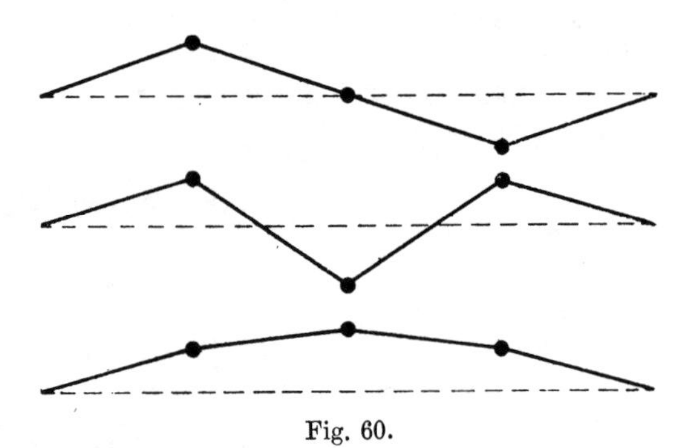

Fig. 60.

Assuming that y_1, y_2, y_3 vary as $e^{i\sigma t}$, we have

$$\left.\begin{aligned} (\sigma^2 - 2\mu)\, y_1 + \mu y_2 &= 0, \\ \mu y_1 + (\sigma^2 - 2\mu)\, y_2 + \mu y_3 &= 0, \\ \mu y_2 + (\sigma^2 - 2\mu)\, y_3 &= 0, \end{aligned}\right\} \quad \text{.......................(13)}$$

where $\mu = P/ma$. Eliminating the ratios $y_1 : y_2 : y_3$, we find

$$(\sigma^2 - 2\mu)(\sigma^4 - 4\mu\sigma^2 + 2\mu^2) = 0. \quad \text{.......................(14)}$$

The root $\sigma^2 = 2\mu$ makes $y_2 = 0$ and $y_1 = -y_3$; it was obvious beforehand that this would be a possible mode of vibration. The remaining roots are

$$\sigma^2 = (2 \pm \sqrt{2})\, \mu.$$

The upper sign makes $y_1=y_3=-y_2/\surd 2$, whilst the lower makes $y_1=y_3=y_2/\surd 2$. Both these modes are symmetrical, and the latter type, in which the three deflections have the same sign, has the longest period, as was to be expected. Figure 60 shews the three modes.

The systematic theory of the vibrations of various types of *continuous* systems must be sought elsewhere. In particular, the transverse vibrations of a tense string, which furnish the simplest instance, belong properly to Acoustics. Continuous systems furnish however such interesting illustrations of various points in the present theory that an occasional example will be useful.

Ex. 3. Let us take the small oscillations of a chain hanging vertically from one end.

The origin being taken at the equilibrium position of the lower end, the tension (P) at a height x above this point will be $g\rho x$, where ρ is the line-density, the vertical motion being neglected as of the second order. If y denote the horizontal deflection, the horizontal components of the tension on the lower and upper ends of an element δx will be

$$-P\frac{\partial y}{\partial x} \quad \text{and} \quad P\frac{\partial y}{\partial x}+\frac{\partial}{\partial x}\left(P\frac{\partial y}{\partial x}\right)\delta x,$$

respectively. The equation of motion of this element is accordingly

$$\rho\delta x\frac{\partial^2 y}{\partial t^2}=\frac{\partial}{\partial x}\left(P\frac{\partial y}{\partial x}\right)\delta x,$$

or

$$\frac{\partial^2 y}{\partial t^2}=g\frac{\partial}{\partial x}\left(x\frac{\partial y}{\partial x}\right). \qquad \text{.........................(15)*}$$

To ascertain the normal modes of vibration we assume as usual that y varies as $e^{i\sigma t}$, whence

$$\frac{\partial}{\partial x}\left(x\frac{\partial y}{\partial x}\right)+\frac{\sigma^2}{g}y=0. \qquad \text{.........................(16)}$$

The solution which is finite for $x=0$ can be obtained by a series, but the result assumes a slightly neater form if we replace x by a new independent variable τ, writing

$$x=\tfrac{1}{4}g\tau^2. \qquad \text{.................................(17)†}$$

* This partial differential equation takes the place of the n linear equations of the type (4) of Art. 90 which present themselves in the case of a system of finite freedom.

† The interpretation of τ is as follows. The wave-velocity on a string with uniform tension equal to that which obtains at the point x would be $\surd(P/\rho)$ or $\surd(gx)$. The time which a geometrical point moving always with this local velocity would take to travel from the lower end to the point x is

$$\int_0^x \frac{dx}{\surd(gx)}=\tau.$$

The equation becomes

$$\frac{\partial^2 y}{\partial \tau^2}+\frac{1}{\tau}\frac{\partial y}{\partial \tau}+\sigma^2 y=0. \quad \text{.........................(18)}$$

The solution which is finite for $\tau=0$ is

$$y=CJ_0(\sigma\tau)\cos(\sigma t+\epsilon), \quad \text{.........(19)}$$

where $J_0(\sigma\tau)$ is the 'Bessel's Function' of zero order, viz.

$$J_0(\sigma\tau)=1-\frac{\sigma^2\tau^2}{2^2}+\frac{\sigma^4\tau^4}{2^2\,.\,4^2}-\ldots, \quad \text{...(20)}$$

and C is arbitrary. The value of τ corresponding to the upper end ($x=l$) is

$$\tau_1=2\sqrt{(l/g)}, \quad \text{............(21)}$$

and the condition that this end is fixed gives

$$J_0(\sigma\tau_1)=0. \quad \text{...............(22)}$$

This equation, which replaces the $\Delta(\lambda^2)=0$ of Art. 90, determines the admissible values of σ; and the types of the corresponding normal vibrations are then given by (19).

The roots of (22) are given by

$$\sigma\tau_1/\pi=\cdot 7655,\ 1\cdot 7571,\ 2\cdot 7546,\ \ldots, \quad \text{...(23)}$$

the numbers tending to the form $s-\frac{1}{4}$, where s is integral. The longest period is accordingly

$$\frac{2\pi}{\sigma}=5\cdot 225\sqrt{\left(\frac{l}{g}\right)}. \quad \text{.........(24)}$$

In the modes after the first, the values of τ corresponding to the lower roots give the nodes, or points of rest ($y=0$). Thus in the second mode there is a node at the point determined by

$$\tau/\tau_1=\cdot 7655/1\cdot 7571, \quad \text{or} \quad x/l=\tau^2/\tau_1{}^2=\cdot 190.$$

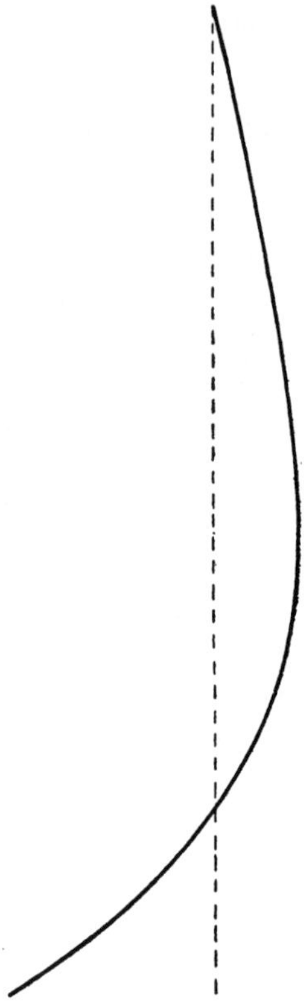

Fig. 61.

The first two modes are represented, on different scales, in the annexed figure. The node represents the point of suspension in the first case.

92. Normal Coordinates. Orthogonal Relations.

If the coordinates $q_1, q_2, \ldots, q_n$ be transformed by a linear substitution with constant coefficients, say

$$q_r=A_r\theta+A_r'\theta'+A_r''\theta''+\ldots, \quad \text{..............(1)}$$

where $\theta, \theta', \theta'', \ldots$ are n new variables, the quadratic character of the expressions for the kinetic and potential energies, and the general nature of the preceding solution, will of course be unchanged.

An important simplification is, however, effected, if we determine the coefficients in (1) so that the formulæ for T and V are both reduced to sums of squares, say

$$2T = a\dot{\theta}^2 + a'\dot{\theta}'^2 + a''\dot{\theta}''^2 + \ldots, \qquad \ldots\ldots\ldots\ldots\ldots(2)$$

$$2V = c\theta^2 + c'\theta'^2 + c''\theta''^2 + \ldots. \qquad \ldots\ldots\ldots\ldots\ldots(3)$$

Algebraically, this is always possible, for in the formulæ (1) of transformation we have n^2 coefficients at our disposal, which can be chosen so that the $\frac{1}{2}n(n-1)$ products (such as $\dot{\theta}\dot{\theta}'$) in the resulting value of T, and the $\frac{1}{2}n(n-1)$ products in the value of V shall vanish. It is moreover still possible to satisfy n additional relations; we might for instance impose the conditions

$$a = a' = a'' = \ldots = 1.$$

Supposing for a moment the above transformation to have been effected, the new variables θ, θ', θ'', ... are called 'principal' or 'normal' coordinates. The coefficients a, a', a'', ... are called the 'principal coefficients of inertia'; they are positive on account of the essentially positive character of T. The coefficients c, c', c'', ... are the 'principal coefficients of stability'; they are positive if V is a minimum in the equilibrium configuration about which the motion is supposed to take place.

In terms of the normal coordinates the equations of motion (Art. 90 (3)) become

$$a\ddot{\theta} + c\theta = 0, \quad a'\ddot{\theta}' + c'\theta' = 0, \quad a''\ddot{\theta}'' + c''\theta'' = 0, \ldots. \quad \ldots(4)$$

These are independent, and the solutions in the case of complete stability, viz.

$$\theta = C\cos(\sigma t + \epsilon), \quad \theta' = C'\cos(\sigma' t + \epsilon'), \quad \theta'' = C''\cos(\sigma'' t + \epsilon''), \ldots, \qquad \ldots\ldots\ldots(5)$$

with $$\sigma^2 = c/a, \quad \sigma'^2 = c'/a', \quad \sigma''^2 = c''/a'', \quad \ldots, \qquad \ldots\ldots\ldots(6)$$

represent the n normal modes of vibration of which the system is capable. In each normal mode one principal coordinate varies alone.

When the roots of $\Delta(\lambda^2)$ are all distinct the requisite transformation is unique, and can be effected by means of certain 'orthogonal' or 'conjugate' relations, which will be useful also later.

Let λ^2 be any root of $\Delta(\lambda^2)$, and let $A_1, A_2, \ldots, A_n$ be chosen to satisfy the n relations (6) of Art. 90, viz.

$$(a_{1r}\lambda^2 + c_{1r}) A_1 + (a_{2r}\lambda^2 + c_{2r}) A_2 + \ldots + (a_{nr}\lambda^2 + c_{nr}) A_n = 0. \quad \ldots\ldots\ldots(7)^*$$

Similarly, if λ'^2 be a *distinct* root, let $A_1', A_2', \ldots, A_n'$ be determined in the same way as functions of λ'^2. Now multiply the n equations (7) by $A_1', A_2', \ldots, A_n'$, in order, and add. The result may be written

$$\lambda^2 T(A, A') + V(A, A') = 0, \quad \ldots\ldots\ldots\ldots\ldots\ldots(8)$$

where

$$2T(A, A') = a_{11} A_1 A_1' + a_{22} A_2 A_2' + \ldots + a_{12}(A_1 A_2' + A_1' A_2) + \ldots, \quad \ldots\ldots\ldots(9)$$

$$2V(A, A') = c_{11} A_1 A_1' + c_{22} A_2 A_2' + \ldots + c_{12}(A_1 A_2' + A_1' A_2) + \ldots. \quad \ldots\ldots\ldots(10)$$

On account of the symmetry of these formulæ we must have also

$$\lambda'^2 T(A, A') + V(A, A') = 0. \quad \ldots\ldots\ldots\ldots\ldots(11)$$

Hence, since by hypothesis $\lambda^2 \neq \lambda'^2$,

$$T(A, A') = 0, \quad V(A, A') = 0, \quad \ldots\ldots\ldots\ldots(12)$$

which are the orthogonal relations referred to.

Again, if we multiply the equations (7) by $A_1, A_2, \ldots, A_n$, respectively, and add, we have

$$\lambda^2 T(A) + V(A) = 0, \quad \ldots\ldots\ldots\ldots\ldots\ldots(13)$$

where

$$2T(A) = 2T(A, A) = a_{11} A_1^2 + a_{22} A_2^2 + \ldots + 2a_{12} A_1 A_2 + \ldots, \quad \ldots(14)$$

$$2V(A) = 2V(A, A) = c_{11} A_1^2 + c_{22} A_2^2 + \ldots + 2c_{12} A_1 A_2 + \ldots \quad \ldots(15)$$

If we now substitute from (1) in the expressions for the kinetic and potential energies, we find that these reduce to sums of squares, since the coefficients of *products* vanish by (12). We obtain in fact the formulæ (2) and (3), with

$$a = T(A), \quad a' = T(A'), \quad a'' = T(A''), \ldots, \quad \ldots\ldots(16)$$

$$c = V(A), \quad c' = V(A'), \quad c'' = V(A''), \ldots. \quad \ldots\ldots(17)$$

Finally, we note that since $a, a', a'', \ldots$ are positive we may write

$$\phi = \sqrt{a}\,\theta, \quad \phi' = \sqrt{a'}\,\theta', \quad \phi'' = \sqrt{a''}\,\theta'', \ldots, \quad \ldots\ldots(18)$$

* This is possible even if the solution of the system is not unique, as in the case of a multiple root.

and so reduce (2) and (3) to the forms

$$2T = \dot{\phi}^2 + \dot{\phi}'^2 + \dot{\phi}''^2 + \dots, \qquad \dots\dots\dots\dots\dots(19)$$

$$2V = \frac{c}{a}\phi^2 + \frac{c'}{a'}\phi'^2 + \frac{c''}{a''}\phi''^2 + \dots. \qquad \dots\dots\dots\dots(20)$$

Ex. Take the problem of three particles attached to a tense string, as in Ex. 2 of Art. 91.

If we write k for ma/P the determinantal equation is

$$\begin{vmatrix} k\lambda^2+2, & -1, & 0 \\ -1, & k\lambda^2+2, & -1 \\ 0, & -1, & k\lambda^2+2 \end{vmatrix} = 0, \qquad \dots\dots\dots\dots\dots\dots\dots(21)$$

the roots of which make

$$k\lambda^2 + 2 = 0, \quad \sqrt{2}, \quad -\sqrt{2},$$

respectively. The minors of the first row of the determinant are

$$(k\lambda^2+2)^2 - 1, \quad k\lambda^2 + 2, \quad 1,$$

respectively. For the values corresponding to the above three roots we have

$$\begin{array}{lll} \alpha_1 = -1, & \alpha_2 = 0, & \alpha_3 = 1, \\ \alpha_1' = 1, & \alpha_2' = \sqrt{2}, & \alpha_3' = 1, \\ \alpha_1'' = 1, & \alpha_2'' = -\sqrt{2}, & \alpha_3'' = 1. \end{array}$$

Hence in accordance with (1) we assume

$$y_1 = -\theta + \theta' + \theta'', \quad y_2 = \sqrt{2}\,(\theta' - \theta''), \quad y_3 = \theta + \theta' + \theta''. \quad \dots\dots\dots(22)$$

This makes

$$T = m\,(\dot{\theta}^2 + 2\dot{\theta}'^2 + 2\dot{\theta}''^2), \qquad \dots\dots\dots\dots\dots\dots\dots\dots\dots(23)$$

$$V = \frac{2P}{a}\{\theta^2 + (2-\sqrt{2})\,\theta'^2 + (2+\sqrt{2})\,\theta''^2\}. \qquad \dots\dots\dots\dots\dots(24)$$

The orthogonal relations (12) can be made to furnish an additional proof that the roots of $\Delta(\lambda^2)$ are all real. For imaginary roots, if any, would occur in pairs, thus

$$\lambda^2 = p + iq, \quad \lambda'^2 = p - iq, \qquad \dots\dots\dots\dots\dots(25)$$

and the corresponding values of A, A' would be conjugate imaginaries, say

$$A_r = M_r + iN_r, \quad A_r' = M_r' - iN_r', \quad \dots\dots\dots\dots(26)$$

which make

$$A_r A_r' = M_r^2 + N_r^2, \quad A_r A_s' + A_r' A_s = 2M_r M_s + 2N_r N_s. \dots(27)$$

Hence, from (9), the relation

$$T(A, A') = 0 \qquad \dots\dots\dots\dots\dots\dots\dots\dots(28)$$

becomes

$$T(M) + T(N) = 0, \qquad \dots\dots\dots\dots\dots\dots\dots(29)$$

which is impossible, since both terms are essentially positive.

93. Theory of Multiple Roots.

In the process of solution of the equations of vibratory motion, explained in Art. 90, it was assumed that the first minors of the determinant $\Delta(\lambda^2)$ do not all vanish simultaneously with $\Delta(\lambda^2)$. This excluded case can only occur when λ^2 is a multiple root. For, differentiating column by column, we have

$$\frac{d\Delta(\lambda^2)}{d(\lambda^2)} = \begin{vmatrix} a_{11}, & a_{21}\lambda^2 + c_{21}, \ldots\ldots, & a_{n1}\lambda^2 + c_{n1} \\ a_{12}, & a_{22}\lambda^2 + c_{22}, \ldots\ldots, & a_{n2}\lambda^2 + c_{n2} \\ \ldots & \ldots & \ldots \\ a_{1n}, & a_{2n}\lambda^2 + c_{2n}, \ldots\ldots, & a_{nn}\lambda^2 + c_{nn} \end{vmatrix} + \text{etc.}, \quad \ldots(1)$$

the number of determinants being n. If α_{rs} be the minor of the constituent in the rth column and sth row of $\Delta(\lambda^2)$, the specimen determinant here written is equal to

$$a_{11}\alpha_{11} + a_{12}\alpha_{12} + \ldots + a_{1n}\alpha_{1n}. \quad \ldots\ldots\ldots\ldots\ldots(2)$$

Hence if the first minors all vanish

$$\frac{d\Delta(\lambda^2)}{d(\lambda^2)} = 0, \quad \ldots\ldots\ldots\ldots\ldots\ldots\ldots\ldots(3)$$

which is the condition for a repeated root.

The converse also holds. For if we add up the n expressions of type (2), and make use of the relations $a_{rs} = a_{sr}$, $\alpha_{rs} = \alpha_{sr}$, the result can be put in the form

$$\frac{d\Delta(\lambda^2)}{d(\lambda^2)} = a_{11}\alpha_{11} + a_{22}\alpha_{22} + \ldots + 2a_{12}\alpha_{12} + \ldots. \quad \ldots\ldots(4)$$

But if $\Delta(\lambda^2) = 0$ we have

$$\alpha_{rr}\alpha_{ss} = \alpha_{rs}^2, \quad \alpha_{pr}\alpha_{qs} = \alpha_{pq}\alpha_{rs}. \quad \ldots\ldots\ldots\ldots\ldots(5)$$

Hence if the first minors do not all vanish there must be at least one minor of the type α_{rr}, say α_{11}, which is not zero; otherwise all the remaining minors (of type α_{rs}) would also vanish. The equation (4) may therefore be written

$$\frac{d\Delta(\lambda^2)}{d(\lambda^2)} = \frac{1}{\alpha_{11}}(a_{11}\alpha_{11}^2 + a_{12}\alpha_{12}^2 + \ldots + 2a_{11}\alpha_{11}\alpha_{12} + \ldots). \quad \ldots(6)$$

The expression in brackets is a quadratic function of the n quantities $\alpha_{11}, \alpha_{12}, \ldots, \alpha_{1n}$, which cannot vanish, by Art. 73.

As already remarked, the occurrence of a multiple root is, from a physical point of view, an accidental property of a dynamical system, which can be made to disappear by the slightest alteration in its constitution. A simple case is that of a particle oscillating under gravity in a smooth concave bowl (Art. 91, Ex. 1). So long as there is the slightest inequality in the principal curvatures at the lowest point, the normal modes are definite and the periods distinct. But if the curvatures are equal, the particle can oscillate in *any* vertical plane through the lowest point, and two such oscillations can be superposed into an elliptic harmonic vibration. Instead of a solution of the type

$$\left.\begin{aligned} x &= A_1 \cos(\sigma_1 t + \epsilon_1) + A_2 \cos(\sigma_2 t + \epsilon_2), \\ y &= k_1 A_1 \cos(\sigma_1 t + \epsilon_1) + k_2 A_2 \cos(\sigma_2 t + \epsilon_2), \end{aligned}\right\} \quad \ldots\ldots(7)$$

where k_1, k_2 are definite constants depending on the curvatures, we have a solution

$$x = F\cos(\sigma t + \alpha), \quad y = G\cos(\sigma t + \beta), \quad \ldots\ldots\ldots(8)$$

involving, but in a different way, the same number of arbitrary constants.

Since the interest of the matter is algebraical rather than dynamical, it must suffice to indicate briefly how the investigation of Art. 90 is to be modified in the case of a multiple root of any order, and in particular how the generality of the solution, with the full number of arbitrary constants, is secured.

Let $\Delta(\lambda^2)$ and the subsidiary determinants derived from it as in Art. 90 be denoted as before by

$$\Delta_n, \quad \Delta_{n-1}, \quad \Delta_{n-2}, \quad \ldots.$$

We have seen that an r-fold root of Δ_n will be an $(r-1)$-fold root of Δ_{n-1}, and so on. Now consider the partial solution

$$q_1 = A_1 e^{\lambda t}, \quad q_2 = A_2 e^{\lambda t}, \quad \ldots, \quad q_n = A_n e^{\lambda t}. \quad \ldots\ldots\ldots(9)$$

If the value of λ^2 makes $\Delta_n = 0$, but not $\Delta_{n-1} = 0$, we can assign an arbitrary value to *one* of the coefficients, and the rest are then uniquely determined. This is the case discussed in Art. 90.

If $\Delta_n = 0$, $\Delta_{n-1} = 0$, but not $\Delta_{n-2} = 0$, we can assign arbitrary values to *two* of the coefficients, and the rest are then determinate. And, generally, in the case of an r-fold root, r of the coefficients, say $A_1, A_2, \ldots, A_r$, may have arbitrary values, the rest being definite linear functions of these.

A similar statement applies to the coefficients in the solution

$$q_1 = B_1 e^{-\lambda t}, \quad q_2 = B_2 e^{-\lambda t}, \quad \ldots, \quad q_n = B_n e^{-\lambda t}. \quad \ldots\ldots(10)$$

Hence, corresponding to an r-fold value of λ^2, we obtain partial solutions involving $2r$ arbitrary constants. And, altogether, allowing for both simple and multiple roots of $\Delta(\lambda^2)$, we have $2n$ arbitrary constants in the complete solution. This is the proper number requisite to adapt the solution to arbitrary initial values of the n coordinates q_n, and n velocities $\dot{q}_r$.

To illustrate the preceding algebraical statements, let $n=5$, and suppose we have a triple root, making $\Delta_5=0$, $\Delta_4=0$, $\Delta_3=0$, but $\Delta_2 \neq 0$. For brevity write

$$e_{rs}=a_{rs}\lambda^2+c_{rs}=e_{sr}.$$

The equations to be solved are then

$$e_{11}A_1+e_{21}A_2+e_{31}A_3+e_{41}A_4+e_{51}A_5=0, \quad \text{.................(i)}$$

$$e_{12}A_1+e_{22}A_2+e_{32}A_3+e_{42}A_4+e_{52}A_5=0, \quad \text{.................(ii)}$$

$$e_{13}A_1+e_{23}A_2+e_{33}A_3+e_{43}A_4+e_{53}A_5=0, \quad \text{.................(iii)}$$

$$e_{14}A_1+e_{24}A_2+e_{34}A_3+e_{44}A_4+e_{54}A_5=0, \quad \text{.................(iv)}$$

$$e_{15}A_1+e_{25}A_2+e_{35}A_3+e_{45}A_4+e_{55}A_5=0. \quad \text{.................(v)}$$

We first shew that there is a solution with $A_1=0$, $A_2=0$, and A_3 arbitrary. Since $\Delta_2 \neq 0$, the equations (iv) and (v) determine A_4 and A_5 in terms of A_3. Since $\Delta_3=0$, the equations (iii), (iv), and (v) are consistent, so that our values of the coefficients will also satisfy (iii). The condition $\Delta_4=0$ ensures in like manner that (ii) is also satisfied, and finally the condition $\Delta_5=0$ shews that (i) will also hold.

We have thus found a solution in which $A_1=0$, $A_2=0$, whilst A_4 and A_5 are uniquely determined by A_3.

If the rows and columns in the above scheme of equations be rearranged, but so as to preserve symmetry about the leading diagonal, the conditions for a triple root, viz. $\Delta_5=0$, $\Delta_4=0$, $\Delta_3=0$, $\Delta_2 \neq 0$, will of course hold in the new arrangement. In this way we can obtain a solution in which $A_1=0$, $A_3=0$, whilst A_2 is arbitrary. And in like manner a solution in which $A_2=0$, $A_3=0$, with A_1 arbitrary. Finally, by superposition, we can build up a solution in which A_1, A_2, A_3 are all arbitrary.

94. Alternative Investigation of Normal Modes.

The theory of normal modes and natural frequencies can be approached in another way. For simplicity of statement we will postulate complete stability, so that V is essentially positive. The more general case would only require a few minor alterations.

The method consists in ascertaining values of $A_1, A_2, \ldots, A_n$ for which the function

$$\frac{V(A)}{T(A)} \quad \text{.............................. (1)}$$

has, under certain conditions, a stationary value, and in shewing that these values are in fact the squares (σ^2) of the natural frequencies, whether these are all distinct or not.

Owing to the essentially positive character of T and V, the function (1) cannot vanish, and cannot become infinite. It has therefore at least two stationary values, at its upper and lower limits.

Now if we apply the ordinary method of the Calculus to the search for a stationary (say a least) value of (1), we are led precisely to the equations (6) and (9) of Art. 90, with σ^2 for the stationary value in question. We have thus at any rate one real solution of the equations, with a positive value of σ^2. Let

$$A_1 : A_2 : \ldots : A_n$$

be the ratios of the coefficients, as thus determined.

We next seek values of $A_1', A_2', \ldots, A_n'$ which shall render the function

$$\frac{V(A')}{T(A')} \qquad \text{.............................(2)}$$

stationary, subject to the condition

$$T(A, A') = 0. \qquad \text{.........................(3)}$$

It is obvious, *à fortiori*, that such a stationary value exists; and if we form the necessary conditions, it will be found that they can be reduced, again, to the type (6) of Art. 90. We have thus a second solution, and the condition (3) shews that it is distinct from the former one, even if the resulting value of σ^2 happens to be the same.

The next step is to ascertain a stationary value of

$$\frac{V(A'')}{T(A'')} \qquad \text{.............................(4)}$$

subject to the conditions

$$T(A, A'') = 0, \quad T(A', A'') = 0, \qquad \text{..............(5)}$$

and so on, until finally the conditions of the type (5) leave only one determination possible.

The method may be illustrated by the special case of $n = 3$. There is here an analogy with a familiar problem of Solid Geometry, and it will be convenient to adopt the notation of that subject. We denote the coordinates by x, y, z, and we will suppose, further, that by a linear transformation the expression for the kinetic energy has been reduced to a sum of squares with unit coefficients. This is always possible in an infinite number of ways. We write, then,

$$2T = \dot{x}^2 + \dot{y}^2 + \dot{z}^2, \quad \text{.......................(6)}$$

and for the potential energy

$$2V = ax^2 + by^2 + cz^2 + 2fyz + 2gzx + 2hxy. \quad \text{......(7)}$$

Forming the equations of motion, and assuming that x, y, z vary as $e^{i\sigma t}$, we have

$$\left.\begin{array}{l} (a - \sigma^2)\,x + hy + gz = 0, \\ hx + (b - \sigma^2)\,y + fz = 0, \\ gx + fy + (c - \sigma^2)\,z = 0, \end{array}\right\} \quad \text{..................(8)}$$

which is a particular case of Art. 90 (6), and thence

$$\begin{vmatrix} a - \sigma^2, & h, & g \\ h, & b - \sigma^2, & f \\ g, & f, & c - \sigma^2 \end{vmatrix} = 0. \quad \text{.........(9)}$$

For a reason given the function

$$u = \frac{ax^2 + by^2 + cz^2 + 2fyz + 2gzx + 2hxy}{x^2 + y^2 + z^2} \quad \text{......(10)}$$

has necessarily a least value. It is a function of the two independent ratios $x:y:z$, and nothing is affected if we take one of these quantities, say z, as fixed, and the others as the independent variables. But if we make $\partial u/\partial x = 0$, $\partial u/\partial y = 0$, it will also follow that $\partial u/\partial z = 0$, since u is of zero dimensions and therefore

$$x\frac{\partial u}{\partial x} + y\frac{\partial u}{\partial y} + z\frac{\partial u}{\partial z} = 0. \quad \text{..................(11)}$$

Now from (10) we have

$$ux + \tfrac{1}{2}(x^2 + y^2 + z^2)\frac{\partial u}{\partial x} = ax + hy + gz, \quad \text{........(12)}$$

with two similar relations. The conditions for a stationary value therefore take the form (8), with $u = \sigma_1^2$, say, a root of (9). Hence there is at any rate one real solution of (8), of the form

$$\frac{x}{x_1} = \frac{y}{y_1} = \frac{z}{z_1}, \quad \text{.......................(13)}$$

where x_1, y_1, z_1 are definite functions of σ_1^2.

The next step is to find a stationary (e.g. a least) value of u, subject to the condition

$$xx_1 + yy_1 + zz_1 = 0. \quad\text{.....................(14)}$$

As in the ordinary method of the Calculus we assume

$$\frac{\partial u}{\partial x} = kx_1, \quad \frac{\partial u}{\partial y} = ky_1, \quad\text{.....................(15)}$$

where k is an undetermined multiplier; and it follows from (11) and (14) that we have then

$$\frac{\partial u}{\partial z} = kz_1, \quad\text{..........................(16)}$$

also. Hence, from (12) and the two parallel equations,

$$\left.\begin{aligned}(a-u)\,x + hy + gz &= \lambda x_1,\\ hx + (b-u)\,y + fz &= \lambda y_1,\\ gx + fy + (c-u)\,z &= \lambda z_1,\end{aligned}\right\} \quad\text{..............(17)}$$

where $$\lambda = \tfrac{1}{2}k\,(x^2 + y^2 + z^2). \quad\text{.................(18)}$$

But if we multiply (17) by x_1, y_1, z_1 in order, and add, we have

$$x\,(ax_1 + hy_1 + gz_1) + y\,(hx_1 + by_1 + fz_1) + z\,(gx_1 + fy_1 + cz_1)$$
$$= \lambda\,(x_1^2 + y_1^2 + z_1^2), \quad\text{......(19)}$$

by (14). Since x_1, y_1, z_1 is a solution of (8), with $\sigma^2 = \sigma_1^2$, the left-hand member reduces to

$$\sigma_1^2\,(xx_1 + yy_1 + zz_1) \quad\text{or}\quad 0.$$

Hence $\lambda = 0$, and the equations (17) reduce to form (8). We thus obtain a second solution of the type

$$\frac{x}{x_2} = \frac{y}{y_2} = \frac{z}{z_2}, \quad\text{........................(20)}$$

and the condition (14) shews that this is distinct from the former, even if the corresponding value (σ_2^2, say) of σ^2 should be the same.

If there were more than three variables, we should have to continue in the same manner. In the present case we have only to determine the ratios $x : y : z$ by means of (14) and the new condition

$$xx_2 + yy_2 + zz_2 = 0, \quad\text{....................(21)}$$

and to shew that they satisfy (8).

Now when x, y, z have these ratios, the determinant

$$\begin{vmatrix} x, & x_1, & x_2 \\ y, & y_1, & y_2 \\ z, & z_1, & z_2 \end{vmatrix} \quad\text{........................(22)}$$

does not vanish, and it is therefore possible to determine three quantities σ^2, λ, μ such that

$$\left.\begin{aligned} \sigma^2 x + \lambda x_1 + \mu x_2 &= ax + hy + gz, \\ \sigma^2 y + \lambda y_1 + \mu y_2 &= hx + by + fz, \\ \sigma^2 z + \lambda z_1 + \mu z_2 &= gx + fy + cz. \end{aligned}\right\} \quad \text{............(23)}$$

If we multiply these by x_1, y_1, z_1 in order, and add, we find, by the same method as in the case of (17), that $\lambda = 0$. Similarly, multiplying by x_2, y_2, z_2 and adding, we find $\mu = 0$. The equations (8), with the present forms of x, y, z, are therefore satisfied.

From a geometrical point of view the above is merely a rather circumstantial method of finding the principal axes of an ellipsoid

$$ax^2 + by^2 + cz^2 + 2fyz + 2gzx + 2hxy = \text{const.},$$

and if we were interested only in the case of $n = 3$ we might have appealed at once to our geometrical knowledge. But it seemed worth while to indicate a systematic procedure available for higher values of n, where geometrical intuitions fail us.

We have been led to three distinct solutions of the equations (8). Including the terms in $e^{-i\sigma t}$ these may be written

$$\left.\begin{aligned} \frac{x}{x_1} = \frac{y}{y_1} = \frac{z}{z_1} = \theta_1, \\ \frac{x}{x_2} = \frac{y}{y_2} = \frac{z}{z_2} = \theta_2, \\ \frac{x}{x_3} = \frac{y}{y_3} = \frac{z}{z_3} = \theta_3, \end{aligned}\right\} \quad \text{......................(24)}$$

where

$$\theta_1 = A_1 \cos \sigma_1 t + B_1 \sin \sigma_1 t, \quad \theta_2 = A_2 \cos \sigma_2 t + B_2 \sin \sigma_2 t,$$
$$\theta_3 = A_3 \cos \sigma_3 t + B_3 \sin \sigma_3 t. \quad \text{......(25)}$$

The quantities θ_1, θ_2, θ_3 are normal coordinates. In terms of them we find

$$2T = (x_1^2 + y_1^2 + z_1^2)\,\dot{\theta}_1^2 + (x_2^2 + y_2^2 + z_2^2)\,\dot{\theta}_2^2 + (x_3^2 + y_3^2 + z_3^2)\,\dot{\theta}_3^2, \quad \text{.........(26)}$$

$$2V = \sigma_1^2(x_1^2 + y_1^2 + z_1^2)\,\theta_1^2 + \sigma_2^2(x_2^2 + y_2^2 + z_2^2)\,\theta_2^2 + \sigma_3^2(x_3^2 + y_3^2 + z_3^2)\,\theta_3^2. \quad \text{.........(27)}$$

95. Stationary Property of the Normal Modes.

Confining ourselves to the case of complete stability, as the most interesting, the preceding investigation shews that the values of σ^2 are stationary for slight variations in the values of the ratios $A_1 : A_2 : \ldots : A_n$. Also that if we imagine the system to be reduced by frictionless constraints to one degree of freedom, so that the above ratios have prescribed values, the frequency will be intermediate between the greatest and least natural frequencies of the system.

The stationary property was noticed by Lagrange. The additional remark just made is due to Rayleigh, who gives the following proof in terms of the normal coordinates of Art. 92. If in the notation of that Art. we put

$$\theta = \mu\phi, \quad \theta' = \mu'\phi, \quad \theta'' = \mu''\phi, \ \ldots, \qquad \text{(1)}$$

the expressions for the kinetic and potential energies become

$$2T = (a\mu^2 + a'\mu'^2 + a''\mu''^2 + \ldots)\,\dot{\phi}^2, \qquad \text{(2)}$$

$$2V = (c\mu^2 + c'\mu'^2 + c''\mu''^2 + \ldots)\,\phi^2, \qquad \text{(3)}$$

and the frequency of the constrained mode is accordingly given by

$$\sigma^2 = \frac{c\mu^2 + c'\mu'^2 + c''\mu''^2 + \ldots}{a\mu^2 + a'\mu'^2 + a''\mu''^2 + \ldots}. \qquad \text{(4)}$$

This is intermediate in value between the greatest and least of the quantities c/a, c'/a', c''/a'', ... proper to the several normal modes. Moreover, if the constrained mode differs little from a normal mode, e.g. if μ', μ'', ... are small compared with μ, the change in the frequency is of the second order.

This stationary property gives a valuable means of estimating the frequency of a natural mode of a system, by means of an assumed approximate type, when the exact determination would be difficult or even impracticable. Many interesting examples of this procedure are given in Rayleigh's *Theory of Sound.*

The effect of a *partial* constraint may be inferred from a remark made in Art. 90. We may suppose the coordinates transformed so that the constraint in question is defined by the vanishing of one of them, say q_1. The frequencies of the altered modes are then determined by the roots of Δ_{n-1}, which were shewn to separate those of Δ_n. In particular the gravest frequency is raised. The effect of an additional constraint may be illustrated by putting $q_2 = 0$. The frequencies are then determined by the roots of Δ_{n-2}. It appears that every additional constraint raises the lowest frequency of the system. This is in accordance with a remark made in Art. 89, that the effect of constraints is to increase the 'stiffness' of the system.

Ex. 1. Take the case of three equal particles attached at equal intervals to a tense string (Art. 91, Ex. 2), and consider an assumed type of symmetrical vibration in which $y_1 = y_3 = \beta y_2$. Then

$$\left.\begin{aligned} T &= \tfrac{1}{2} m (2\beta^2 + 1) \dot{y}_2^2, \\ V &= \tfrac{1}{2} \frac{P}{a} (4\beta^2 - 4\beta + 2) y_2^2. \end{aligned}\right\} \qquad \text{.........(5)}$$

The method explained makes

$$\sigma^2 = \mu \,.\, \frac{4\beta^2 - 4\beta + 2}{2\beta^2 + 1}, \qquad \text{.........(6)}$$

where $\mu = P/ma$. The fraction is stationary for $\beta = \pm 1/\surd 2$, and the corresponding values of σ^2 are as in Art. 91. In this case it was evident beforehand that the assumed type would include the true natural modes of symmetrical vibration.

Ex. 2. As an example, where there is infinite freedom, take the problem of a chain hanging vertically from one end.

If y be the horizontal deflection at a distance s (measured along the chain) from the upper end, we have

$$2T = \rho \int_0^l \dot{y}^2 \, ds, \qquad \text{.........(7)}$$

where l is the total length. The height of any point above its equilibrium position is increased by

$$s - \int_0^s \cos \psi \, ds = 2 \int_0^s \sin^2 \tfrac{1}{2} \psi \, ds,$$

where ψ denotes inclination to the vertical. Since $\sin \psi = \partial y / \partial s$, $= y'$, say, we have, to the second order,

$$\begin{aligned} 2V &= g\rho \int_0^l ds \int_0^s y'^2 \, ds \\ &= g\rho \left[s \int_0^s y'^2 \, ds \right]_0^l - g\rho \int_0^l s y'^2 \, ds \\ &= g\rho \int_0^l (l - s) \, y'^2 \, ds. \qquad \text{.........(8)} \end{aligned}$$

Let us choose as an assumed type

$$y = \eta \frac{s}{l} \left(1 + \beta \frac{s}{l} \right), \qquad \text{.........(9)}$$

where η involves t only. This makes

$$\left.\begin{aligned} 2T &= \rho l \left(\tfrac{1}{3} + \tfrac{1}{2}\beta + \tfrac{1}{5}\beta^2 \right) \dot{\eta}^2, \\ 2V &= g\rho \left(\tfrac{1}{2} + \tfrac{2}{3}\beta + \tfrac{1}{3}\beta^2 \right) \eta^2, \end{aligned}\right\} \qquad \text{.........(10)}$$

and therefore

$$\sigma^2 = \frac{g}{l} \cdot \frac{15 + 20\beta + 10\beta^2}{10 + 15\beta + 6\beta^2}. \qquad \text{.........(11)}$$

We know that, whatever the value of β, this will give us an over-estimate of the frequency of the slowest mode. If we put $\beta=0$ we get the period of a rigid bar, viz.

$$\frac{2\pi}{\sigma}=5{\cdot}130\sqrt{\frac{l}{g}}. \qquad \text{.........(12)}$$

The minimum value of the fraction in (11) is 1·4460, whence

$$\frac{2\pi}{\sigma}=5{\cdot}225\sqrt{\frac{l}{g}}. \qquad \text{.........(13)}$$

This agrees with the correct value (Art. 91 (24)), so far as the digits go.

The comparison of (12) with (13) illustrates the principle that the introduction of any constraint into a system *raises* the frequency of the gravest mode.

96. Forced Oscillations.

We proceed to consider the *forced* vibrations of a system under prescribed *external* forces Q_r, the typical equation being

$$a_{1r}\ddot{q}_1 + a_{2r}\ddot{q}_2 + \ldots + a_{nr}\ddot{q}_n + c_{1r}q_1 + c_{2r}q_2 + \ldots + c_{nr}q_n = Q_r. \qquad \text{.........(1)}$$

The most important case is where the forces Q_r are of the simple-harmonic type $\cos(\sigma t+\epsilon)$. The most general law of variation with the time can be based on this by superposition. Analytically, it is simplest to assume that Q_r varies as $e^{i\sigma t}$, with a complex coefficient. Owing to the linearity of the equations the 'time-factor' $e^{i\sigma t}$ will run through all the terms, and need not always be exhibited.

The equation (1) now takes the form

$$(c_{1r}-\sigma^2 a_{1r})\,q_1 + (c_{2r}-\sigma^2 a_{2r})\,q_2 + \ldots + (c_{nr}-\sigma^2 a_{nr})\,q_n = Q_r. \qquad \text{.........(2)}$$

Solving we find

$$\Delta(\sigma^2)\,.\,q_r = \alpha_{1r}Q_1 + \alpha_{2r}Q_2 + \ldots + \alpha_{nr}Q_n, \qquad \text{.........(3)}$$

where $\Delta(\sigma^2)$ is the determinant of the coefficients in (2), and α_{1r}, α_{2r}, ..., α_{nr} are the minors of its rth row. Every particle of the system executes in general a simple vibration of the imposed period $2\pi/\sigma$; and all the particles pass simultaneously through their equilibrium positions.

The amplitude becomes very great when σ^2 approximates to a root of

$$\Delta(\sigma^2)=0, \qquad \text{.........(4)}$$

i.e. when the imposed period nearly coincides with one of the natural periods. This is the principle of 'resonance,' which is of great importance in Acoustics, and in the Dynamics of Machinery.

Since the determinant $\Delta(\sigma^2)$ is symmetrical, $\alpha_{rs} = \alpha_{sr}$. The coefficient of Q_r in the expression for q_s is therefore identical with the coefficient of Q_s in the value of q_r. This is the basis of an important 'reciprocal theorem' formulated by Helmholtz, and afterwards extended by Rayleigh. As in the case of some of our previous theorems, the most important applications are to systems of infinite freedom, as in Acoustics.

The solution (3) is of course only a 'particular integral' of the equations (1). For a complete solution we must superpose the expressions for the free vibrations (Art. 90) with their $2n$ arbitrary constants. This enables us to adapt the solution to arbitrary initial conditions of displacement and velocity.

If the system be referred to its normal coordinates (Art. 92), the equations of type (2) are replaced by

$$(c - \sigma^2 a)\,\theta = Q, \quad (c' - \sigma^2 a')\,\theta' = Q', \quad (c'' - \sigma^2 a'')\,\theta'' = Q'', \; \ldots, \qquad \text{.........(5)}$$

the notation being as in the Art. referred to. If $\sigma_0, \sigma_0', \sigma_0'', \ldots$ be the values of σ in the various normal modes, viz.

$$\sigma_0^2 = c/a, \quad \sigma_0'^2 = c'/a', \quad \sigma_0''^2 = c''/a'', \; \ldots, \qquad \text{.........(6)}$$

we have

$$\theta = \frac{Q}{c\,(1 - \sigma^2/\sigma_0^2)}, \quad \theta' = \frac{Q'}{c'\,(1 - \sigma^2/\sigma_0'^2)}, \quad \theta'' = \frac{Q''}{c''\,(1 - \sigma^2/\sigma_0''^2)}, \ldots. \qquad \text{.........(7)}$$

Now Q/c, Q'/c', Q''/c'', ... are the statical displacements which would be produced by *constant* forces of the respective normal types, equal to the instantaneous values of these forces in the actual case. They may be called the 'equilibrium values' of the displacements. The formulæ shew that if the period $(2\pi/\sigma)$ of the imposed vibration is long compared with those of the normal modes of free vibration, the system is practically, at any instant, in the 'equilibrium' configuration corresponding to the disturbing forces. This is the assumption virtually made in Bernoulli's 'equilibrium theory' of the tides; but the fundamental condition as to the relation of the periods is not fulfilled in the case of the tides of short period (semi-diurnal and diurnal).

If the frequency of the imposed vibration is very nearly equal to one of the natural frequencies, the amplitude of the corresponding normal coordinate tends to become great in comparison with the rest, and the forced vibration of the system therefore approximates to the type of a normal mode.

97. Effect of Dissipative Forces.

In practice the vibrations of a dynamical system are more or less affected by dissipative forces of various kinds. In order to obtain at all events a qualitative representation of these it is usual to introduce into the equations frictional forces proportional to the generalized velocities. This procedure will be familiar to the reader in the case of the damped oscillations of a system of one degree of freedom [D. 94].

In many cases where the origin of these forces is known, the representative terms in the generalized equations are found to be of a special type, depending on a certain function of the velocities. Suppose, for instance, that a particle m is resisted by a force proportional to its velocity, so that its equations of motion are

$$m\ddot{x} + k\dot{x} = X, \quad m\ddot{y} + k\dot{y} = Y, \quad m\ddot{z} + k\dot{z} = Z. \quad \text{.........(1)}$$

In the process of forming the general equations of motion we multiply (1) by $\partial x/\partial q_r$, $\partial y/\partial q_r$, $\partial z/\partial q_r$, respectively, and add. It follows from Art. 73 (1) that these factors are respectively equal to $\partial\dot{x}/\partial\dot{q}_r$, $\partial\dot{y}/\partial\dot{q}_r$, $\partial\dot{z}/\partial\dot{q}_r$. Hence, in addition to the terms previously obtained in Art. 77 (5), we have an expression

$$\frac{1}{2}\frac{\partial}{\partial\dot{q}_r}\Sigma k\,(\dot{x}^2 + \dot{y}^2 + \dot{z}^2). \quad \text{.....................(2)}$$

Again, suppose that between any two particles m_1, m_2 there is a force varying as the *relative* velocity. In the equations of motion of m_1 we have terms

$$k\,(\dot{x}_1 - \dot{x}_2), \quad k\,(\dot{y}_1 - \dot{y}_2), \quad k\,(\dot{z}_1 - \dot{z}_2), \quad \text{...........(3)}$$

and, in the equations of m_2, terms

$$k\,(\dot{x}_2 - \dot{x}_1), \quad k\,(\dot{y}_2 - \dot{y}_1), \quad k\,(\dot{z}_2 - \dot{z}_1). \quad \text{...........(4)}$$

If we multiply the expressions (3) by $\partial\dot{x}_1/\partial\dot{q}_r$, $\partial\dot{y}_1/\partial\dot{q}_r$, $\partial\dot{z}_1/\partial\dot{q}_r$, and the expressions (4) by $\partial\dot{x}_2/\partial\dot{q}_r$, $\partial\dot{y}_2/\partial\dot{q}_r$, $\partial\dot{z}_2/\partial\dot{q}_r$, and add, we obtain, finally, a result

$$\frac{1}{2}\frac{\partial}{\partial\dot{q}_r}\Sigma k\,\{(\dot{x}_1 - \dot{x}_2)^2 + (\dot{y}_1 - \dot{y}_2)^2 + (\dot{z}_1 - \dot{z}_2)^2\}. \quad \text{.........(5)}$$

Hence, in all such cases, the generalized equations of (small) motion take the form

$$\frac{d}{dt}\frac{\partial T}{\partial \dot{q}_r} + \frac{\partial F}{\partial \dot{q}_r} + \frac{\partial V}{\partial q_r} = 0, \quad \text{......................(6)}$$

where F is a homogeneous quadratic function of the generalized velocities, say

$$2F = b_{11}\dot{q}_1^2 + b_{22}\dot{q}_2^2 + \ldots + 2b_{12}\dot{q}_1\dot{q}_2 + \ldots; \quad \text{.........(7)}$$

this function is moreover essentially positive.

If we multiply (6) by $\dot{q}_r$, and add up the n equations thus obtained, we find

$$\frac{d}{dt}(T+V) = -2F. \quad \text{......................(8)}$$

Hence the function $2F$ measures the rate at which the mechanical energy of the system is being diminished by friction. It was introduced into the theory of vibrations by Rayleigh*, and termed by him the 'dissipation function.' Kelvin proposed the name 'dissipativity.'

98. Free Oscillations with Friction.

The equations of free motion are now of the type

$$\begin{aligned} a_{1r}\ddot{q}_1 + a_{2r}\ddot{q}_2 + \ldots + a_{nr}\ddot{q}_n & \\ + b_{1r}\dot{q}_1 + b_{2r}\dot{q}_2 + \ldots + b_{nr}\dot{q}_n & \\ + c_{1r}q_1 + c_{2r}q_2 + \ldots + c_{nr}q_n &= 0, \quad \text{...(1)} \end{aligned}$$

where $\quad a_{rs} = a_{sr}, \quad b_{rs} = b_{sr}, \quad c_{rs} = c_{sr}. \quad$ ....(2)

If we assume $\quad q_r = A_r e^{\lambda t}, \quad$ ....(3)

we have n equations of the type

$$\begin{aligned} (a_{1r}\lambda^2 + b_{1r}\lambda + c_{1r}) A_1 + (a_{2r}\lambda^2 + b_{2r}\lambda + c_{2r}) A_2 + \ldots & \\ + (a_{nr}\lambda^2 + b_{nr}\lambda + c_{nr}) A_n &= 0. \quad \text{...(4)} \end{aligned}$$

Eliminating the $n-1$ ratios

$$A_1 : A_2 : \ldots : A_n$$

we have a symmetrical determinantal equation of the $2n$th degree in λ, which we write

$$\Delta(\lambda) = 0. \quad \text{..........................(5)}$$

* *Proc. Lond. Math. Soc.* (1), t. iv (1873); *Theory of Sound*, c. v.

Corresponding to any one root of this we have

$$\frac{A_1}{\alpha_1} = \frac{A_2}{\alpha_2} = \ldots = \frac{A_n}{\alpha_n}, \quad \text{.....................(6)}$$

where $\alpha_1, \alpha_2, \ldots, \alpha_n$ are the minors of any one row in $\Delta(\lambda)$ (cf. Art. 90).

Proceeding as in Art. 92, but replacing the symbols

$$A_1, A_2, \ldots, A_n \text{ by } \alpha_1, \alpha_2, \ldots, \alpha_n,$$

we find

$$\lambda^2 T(\alpha) + \lambda F(\alpha) + V(\alpha) = 0. \quad \text{..............(7)}$$

If, as we will assume, V as well as F and T is essentially positive, this equation shews that real values of λ, which imply real values of the minors α, must be negative.

Again, if λ' be a second root of (5), and if we distinguish the minors of $\Delta(\lambda')$ by accents, we find, with an obvious extension of the notations of Art. 92,

$$\lambda^2 T(\alpha, \alpha') + \lambda F(\alpha, \alpha') + V(\alpha, \alpha') = 0, \quad \text{...........(8)}$$

$$\lambda'^2 T(\alpha, \alpha') + \lambda' F(\alpha, \alpha') + V(\alpha, \alpha') = 0. \quad \text{...........(9)}$$

Hence if λ, λ' are unequal

$$\lambda + \lambda' = -\frac{F(\alpha, \alpha')}{T(\alpha, \alpha')}, \quad \lambda\lambda' = \frac{V(\alpha, \alpha')}{T(\alpha, \alpha')}. \quad \text{..........(10)}$$

If λ, λ' are a conjugate pair of complex roots, say

$$\lambda = \rho + i\sigma, \quad \lambda' = \rho - i\sigma, \quad \text{.................(11)}$$

we may write

$$\alpha_r = \mu_r + i\nu_r, \quad \alpha_r' = \mu_r' - i\nu_r'. \quad \text{...............(12)}$$

We find

$$2\rho = -\frac{F(\mu) + F(\nu)}{T(\mu) + T(\nu)}, \quad \rho^2 + \sigma^2 = \frac{V(\mu) + V(\nu)}{T(\mu) + T(\nu)}. \quad \text{...(13)}$$

The former of these equations shews that complex roots of (5) will have negative real parts.

For each *real* root of (5) we have therefore a solution of the type

$$q_r = C\alpha_r e^{\rho t}, \quad \text{......................(14)}$$

where ρ is negative. The displacements decay without oscillation, and the motion is accordingly styled 'aperiodic.'

When the solution corresponding to a pair of conjugate imaginary roots is put in real form, we obtain

$$q_r = A\{\mu_r \cos(\sigma t + \epsilon) - \nu_r \sin(\sigma t + \epsilon)\} e^{\rho t}, \quad \text{.........(15)}$$

where the constants A, ϵ are arbitrary, but the same for all the coordinates. This represents a simple harmonic oscillation whose amplitude diminishes asymptotically to zero. It was found in Art. 90 that in the absence of friction the phase was at any instant the same for each coordinate. Equation (15) shews that this statement no longer holds. It is easily seen that the motion of any particle will in general be (damped) elliptic harmonic.

When the frictional coefficients b_{rr}, b_{rs} are small, considerations of continuity indicate that the roots of (5) will be imaginary, and that the quantities ν_r will be small. The equations (13) then shew that ρ is small, and that, to the first order of small quantities,

$$\sigma^2 = \frac{V(\mu)}{T(\mu)}. \quad \text{......................(16)}$$

The quantities μ_r will differ only slightly from the values which they have in the case of no friction, and the stationary property of the normal modes, proved in Art. 95, shews that the value of σ^2 given by (16) is affected by friction to the extent only of a small quantity of the *second* order.

The effect of slight friction is therefore mainly on the *amplitudes* of vibration, the natural periods being practically unaltered.

The method to be pursued in the treatment of *forced* oscillations with friction will be understood from what has been said in Art. 96. When the frictional coefficients are small, the modification of previous results is slight, except in the case of exact or approximate coincidence between a natural and an imposed frequency. The general character of the results is perhaps sufficiently exemplified by the case of one degree of freedom [D. 95]. The theory is fully developed in Rayleigh's *Theory of Sound.*

99. Oscillations of a Cyclic System.

The most general form which the equations of small motion of a dynamical system can assume, when modified by the introduction of terms proportional to the velocities, is

$$\frac{d}{dt}\frac{\partial T}{\partial \dot{q}_r} + B_{1r}\dot{q}_1 + B_{2r}\dot{q}_2 + \ldots + B_{nr}\dot{q}_n + \frac{\partial U}{\partial q_r} = Q_r, \quad \text{......(1)}$$

where the relation $B_{rs} = B_{sr}$ is *not* assumed to hold. This will apply to both cyclic and rotating systems, provided in each case we assign the proper form to the function U (see below). If we put

$$b_{rs} = b_{sr} = \tfrac{1}{2}(B_{rs} + B_{sr}), \quad \text{.................}(2)$$

$$\beta_{rs} = -\beta_{sr} = \tfrac{1}{2}(B_{rs} - B_{sr}), \quad \text{.................}(3)$$

the equation may be written

$$\frac{d}{dt}\frac{\partial T}{\partial \dot{q}_r} + \frac{\partial F}{\partial \dot{q}_r} + \beta_{1r}\dot{q}_1 + \beta_{2r}\dot{q}_2 + \ldots + \beta_{nr}\dot{q}_n + \frac{\partial U}{\partial q_r} = Q_r, \ \ldots(4)$$

where F has the same form as in Art. 97 (7).

The terms arising from F are of the type already considered, and may therefore be classed as representing 'frictional' or 'dissipative' forces. If we multiply (4) by $\dot{q}_r$, and add the results for all the equations, we find

$$\frac{d}{dt}(T + U) = -2F + Q_1\dot{q}_1 + Q_2\dot{q}_2 + \ldots + Q_n\dot{q}_n, \quad \text{......}(5)$$

since the terms involving β's cancel by (3).

The terms in (4) containing β's, on the other hand, are such as occur in 'cyclic' systems with latent motion (Art. 84), and are called 'gyrostatic' terms. To make the correspondence exact, we must put $U = V + K$, in (1), where K denotes the kinetic energy of the latent motions alone. The symbol T must also be replaced by $\mathfrak{T}$, which denotes the kinetic energy which would remain if the cyclic motions were annulled.

Equations of the same type apply also to the case of motion relative to a rotating body (Art. 80), provided U now stand for $V - T_0$, where T_0 is the kinetic energy of the system when rotating in relative 'rest' in the configuration $(q_1, q_2, \ldots, q_n)$.

In either case, the conditions of steady motion, or of relative equilibrium, are of the same type

$$\frac{\partial U}{\partial q_r} = 0, \quad \text{............................}(6)$$

as in Art. 86.

The argument of Dirichlet applies without modification to shew that a *sufficient* condition of stability is that U should be a minimum in the equilibrium configuration. This is in virtue of the equations

$$\mathfrak{T} + V + K = \text{const.}, \quad \text{.......................}(7)$$

$$\mathfrak{T} + V - T_0 = \text{const.}, \quad \text{.......................}(8)$$

proved in Arts. 84, 80 for the respective cases. The same condition is also *necessary*, if the influence of dissipative forces affecting the coordinates $q_1, q_2, \ldots, q_n$ is taken into account. This follows from Kelvin's argument, reproduced in Art. 86. We shall therefore assume in what follows that U is a minimum.

To examine the influence of the gyrostatic terms on the small oscillations we will ignore friction and assume, for simplicity, that $F=0$. We will also assume the coordinates $q_1, q_2, \ldots, q_n$ to be adjusted so as to vanish when the conditions (6) are fulfilled. We write therefore

$$2U = c_{11}q_1^2 + c_{22}q_2^2 + \ldots + 2c_{12}q_1q_2 + \ldots. \qquad \text{.........(9)}$$

It is to be remarked, however, that owing to the fact that K, or T_0, is involved in the value of U, the coefficients in this expression may depend in part on the values of the constant momenta corresponding to the ignored coordinates, or on the angular velocity of the rotating body, as the case may be.

The equations of motion may be further simplified by a transformation of coordinates which shall reduce the functions $\mathfrak{T}$ and U simultaneously to sums of squares, say

$$2\mathfrak{T} = a_1\dot{q}_1^2 + a_2\dot{q}_2^2 + \ldots + a_n\dot{q}_n^2, \qquad \text{..............(10)}$$

$$2U = c_1q_1^2 + c_2q_2^2 + \ldots + c_nq_n^2. \qquad \text{..............(11)}$$

This is always possible, but it is to be remembered that the particular transformation required may vary with the values of the constant momenta, or constant angular velocity, referred to. The new coordinates may be called the 'principal coordinates' of the system, but it must not be assumed that they retain the mutually independent character of the 'normal coordinates' of an acyclic system. For instance, modes of motion are not as a rule possible in which one principal coordinate varies alone.

The equations (4) now take the forms

$$\left.\begin{array}{l} a_1\ddot{q}_1 + c_1q_1 \qquad\qquad + \beta_{12}\dot{q}_2 + \beta_{13}\dot{q}_3 + \ldots + \beta_{1n}\dot{q}_n = Q_1, \\ a_2\ddot{q}_2 + c_2q_2 + \beta_{21}\dot{q}_1 \qquad\qquad + \beta_{23}\dot{q}_3 + \ldots + \beta_{2n}\dot{q}_n = Q_2, \\ \ldots\ldots\ldots\ldots\ldots\ldots\ldots\ldots\ldots\ldots\ldots\ldots \\ a_n\ddot{q}_n + c_nq_n + \beta_{n1}\dot{q}_1 + \beta_{n2}\dot{q}_2 + \beta_{n3}\dot{q}_3 + \ldots \qquad = Q_n. \end{array}\right\} \quad \text{...(12)}$$

To investigate the free motions, we put $Q_1, Q_2, \ldots, Q_n = 0$, and assume

$$q_1 = A_1e^{\lambda t}, \quad q_2 = A_2e^{\lambda t}, \quad \ldots, \quad q_n = A_ne^{\lambda t}. \qquad \text{.........(13)}$$

Substituting, we find

$$\left.\begin{array}{llll}(a_1\lambda^2+c_1)A_1 & +\beta_{12}\lambda A_2+\ldots & & +\beta_{1n}\lambda A_n=0,\\ \beta_{21}\lambda A_1 & +(a_2\lambda^2+c_2)A_2+\ldots & & +\beta_{2n}\lambda A_n=0,\\ \ldots\ldots & \ldots\ldots & & \ldots\ldots\\ \beta_{n1}\lambda A_1 & +\beta_{n2}\lambda A_2+\ldots & & +(a_n\lambda^2+c_n)A_n=0.\end{array}\right\} \quad \ldots(14)$$

Eliminating the ratios $A_1 : A_2 : \ldots : A_n$, we get the equation

$$\begin{vmatrix} a_1\lambda^2+c_1, & \beta_{12}\lambda, & \ldots & \beta_{1n}\lambda \\ \beta_{21}\lambda, & a_2\lambda^2+c_2, & \ldots & \beta_{2n}\lambda \\ \ldots & \ldots & \ldots & \ldots \\ \beta_{n1}\lambda, & \beta_{n2}\lambda, & \ldots & a_n\lambda^2+c_n \end{vmatrix} = 0, \quad \ldots\ldots(15)$$

or, as we may write it for shortness,

$$\Delta(\lambda)=0. \quad \ldots\ldots(16)$$

The solution of (14) corresponding to any particular root λ is

$$\frac{A_1}{\alpha_1}=\frac{A_2}{\alpha_2}=\ldots=\frac{A_n}{\alpha_n}=C, \quad \ldots\ldots(17)$$

where $\alpha_1, \alpha_2, \ldots, \alpha_n$ are the minors of any row in (15), and C is arbitrary.

The determinant $\Delta(\lambda)$ is a 'skew' determinant, in virtue of the relations $\beta_{rs}=-\beta_{sr}$. If we reverse the sign of λ, the rows and columns are simply interchanged, and the value of the determinant is therefore unaltered. The roots of (15) will therefore occur in pairs of the form

$$\lambda=\pm(\rho+i\sigma), \quad \ldots\ldots(18)$$

but it is evident from Kelvin's criterion that if U is a minimum in the zero configuration we must have $\rho=0$, so that the motion is strictly periodic.

To obtain an analytical proof of this result we adapt the procedure of Art. 92. If in (14) we replace $A_1, A_2, \ldots, A_n$ by $\alpha_1, \alpha_2, \ldots, \alpha_n$, and multiply the resulting equations by $\alpha_1, \alpha_2, \ldots, \alpha_n$ respectively, we find on addition

$$\lambda^2\mathfrak{T}(\alpha)+U(\alpha)=0. \quad \ldots\ldots(19)$$

This shews that the values of λ^2, if real, must be negative.

Again, let us denote by $\alpha_1', \alpha_2', \ldots, \alpha_n'$ the values of the minors corresponding to any other root of (16) which is not such that $\lambda+\lambda'=0$. We find

$$\lambda^2\mathfrak{T}(\alpha,\alpha')+\lambda\Sigma\beta_{rs}(\alpha_r'\alpha_s-\alpha_r\alpha_s')+U(\alpha,\alpha')=0, \quad \ldots(20)$$

and

$$\lambda'^2\mathfrak{T}(\alpha,\alpha')+\lambda'\Sigma\beta_{rs}(\alpha_r\alpha_s'-\alpha_r'\alpha_s)+U(\alpha,\alpha')=0. \quad \ldots(21)$$

Eliminating the middle terms, and dividing by $\lambda + \lambda'$, we have

$$\lambda\lambda' = -\frac{U(\alpha, \alpha')}{\mathfrak{T}(\alpha, \alpha')}. \quad \text{.....................(22)}$$

If complex roots were admissible, we might put

$$\lambda = \rho + i\sigma, \quad \lambda' = \rho - i\sigma. \quad \text{.................(23)}$$

Hence writing $\quad \alpha_r = \mu_r + i\nu_r, \quad \alpha_r' = \mu_r - i\nu_r, \quad$ ....(24)

we should have $\quad \rho^2 + \sigma^2 = -\dfrac{U(\mu) + U(\nu)}{\mathfrak{T}(\mu) + \mathfrak{T}(\nu)}, \quad$ ....(25)

which is impossible if U be essentially positive. We conclude that the roots of (16) occur in pairs of the form

$$\lambda = \pm i\sigma. \quad \text{........................(26)}$$

It will appear presently that the converse statement, viz. that the roots cannot be all of this type unless U is a minimum in the zero configuration, does not hold in the present case, unless frictional forces be taken into account (see Art. 100).

To examine the character of a small oscillation, we note that the minors $\alpha_1, \alpha_2, \ldots, \alpha_n$ will in general involve odd as well as even powers of λ, and will therefore be complex. Putting

$$\lambda = i\sigma, \quad \alpha_r = \mu_r + i\nu_r, \quad \text{.................(27)}$$

we have $\quad q_r = C(\mu_r + i\nu_r)e^{i\sigma t}. \quad$ ....(28)

Writing $C = He^{i\epsilon}$, and taking the real part, we get

$$q_r = H\{\mu_r \cos(\sigma t + \epsilon) - \nu_r \sin(\sigma t + \epsilon)\}, \quad \text{.........(29)}$$

where the constants H and ϵ are arbitrary. This represents what may be called a 'natural mode' of oscillation of the system. The number of such modes is equal to that of the degrees of freedom.

It will be observed that the phase at any instant is not uniform throughout the system, being different for the different coordinates. It will be found that in any natural mode the motion of any particle is in general elliptic harmonic.

Examples of the small vibrations of a cyclic or of a rotating system of finite freedom have occurred incidentally in various parts of this book*. The following is a comparatively simple instance of a continuous system.

Ex. A uniform endless chain revolves in its own plane in the form of a circle; to investigate the small oscillations about the steady motion.

* See Arts. 47, 57, 79.

Let a be the radius, ρ the line-density, ω the angular velocity. The polar coordinates of an element of the chain when at rest being (a, θ), the coordinates in the disturbed state may be written

$$r = a + u, \quad \theta_1 = \theta + \omega t + \frac{v}{a}, \qquad \text{.........................(30)}$$

where u, v are small. If ϕ be the angle which the tangent makes with the radius vector, we have, since the chain is assumed to be inextensible,

$$\cos\phi = \frac{\partial r}{\partial s} = \frac{\partial u}{a\partial\theta} = \frac{u'}{a}, \qquad \text{.........................(31)}$$

where the accent denotes differentiation with respect to θ. Hence

$$\phi = \tfrac{1}{2}\pi - \frac{u'}{a}, \qquad \text{.........................(32)}$$

to the first order. Again

$$\sin\phi = \frac{r d\theta_1}{ds} = \left(1 + \frac{u}{a}\right)\left(1 + \frac{v'}{a}\right). \qquad \text{.........................(33)}$$

Since $\sin\phi = 1$, to the first order, we have

$$u + v' = 0. \qquad \text{.........................(34)}$$

The radial and transverse accelerations are

$$\ddot{r} - r\dot{\theta}_1^2 = \ddot{u} - (a+u)\left(\omega + \frac{\dot{v}}{a}\right)^2 = \ddot{u} - 2\omega\dot{v} - \omega^2 u - \omega^2 a, \qquad \text{.........(35)}$$

$$r\ddot{\theta}_1 + 2\dot{r}\dot{\theta}_1 = (a+u)\frac{\ddot{v}}{a} + 2\dot{u}\left(\omega + \frac{\dot{v}}{a}\right) = \ddot{v} + 2\omega\dot{u}, \qquad \text{.........................(36)}$$

with the same approximation.

If ψ be the inclination of the tangent to a fixed line, we may write

$$\psi = \theta_1 + \phi, \qquad \text{.........................(37)}$$

and therefore

$$\delta\psi = \left(1 + \frac{v' - u''}{a}\right)\delta\theta, \qquad \text{.........................(38)}$$

from (30) and (32).

We now calculate the forces on an element $PQ\,(=a\delta\theta)$. If T be the tension, these reduce to δT along the tangent, and $T\delta\psi$ along the normal [S. 80]. Resolving along and at right angles to the radius vector OP, we have

$$\delta T\cos\phi - T\delta\psi\sin\phi = -T\left(1 + \frac{v' - u''}{a}\right)\delta\theta, \qquad \text{...............(39)}$$

and

$$\delta T\sin\phi + T\delta\psi\cos\phi = \delta T + T\frac{u'}{a}\delta\theta, \qquad \text{.........................(40)}$$

respectively. The equations of motion of the element $\rho a\delta\theta$ are therefore

$$\rho a(\ddot{u} - 2\omega\dot{v} - \omega^2 u - \omega^2 a) = -T\left(1 + \frac{v' - u''}{a}\right), \qquad \text{...........(41)}$$

$$\rho a(\ddot{v} + 2\omega\dot{u}) = \frac{\partial T}{\partial\theta} + \frac{Tu'}{a}. \qquad \text{.........................(42)}$$

As a first approximation we have $T=\rho\omega^2 a^2$, as is otherwise known [S. 80]. Putting this value of T in the small terms, and eliminating T, we find

$$\ddot{u}' - 2\omega\dot{v}' - \omega^2 u' + \ddot{v} + 2\omega\dot{u} = -\omega^2 v'' + \omega^2 u''' + \omega^2 u', \qquad (43)$$

or, on eliminating u by means of (34),

$$\ddot{v}'' - \ddot{v} + 4\omega\dot{v}' - \omega^2 v^{\text{iv}} - 3\omega^2 v'' = 0. \qquad (44)$$

This might be treated on the usual plan, assuming that v varies as $e^{i\sigma t}$, but the work is shortened if we assume at once

$$v = Ce^{i(\sigma t - s\theta)}, \qquad (45)$$

where s is any positive or negative integer. This requires

$$\sigma^2(s^2+1) + 4\sigma\omega s - \omega^2 s^2(s^2-3) = 0, \qquad (46)$$

the roots of which are

$$\sigma = -s\omega, \quad \frac{s(s^2-3)}{s^2+1}\omega. \qquad (47)$$

The formula (45) may be interpreted as representing a wave travelling round the circle with either of the two angular velocities

$$\frac{\sigma}{s} = -\omega, \quad \text{or} \quad \frac{s^2-3}{s^2+1}\omega. \qquad (48)$$

These are the velocities relative to the chain. To get the velocities *in space* we must add ω to the results, which become

$$0, \text{ and } \frac{2(s^2-1)}{s^2+1}\omega, \qquad (49)$$

respectively.

That a disturbed form would be possible which is stationary in space was known beforehand [D. 50]. That *both* waves should be stationary for $s=\pm 1$ is also obvious, since the disturbed form is then circular, to our order of approximation.

100. Kinetic Stability.

It has been objected to the Lagrangian theory of small oscillations about absolute equilibrium, as developed in Art. 90 with the neglect of small quantities of the second order, that it cannot furnish a decisive test of stability, since the neglected terms might conceivably in the course of the motion become important.

As a matter of strict logic the objection is valid, but since the criterion arrived at, viz. that a necessary and sufficient condition of equilibrium is that the potential energy in the zero configuration should be a minimum, is confirmed by the argument of Dirichlet and Kelvin (Art. 86), the point is hardly important.

The argument referred to, when extended to gyrostatic systems, shews that a *sufficient* condition of stability, even when dissipative forces are ignored, is that $V+K$, or $V-T_0$, as the case may be, should be a minimum. And so far there is agreement with the approximate treatment based on linear equations (Art. 99). On the *analytical* theory, however, this condition is no longer essential, and we may have periodic solutions, apparently indicating stability, even when the function in question is a maximum.

To prove this statement it will be sufficient to consider the case of two degrees of freedom. If friction be neglected the equations of small motion, in terms of principal coordinates, have the form

$$\left.\begin{array}{l} a_1\ddot{q}_1 - \beta\dot{q}_2 + c_1 q_1 = 0, \\ a_2\ddot{q}_2 + \beta\dot{q}_1 + c_2 q_2 = 0, \end{array}\right\} \quad \text{.....................(1)}$$

where a_1, a_2 are necessarily positive. Assuming that q_1 and q_2 vary as $e^{\lambda t}$ we find

$$a_1 a_2 \lambda^4 + (a_1 c_2 + a_2 c_1 + \beta^2)\lambda^2 + c_1 c_2 = 0. \quad \text{.........(2)}$$

The two values of λ^2 will be real provided

$$\beta^4 + 2\beta^2(a_1 c_2 + a_2 c_1) + (a_2 c_1 - a_1 c_2)^2 > 0. \quad \text{.........(3)}$$

If c_1, c_2 are both positive this condition is fulfilled, and the two values of λ^2 are moreover negative (cf. Art. 99). We have then

$$\lambda = \pm i\sigma_1, \quad \pm i\sigma_2, \quad \text{.......................(4)}$$

where σ_1^2, σ_2^2 are the roots of

$$a_1 a_2 \sigma^4 - (a_1 c_2 + a_2 c_1 + \beta^2)\sigma^2 + c_1 c_2 = 0. \quad \text{.........(5)}$$

The solution therefore consists of simple harmonic vibrations, so that the zero configuration is stable from the analytical point of view, as well as in virtue of Kelvin's criterion.

It is evident, however, that the condition (3) may be fulfilled even if c_1 or c_2, or both, be negative, provided the gyrostatic constant β be sufficiently great. If c_1, c_2 have opposite signs one value of λ^2 is then positive and the other negative. The positive root gives solutions of the types $e^{\lambda t}$, $e^{-\lambda t}$, and so indicates instability.

If, however, c_1 and c_2 are both negative, whilst $a_1 c_2 + a_2 c_1 + \beta^2$ is positive, as may evidently happen, the two values of λ^2 (when real) are negative, indicating stability, although the function $V + K$ or $V - T_0$, is now a *maximum* in the zero configuration.

This is not of course in contradiction with the Kelvin criterion, which contemplates the influence of dissipative forces on the palpable coordinates. If we introduce frictional terms into our approximate equations, the results become concordant. We have now

$$\left.\begin{aligned} a_1\ddot{q}_1 + c_1 q_1 + b_{11}\dot{q}_1 + (b_{12} - \beta)\,\dot{q}_2 = 0,\\ a_2\ddot{q}_2 + c_2 q_2 + (b_{21} + \beta)\,\dot{q}_1 + b_{22}\dot{q}_2 = 0,\end{aligned}\right\} \quad \text{............(6)}$$

where b_{11}, b_{12}, b_{22} are the coefficients in the dissipation function

$$2F = b_{11}\dot{q}_1^2 + 2b_{12}\dot{q}_1\dot{q}_2 + b_{22}\dot{q}_2^2 \text{..................(7)}$$

Since this is assumed to be essentially positive we have

$$b_{11} > 0, \quad b_{22} > 0, \quad b_{11}b_{22} - b_{12}^2 > 0 \text{................(8)}$$

The equation (2) is now replaced by

$$a_1 a_2 \lambda^4 + (a_2 b_{11} + a_1 b_{22})\,\lambda^3 + (a_2 c_1 + a_1 c_2 + \beta^2 + b_{11}b_{22} - b_{12}^2)\,\lambda^2 + (b_{11}c_2 + b_{22}c_1)\,\lambda + c_1 c_2 = 0 \text{....(9)}$$

We will suppose that the roots of (2) in λ^2 are real and negative. This implies that c_1 and c_2 have the same sign. Further, we will for simplicity suppose the frictional coefficients to be small. We write, therefore, for the roots of (9),

$$\lambda = \rho_1 \pm i\sigma_1, \quad \rho_2 \pm i\sigma_2, \quad \text{.................(10)}$$

where ρ_1, ρ_2 are small, and σ_1, σ_2 are determined by (5). Neglecting terms of the second order in b_{11}, b_{12}, b_{22}, we find

$$\rho_1 + \rho_2 = -\tfrac{1}{2}\left(\frac{b_{11}}{a_1} + \frac{b_{22}}{a_2}\right), \quad \text{..............(11)}$$

and

$$\frac{\rho_1}{\sigma_1^2} + \frac{\rho_2}{\sigma_2^2} = -\tfrac{1}{2}\left(\frac{b_{11}}{c_1} + \frac{b_{22}}{c_2}\right). \quad \text{..............(12)}$$

Hence either ρ_1 or ρ_2 must be negative, but both cannot be negative unless the sign of c_1 and c_2 (assumed to be the same) be positive. If c_1 and c_2 be both negative, ρ_1 and ρ_2 must have opposite signs. Since

$$\rho_1\left(\frac{1}{\sigma_2^2} - \frac{1}{\sigma_1^2}\right) = -\frac{1}{2\sigma_2^2}\left(\frac{b_{11}}{a_1} + \frac{b_{22}}{a_2}\right) + \tfrac{1}{2}\left(\frac{b_{11}}{c_1} + \frac{b_{22}}{c_2}\right), \quad \text{...(13)}$$

the positive value of ρ will correspond to the smaller value of σ, i.e. to the slower of the two fundamental oscillations. The amplitude of this oscillation will therefore gradually increase whilst that of the other will decay*. We may refer for illustration to a problem discussed in Art. 41.

* The above investigation is repeated, with merely verbal alterations, from the author's *Hydrodynamics*, 2nd ed., 1895.

We are thus led to distinguish, in the case of rotational and cyclic systems, between stability as indicated by the classical Lagrangian method of small oscillations where friction is neglected, and stability as determined by the Dirichlet-Kelvin test. The distinction was first pointed out by Kelvin, and was further insisted upon by Poincaré in his researches on the possible forms of equilibrium of masses of liquid rotating under the mutual gravitation of their parts. The two types have been designated as 'ordinary' or 'temporary,' and 'practical' or 'permanent' or 'secular' stability, respectively, the latter name being doubtless suggested in relation to cosmical applications.

The criteria of 'ordinary' stability, i.e. the conditions under which the equation for the determination of λ in the general case should have roots which are either pure imaginaries, or have their real parts negative, have been discussed by Routh*. They are naturally of a somewhat elaborate character.

Ex. 1. In the theory of central forces in two dimensions we have, taking the mass to be unity,

$$2T = \dot{r}^2 + r^2\dot{\theta}^2. \qquad (14)$$

If the force vary as $1/r^n$ we have

$$V = -\frac{\mu}{(n-1)\,r^{n-1}}. \qquad (15)$$

Putting $r^2\dot{\theta} = \kappa$, where κ is the constant angular momentum about the centre of force, we have in the notation of Art. 84

$$K = \tfrac{1}{2}\kappa^2/r^2, \qquad (16)$$

and therefore

$$U = V + K = \frac{1}{2}\frac{\kappa^2}{r^2} - \frac{\mu}{(n-1)\,r^{n-1}}, \qquad \frac{dU}{dr} = -\frac{\kappa^2}{r^3} + \frac{\mu}{r^n}. \qquad (17)$$

The condition for a circular orbit is that U should be stationary, or

$$\kappa^2 = \mu/r^{n-3}. \qquad (18)$$

This makes

$$\frac{d^2U}{dr^2} = \frac{3\kappa^2}{r^4} - \frac{n\mu}{r^{n+1}} = \frac{\kappa^2}{r^4}(3-n). \qquad (19)$$

Hence U is a minimum, i.e. the circular orbit is permanently stable, only if $n < 3$.

It is known that this is also the necessary condition for 'temporary' stability [D. 87].

Ex. 2. A particle moves on the inner surface of a spherical bowl which rotates with constant angular velocity ω about the vertical diameter. In this

* In his essay on *Stability of Motion*. Particular cases have been met with in Arts. 71, 78, above.

case the equilibrium of the particle when in the lowest position is 'ordinarily' stable since the rotation is irrelevant if the bowl be smooth.

If a be the radius, m the mass of the particle, and θ its angular distance from the lowest point, we have

$$U = V - T_0 = -mga\cos\theta - \tfrac{1}{2}m\omega^2 a^2 \sin^2\theta. \qquad (20)$$

Hence

$$\frac{dU}{d\theta} = mga\sin\theta\left(1 - \frac{\omega^2 a}{g}\cos\theta\right), \qquad (21)$$

$$\frac{d^2U}{d\theta^2} = mga\cos\theta\left(1 - \frac{\omega^2 a}{g}\cos\theta\right) + m\omega^2 a^2\sin^2\theta. \qquad (22)$$

For relative equilibrium we must have $dU/d\theta = 0$. If $\omega^2 < g/a$ this is only possible when $\theta = 0$, and (22) shews that U is then a minimum, indicating permanent stability. If $\omega^2 > g/a$, there is a second position of relative equilibrium given by

$$\cos\theta = g/\omega^2 a. \qquad (23)$$

The formula (22) shews that this is stable, whilst the lowest position is now unstable. In the position given by (23) the particle rotates with the bowl like the bob of a conical pendulum.

To examine the initial stages when the particle is slightly disturbed from its position at the lowest point, we employ horizontal axes Ox, Oy passing through the lowest point, and rotating with the bowl. Introducing a frictional force varying as the relative velocity, we have

$$\left.\begin{aligned} \ddot{x} - 2\omega\dot{y} - \omega^2 x &= -k\dot{x} - n^2 x, \\ \ddot{y} + 2\omega\dot{x} - \omega^2 y &= -k\dot{y} - n^2 y, \end{aligned}\right\} \qquad (24)$$

where $n^2 = g/a$. These equations combine into

$$\ddot{\zeta} + (2i\omega + k)\dot{\zeta} + (n^2 - \omega^2)\zeta = 0, \qquad (25)$$

where $\zeta = x + iy$. Assuming that ζ varies as $e^{\lambda t}$, we find

$$(\lambda + i\omega)^2 = -n^2\left(1 + \frac{k\lambda}{n^2}\right). \qquad (26)$$

If k be small we have, for a first approximation, $\lambda = -i\omega \pm in$, and consequently, for the next,

$$\lambda = -i(\omega \mp n) - \tfrac{1}{2}k\left(1 \mp \frac{\omega}{n}\right). \qquad (27)$$

If ξ, η be Cartesian coordinates referred to *fixed* axes through O, the complete solution is

$$\xi + i\eta = \zeta e^{i\omega t} = C_1 e^{-\nu_1 t + int} + C_2 e^{-\nu_2 t - int}, \qquad (28)$$

where

$$\nu_1 = \tfrac{1}{2}k\left(1 - \frac{\omega}{n}\right), \quad \nu_2 = \tfrac{1}{2}k\left(1 + \frac{\omega}{n}\right). \qquad (29)$$

This solution, when put in real form, shews that the motion is made up of two circular vibrations in opposite directions, of period $2\pi/n$; moreover that if $\omega^2 > n^2$, i.e. $\omega^2 a > g$, that circular vibration whose sense agrees with ω continually increases in amplitude. The particle works its way outwards in an ever widening spiral path, approximating to the stable position of relative equilibrium given by (23).

Ex. 3. A pendulum symmetrical about its longitudinal axis hangs by a universal flexure joint from a vertical spindle which is made to rotate with the constant angular velocity ω.

This problem has been discussed by the Lagrangian method in Art. 79, Ex. 4. In the steady motion we have $\dot{\theta}=0$, $\dot{\psi}=\omega$, and therefore

$$V-T_0=-Mgh\cos\theta-\tfrac{1}{2}(A\sin^2\theta+C\cos^2\theta)\,\omega^2. \quad\text{..........(30)}$$

Hence

$$\frac{d}{d\theta}(V-T_0)=\{Mgh-(A-C)\,\omega^2\cos\theta\}\sin\theta, \quad\text{..........(31)}$$

$$\frac{d^2}{d\theta^2}(V-T_0)=(A-C)\,\omega^2\sin^2\theta+\{Mgh-(A-C)\,\omega^2\cos\theta\}\cos\theta. \quad\text{...(32)}$$

The vertical position $\theta=0$ is one of relative equilibrium, but is unstable if

$$\omega^2>\frac{Mgh}{A-C}. \quad\text{..........(33)}$$

When this condition is fulfilled there is a second position of relative equilibrium given by

$$\cos\theta=\frac{Mgh}{(A-C)\,\omega^2}. \quad\text{..........(34)}$$

This is stable if (as in the usual form of the experiment) $A>C$.

EXAMPLES. XIV.

1. Shew how the solution (14) of Art. 90 verifies the equation of energy.

2. A uniform horizontal rectangular plate rests on four similar springs at the corners. Find the normal modes of vibration, and prove that their frequencies are as $1:\sqrt{3}:\sqrt{3}$.

3. A uniform rectangular plate whose edges are $2a$, $2b$, hangs by two equal vertical strings of length l attached to the ends of the upper edge ($2a$) which is horizontal. Describe the characters of the four normal modes of oscillation.

Prove that the periods are those of simple pendulums of lengths

$$l,\quad \tfrac{1}{3}l,\quad \tfrac{1}{2}l+\tfrac{2}{3}b\pm\tfrac{1}{2}\sqrt{(l^2+\tfrac{4}{3}bl+\tfrac{16}{9}b^2)}.$$

4. A uniform bar of length $2a$ is suspended by a short string of length l. Prove that the period of the slower vibration is greater than that of the bar when swinging about one end, in the ratio

$$1+\frac{9}{32}\frac{l}{a},$$

approximately. Also that the period of the quicker vibration is $\pi\sqrt{(l/g)}$, nearly.

5. A heavy plate hangs in a horizontal position by three vertical strings of unequal lengths. Shew that the normal modes of vibration are (1) a rotation about either of two vertical lines in a plane through the centre of mass, and (2) a swing parallel to this plane.

6. A thin cylindrical shell of radius a and mass M rests with its axis horizontal on a horizontal plane. Inside it there is placed a circular cylinder of mass m, radius b, and radius of gyration κ. If the system is set rocking,

form the equations of finite motion. If the motion be small, prove that the length of the equivalent simple pendulum is

$$\frac{2M(1+\kappa^2/b^2)(a-b)}{2M+m(1+\kappa^2/b^2)}.$$

7. An elliptic hole of semi-axes a, b is cut in a thin horizontal sheet of metal, and a sphere of radius c ($>b$) rests on it, its centre being therefore at a height h, $=\sqrt{(c^2-b^2)}$. If the sphere be set rocking through a small angle, the length of the equivalent simple pendulum is

$$\frac{(a^2-b^2)(\kappa^2+h^2)}{b^2h},$$

where κ is the radius of gyration of the sphere about a diameter.

8. A string of length $4a$ is weighted at equal intervals with three weights m, M, m, respectively, and is suspended from two points A, B symmetrically. Prove that if M performs small vertical oscillations, the length of the equivalent simple pendulum is

$$\frac{a\cos\alpha\cos\beta\sin(\alpha-\beta)\cos(\alpha-\beta)}{\sin\alpha\cos^2\alpha+\sin\beta\cos^2\beta},$$

where α, β are the inclinations of the parts of the string to the vertical.

9. A solid free to turn about the origin O is in equilibrium in a field of force such that the potential energy due to small rotations ξ, η, ζ about the coordinate axes is

$$V=\tfrac{1}{2}(a\xi^2+b\eta^2+c\zeta^2+2f\eta\zeta+2g\zeta\xi+2h\xi\eta);$$

prove that the normal modes of vibration consist of rotations about the common conjugate diameters of the momental ellipsoid at O and the quadric

$$ax^2+\dots+2fyz+\dots=1.$$

Shew that this includes the case of ordinary gravity, the centre of gravity not being restricted to lie in one of the principal axes of inertia at O. (Ball.)

10. A string of length l, $=(n+1)a$, loaded at intervals a with equal masses m, is fixed at both ends and subject to a tension T_1. If the particles receive small lateral displacements y_r, the potential energy is

$$V=\frac{1}{2}\frac{T_1}{a}\{y_1^2+(y_2-y_1)^2+(y_3-y_2)^2+\dots+y_n^2\}.$$

Form the equations of motion, and prove that the solution is

$$y_r=C\sin\frac{rs\pi a}{l}\cos(\sigma t+\epsilon),$$

where $s=1, 2, 3, \dots, n$, and

$$\sigma=2\sqrt{\left(\frac{T_1}{ma}\right)}\sin\frac{s\pi a}{2l}.$$

Find the limiting form of the result when $m/a=\rho$, $n\to\infty$. (Lagrange.)

11. A string of length na hangs vertically from one end, and is loaded at equal intervals a with n particles each of mass m. If this position be slightly disturbed, prove that the gain of potential energy is

$$V=\Sigma_r\frac{mg}{2a}\{ny_1^2+(n-1)(y_2-y_1)^2+(n-2)(y_3-y_2)^2+\dots+(y_n-y_{n-1})^2\},$$

where y_r is the lateral displacement of the rth particle from the top.

If $n=3$, prove that the lengths of the equivalent pendulums in the three normal modes are

$$2{\cdot}405\,a, \quad {\cdot}436\,a, \quad {\cdot}159\,a.$$

Find the limiting value of the above expression for V when

$$m/a=\rho, \quad na=l, \quad n\to\infty.$$

12. If a tense string be fixed at the ends ($x=0, l$), the potential energy of a small lateral displacement is

$$\tfrac{1}{2}T_1\int_0^l\left(\frac{\partial y}{\partial x}\right)^2 dx,$$

where T_1 is the tension, and y the displacement at the point x.

Assuming as an approximate form of vibration

$$y=Cx\,(l-x),$$

prove that the resulting estimate of the period is

$${\cdot}9935\times\frac{2l}{\sqrt{(T_1/\rho)}},$$

where ρ is the line-density.

13. A circular membrane of radius a is fixed at the edge and is subject to a uniform tension T_1. Prove that in a normal displacement z, symmetrical about the centre, the potential energy is

$$V=\tfrac{1}{2}T_1\int_0^a\left(\frac{\partial z}{\partial r}\right)^2 . \, 2\pi r\,dr.$$

Assuming

$$z=C\left(1-\frac{r^2}{a^2}\right),$$

obtain an approximate estimate of the longest period.

[$2{\cdot}56492\times\sqrt{(\rho a^2/T_1)}$, where ρ is the surface-density, the correct factor being $2{\cdot}61253$.]

14. Repeat the calculation of Ex. 13, assuming

$$z=C\left(1-\frac{r^2}{a^2}\right)\left(1+\beta\frac{r^2}{a^2}\right),$$

and adjusting the value of β so as to make the period a maximum.

[$2{\cdot}61235\sqrt{(\rho a^2/T_1)}$.]

15. Prove that the equations of motion of a gyrostatic system with three degrees of freedom can be reduced to the same form as for a particle attached by springs to a body rotating about a fixed axis.

16. A particle is subject to a centre of attractive force varying as $e^{-r/a}/r^2$. Prove that a circular orbit will be stable or unstable according as its radius is less or greater than a.

17. Four equal particles, connected by equal strings forming a rhombus, rotate in their own plane about the centre, under no external forces. Prove that the relative configuration is neutral for all forms of the rhombus.

18. Two particles M, m are attached to a string at distances a and $a+b$ from one end O, which is fixed. The system revolves about O in one plane, the angular momentum about O being μ. Prove that

$$2K = \frac{\mu^2}{Ma^2 + m(a^2 + 2ab\cos\chi + b^2)},$$

where χ is the angle between the directions of the two portions of the string, and thence that the straight form of the string is stable.

Prove that the period of a small oscillation is $2\pi/\sigma$, where

$$\frac{\sigma^2}{\omega^2} = \frac{Ma^2 + m(a+b)^2}{Mab},$$

ω being the angular velocity of the steady rotation.

19. A body of mass M can turn freely about a vertical axis O, and a second body of mass m can turn freely about a parallel axis O' carried by the former body. The mass-centre G of the first body is assumed to lie in the plane of the axes O, O'. The radius of gyration of the first body about O is k, that of the second body about a vertical through its mass-centre G' is κ, the distance between O and O' is a, and the distance of G' from O' is b. Prove that the kinetic energy is given by

$$2T = (A + 2H\cos\chi)\dot{\theta}^2 + 2(H\cos\chi + B)\dot{\theta}\dot{\chi} + B\dot{\chi}^2,$$

where θ is the angle turned through by the first body, χ is the angle which the plane of G' and O' makes with that of O and O' (vanishing when G' is farthest from O), and

$$A = Mk^2 + m(a^2 + b^2 + \kappa^2), \quad H = mab, \quad B = m(b^2 + \kappa^2).$$

If there are no external forces, and μ be the constant angular momentum about O, prove that, in the notation of Art. 84,

$$2K = \frac{\mu^2}{A + 2H\cos\chi},$$

and thence that the configurations $\chi = 0$, $\chi = \pi$ are respectively stable and unstable.

If the first body be maintained in uniform rotation ω, prove that

$$2T_0 = (A + 2H\cos\chi)\,\omega^2,$$

leading to the same results as to stability.

Prove that the frequency of a small oscillation about the stable configuration is given by

$$\frac{\sigma^2}{\omega^2} = \frac{(A+2H)H}{AB - H^2 - B^2}.$$

20. A uniform chain of length l is rotating with angular velocity ω about one end, which is fixed, the other end being free. Prove that for small deviations z normal to the plane of rotation the equation of motion is

$$\ddot{z} = \tfrac{1}{2}\omega^2(l^2 - x^2)\,z'' - \omega^2 x z',$$

where the accents denote differentiation with respect to the distance x from the fixed end.

Prove that the displacement in a normal mode is of the type

$$z = C\left\{\frac{x}{l} - \frac{(n-1)(n+2)}{2.3}\frac{x^3}{l^3} + \frac{(n-3)(n-1)(n+2)(n+4)}{2.3.4.5}\frac{x^5}{l^5} - \dots\right\}.$$

Shew that the series does not converge for $x = l$ unless n is an odd integer; and that the corresponding frequency is then given by

$$\frac{\sigma^2}{\omega^2} = \tfrac{1}{2}n(n+1). \qquad \text{(Southwell.)}$$

21. Prove that for transverse displacements y in the plane of rotation the equation is

$$\ddot{y} = \tfrac{1}{2}\omega^2(l^2 - x^2)\,y'' - \omega^2 x y' + \omega^2 y,$$

and that the frequencies of the various modes of vibration are given by

$$\frac{\sigma^2}{\omega^2} = \tfrac{1}{2}(n-1)(n+2)$$

where n is odd.

CHAPTER XII

VARIATIONAL METHODS

101. Lagrange's Variational Equation.

Starting as in Art. 77 from the equations of motion of any individual particle of a system, viz.

$$m\ddot{x} = X, \quad m\ddot{y} = Y, \quad m\ddot{z} = Z, \quad \text{..............(1)}$$

we denote by $x + \delta x$, $y + \delta y$, $z + \delta z$ the coordinates at time t of the same particle in any arbitrary configuration of the system (consistent with its connections) which differs infinitely little from the actual configuration. Multiplying the above equations by δx, δy, δz, respectively, and summing for all the particles of the system, we have

$$\Sigma m (\ddot{x}\,\delta x + \ddot{y}\,\delta y + \ddot{z}\,\delta z) = \Sigma (X\,\delta x + Y\delta y + Z\delta z). \text{......(2)}$$

This is 'Lagrange's variational equation.'

The special advantage of this form is that in calculating the right-hand side we may omit all forces which do no work in any infinitesimal change of configuration. Thus in the case of a conservative system free from external force we have by Art. 77 (12)

$$\Sigma m (\ddot{x}\,\delta x + \ddot{y}\,\delta y + \ddot{z}\,\delta z) + \delta V = 0. \quad \text{...........(3)}$$

The formula (2) or (3) is most useful in the theory of the small motions of continuous systems. It is freely employed, for instance, in Rayleigh's *Theory of Sound*. The following example may serve to illustrate the procedure.

Ex. In the case of a chain hanging vertically from one end treated in Art. 91, Ex. 3, we have

$$V = \tfrac{1}{2} g\rho \int_0^l (l - s)\, y'^2\, ds, \quad \text{..........................(4)}$$

where s denotes distance from the fixed end, and $y' = \partial y/\partial s$. Hence

$$\delta V = g\rho \int_0^l (l - s) y'\, \delta y'\, ds$$

$$= g\rho \Big[(l - s)\, y'\, \delta y \Big]_0^l - g\rho \int_0^l \frac{\partial}{\partial s} \{(l - s)\, y'\}\, \delta y\, ds. \quad \text{...........(5)}$$

The integrated term vanishes for $s=l$, and also for $s=0$, since $\delta y=0$ at the upper end. The equation (3) therefore takes the form

$$\int_0^l \left[\ddot{y} - g\frac{\partial}{\partial s}\{(l-s)\,y'\}\right]\delta y\, ds = 0. \quad\text{.....................(6)}$$

Since the function δy is arbitrary, subject only to the condition that it must vanish for $s=0$, the remaining factor of the integrand must vanish for all values of s. Hence

$$\ddot{y} = g\frac{\partial}{\partial s}\{(l-s)\,y'\}. \quad\text{.............................(7)}$$

This is equivalent to (15) of Art. 91, where x was measured from the *lower* end.

102. Transformation to General Coordinates.

We contemplate in the first instance a system such that any possible configuration can be completely specified by means of n independent coordinates $q_1, q_2, \ldots, q_n$.

The quantities δx, δy, δz vary by definition with the time, and it is important to notice that the operations δ and d/dt are commutative. For instance,

$$\delta\dot{x} = \frac{d}{dt}(x + \delta x) - \frac{dx}{dt} = \frac{d}{dt}\,\delta x. \quad\text{...............(1)}$$

Hence

$$\Sigma m\,(\ddot{x}\,\delta x + \ddot{y}\,\delta y + \ddot{z}\,\delta z)$$
$$= \frac{d}{dt}\Sigma m\,(\dot{x}\,\delta x + \dot{y}\,\delta y + \dot{z}\,\delta z) - \Sigma m\,(\dot{x}\,\delta\dot{x} + \dot{y}\,\delta\dot{y} + \dot{z}\,\delta\dot{z}).\ldots(2)$$

The formulæ for $\dot{x}$, $\dot{y}$, $\dot{z}$ in terms of the generalized coordinates and velocities have been written out in Art. 73. We have also

$$\left.\begin{aligned}
\delta x &= \frac{\partial x}{\partial q_1}\,\delta q_1 + \frac{\partial x}{\partial q_2}\,\delta q_2 + \ldots + \frac{\partial x}{\partial q_n}\,\delta q_n,\\
\delta y &= \frac{\partial y}{\partial q_1}\,\delta q_1 + \frac{\partial y}{\partial q_2}\,\delta q_2 + \ldots + \frac{\partial y}{\partial q_n}\,\delta q_n,\\
\delta z &= \frac{\partial z}{\partial q_1}\,\delta q_1 + \frac{\partial z}{\partial q_2}\,\delta q_2 + \ldots + \frac{\partial z}{\partial q_n}\,\delta q_n.
\end{aligned}\right\}\quad\text{............(3)}$$

If we substitute in the expression

$$\Sigma m\,(\dot{x}\,\delta x + \dot{y}\,\delta y + \dot{z}\,\delta z),\quad\text{.....................(4)}$$

the coefficient of δq_r is

$$\Sigma m\left(\dot{x}\frac{\partial x}{\partial q_r} + \dot{y}\frac{\partial y}{\partial q_r} + \dot{z}\frac{\partial z}{\partial q_r}\right) = a_{1r}\dot{q}_1 + a_{2r}\dot{q}_2 + \ldots + a_{nr}\dot{q}_n,\ldots(5)$$

in the notation of Art. 73. Hence, writing as before

$$p_r = \frac{\partial T}{\partial \dot{q}_r}, \quad \text{.........................(6)}$$

we have $$\Sigma m(\dot{x}\delta x + \dot{y}\delta y + \dot{z}\delta z) = \sum_r p_r \delta q_r, \quad \text{..............(7)}$$

where $\sum_r$ indicates a summation for values of r from 1 to n. This formula is fundamental.

Again,

$$\Sigma m(\dot{x}\delta\dot{x} + \dot{y}\delta\dot{y} + \dot{z}\delta\dot{z}) = \delta\Sigma\tfrac{1}{2}m(\dot{x}^2 + \dot{y}^2 + \dot{z}^2)$$

$$= \delta T = \sum_r \left(\frac{\partial T}{\partial \dot{q}_r}\delta\dot{q}_r + \frac{\partial T}{\partial q_r}\delta q_r\right). \quad \text{...(8)}$$

Substituting from (7) and (8) in (2), and omitting the terms which cancel by (6), we have

$$\Sigma m(\ddot{x}\delta x + \ddot{y}\delta y + \ddot{z}\delta z) = \sum_r \left(\frac{dp_r}{dt} - \frac{\partial T}{\partial q_r}\right)\delta q_r. \quad \text{......(9)}$$

With the same notation for generalized components of force as in Art. 77 we have

$$\Sigma(X\delta x + Y\delta y + Z\delta z) = \sum_r P_r \delta q_r. \quad \text{...........(10)}$$

The equation (2) of Art. 101 is thus transformed into

$$\sum_r \left(\frac{dp_r}{dt} - \frac{\partial T}{\partial q_r} - P_r\right)\delta q_r = 0. \quad \text{..............(11)}$$

Since the quantities δq_r are quite independent, this requires

$$\frac{dp_r}{dt} - \frac{\partial T}{\partial q_r} = P_r, \quad \text{.......................(12)}$$

which is the typical Lagrangian equation of Art. 77. The investigation now given is substantially that by which the equations were originally obtained.

When the geometrical relations vary with the time, we must employ the forms of $\dot{x}$, $\dot{y}$, $\dot{z}$ given in Art. 79 (2). We find that the coefficient of δq_r in the transformation of (4) is

$$\alpha_r + a_{1r}\dot{q}_1 + a_{2r}\dot{q}_2 + \ldots + a_{nr}\dot{q}_n. \quad \text{...........(13)}$$

Hence $$\Sigma m(\dot{x}\delta x + \dot{y}\delta y + \dot{z}\delta z) = \sum_r p_r \delta q_r, \quad \text{...........(14)}$$

where $$p_r = \frac{\partial T}{\partial \dot{q}_r}, \quad \text{..........................(15)}$$

since T has now the form given in Art. 79 (3). The rest of the work is as before.

103. Redundant Coordinates.

The advantages of Lagrange's variational method are not confined to the case where the coordinates $q_1, q_2, \ldots, q_n$, in terms of which the configuration is supposed expressed, are all independent.

We may in the first instance suppose them to be connected by m relations, less than n in number, say

$$A(t, q_1, q_2, \ldots, q_n) = 0, \quad B(t, q_1, q_2, \ldots, q_n) = 0, \ldots \qquad (1)$$

These may be interpreted as introducing partial constraints in a system previously free. The variations δq_r in the formula (11) of the preceding Art. are no longer independent, but are connected by the relations

$$\sum_r \frac{\partial A}{\partial q_r} \delta q_r = 0, \quad \sum_r \frac{\partial B}{\partial q_r} \delta q_r = 0. \qquad (2)$$

Introducing undetermined multipliers $\lambda, \mu, \ldots$, one for each of these equations, we have

$$\sum_r \left(\frac{dp_r}{dt} - \frac{\partial T}{\partial q_r} - P_r - \lambda \frac{\partial A}{\partial q_r} - \mu \frac{\partial B}{\partial q_r} - \ldots \right) \delta q_r = 0. \qquad (3)$$

As in the theory of conditional maxima and minima we may suppose $\lambda, \mu, \ldots$, to be adjusted so as to annul the coefficients of m of these variations δq_r. The remaining $n - m$ variations can then be regarded as independent, so that their coefficients separately vanish. Thus we obtain n equations of the type

$$\frac{dp_r}{dt} - \frac{\partial T}{\partial q_r} = P_r + \lambda \frac{\partial A}{\partial q_r} + \mu \frac{\partial B}{\partial q_r} + \ldots. \qquad (4)$$

These equations, in conjunction with (1), determine the n coordinates $q_1, q_2, \ldots, q_n$, and the m multipliers $\lambda, \mu, \ldots$.

Ex. 1. As a simple example, take the case of a particle constrained by a moving surface

$$\phi(x, y, z, t) = 0. \qquad (5)$$

The equations are

$$m\ddot{x} = X + \lambda \frac{\partial \phi}{\partial x}, \quad m\ddot{y} = Y + \lambda \frac{\partial \phi}{\partial y}, \quad m\ddot{z} = Z + \lambda \frac{\partial \phi}{\partial z}. \qquad (6)$$

If the surface is smooth, its reaction on the particle is not included in the forces X, Y, Z, since it does no work. The reaction (R) is in fact represented by the undetermined multiplier λ; thus

$$R = \lambda \left\{ \left(\frac{\partial \phi}{\partial x}\right)^2 + \left(\frac{\partial \phi}{\partial y}\right)^2 + \left(\frac{\partial \phi}{\partial z}\right)^2 \right\}. \qquad (7)$$

Again, it may happen that, although there are no invariable relations between the coordinates $q_1, q_2, \ldots, q_n$, yet from the circumstances of the particular problem certain geometrical relations are imposed on their *variations*, e.g.

$$\sum_r (A_r \delta q_r) = 0, \quad \sum_r (B_r \delta q_r) = 0, \ldots, \qquad \text{.........(8)}$$

where the coefficients A_r, B_r, ... are functions of $q_1, q_2, \ldots, q_n$, and (possibly) of t. It is assumed that these equations are not integrable; otherwise we fall back on the previous case.

The equations of motion are obtained as before by the use of indeterminate multipliers. Thus

$$\frac{d}{dt}\frac{\partial T}{\partial \dot{q}_r} - \frac{\partial T}{\partial q_r} = P_r + \lambda A_r + \mu B_r + \ldots. \qquad \text{.........(9)}$$

The coordinates $q_1, q_2, \ldots, q_n$ and the multipliers λ, μ, ... are determined by these equations and by the velocity conditions corresponding to (7), viz.

$$\sum_r A_r \dot{q}_r = 0, \quad \sum_r B_r \dot{q}_r = 0, \ldots. \qquad \text{.........(10)}$$

Systems in which the restrictions are of the type (1) were designated by Hertz* as 'holonomous,' the meaning being that the constraints are expressed by *integral* as distinguished from *differential* relations. Except for mathematical convenience we might by elimination express any possible configuration in terms of the $n - m$ independent coordinates.

Systems, on the other hand, which are subject to non-integrable conditions of the type (8) are called 'non-holonomous.' Cases of this kind may occur when we have a solid rolling (without sliding) on a fixed or moving surface.

Ex. 2. Suppose we have a solid moving in contact with a smooth horizontal plane. Any position whatever of the solid may be specified by the Cartesian coordinates x, y, z of the mass-centre and Euler's angular coordinates θ, ψ, ϕ. Let θ be measured from the axis of z, which we will suppose drawn vertically upwards. We have then

$$2T = 2T_1 + M(\dot{x}^2 + \dot{y}^2 + \dot{z}^2), \qquad \text{.........(11)}$$

and

$$V = Mgz, \qquad \text{.........(12)}$$

where T_1 is the kinetic energy of the motion relative to the mass-centre, and is therefore a function of the angular coordinates and velocities.

* H. Hertz (1857—94), *Prinzipien der Mechanik*, Leipzig, 1894.

The geometrical relation, expressing contact with the plane, has the form

$$z=f(\theta, \phi). \qquad \text{.................................(13)}$$

This is of the 'holonomous' type; and it is immaterial whether we eliminate z and $\dot{z}$ by means of this equation, and so reduce the number of coordinates to five, or whether we retain all six coordinates and use an indeterminate multiplier. The results obtained in these two ways will be consistent.

Ex. 3. If the plane be 'rough,' and the motion of the solid therefore restricted to be one of pure rolling, the vertical coordinate z can be eliminated as before, but the fact that the point of the solid which is in contact with the plane is momentarily at rest imposes two additional conditions of a non-integrable character.

To take a fairly simple example, let the outer surface of the solid be spherical, of radius a, and let the mass-centre coincide with the centre of figure. Let us further suppose that the (fixed) plane of zx is that from which the angle ψ of Euler's scheme is measured.

In a small displacement the sphere rotates about an instantaneous axis through the point of contact. A small variation in the value of θ alone displaces the centre through a horizontal distance $a\,\delta\theta$ in the plane of ZC (Fig. 32, p. 83). The components of this parallel to x and y are $a\,\delta\theta\cos\psi$ and $a\,\delta\theta\sin\psi$, respectively. A small change in the value of ψ alone does not affect the position of the centre. A rotation $\delta\phi$ about OC has a horizontal component $\delta\phi\sin\theta$, and if this be transferred to an axis through the point of contact, the consequent displacements of the centre are $a\,\delta\phi\sin\theta\sin\psi$ and $-a\,\delta\phi\sin\theta\cos\psi$ parallel to x and y, respectively. The required relations are therefore

$$\left.\begin{aligned} \delta x - a\cos\psi\,\delta\theta - a\sin\theta\sin\psi\,\delta\phi = 0, \\ \delta y - a\sin\psi\,\delta\theta + a\sin\theta\cos\psi\,\delta\phi = 0. \end{aligned}\right\} \qquad \text{..................(14)}$$

Although problems of this kind are generally best treated in other ways, we may proceed to write down the dynamical equations of the present method, on the assumption that the sphere, though not necessarily homogeneous, has kinetic symmetry about an axis.

We have then

$$2T = M(\dot{x}^2+\dot{y}^2) + A(\dot{\theta}^2 + \sin^2\theta\,\dot{\psi}^2) + C(\dot{\phi}+\cos\theta\,\dot{\psi})^2, \qquad \text{.........(15)}$$

whilst V is constant. The method gives

$$M\ddot{x} = \lambda, \quad M\ddot{y} = \mu, \qquad \text{..............................(16)}$$

$$A\ddot{\theta} - A\sin\theta\cos\theta\,\dot{\psi}^2 + C(\dot{\phi}+\cos\theta\,\dot{\psi})\sin\theta\,\dot{\psi} = -\lambda a\cos\psi - \mu a\sin\psi, \quad (17)$$

$$\frac{d}{dt}\{A\sin^2\theta\,\dot{\psi} + C\cos\theta\,(\dot{\phi}+\cos\theta\,\dot{\psi})\} = 0, \qquad \text{..............(18)}$$

$$\frac{d}{dt}C(\dot{\phi}+\cos\theta\,\dot{\psi}) = -\lambda a\sin\theta\sin\psi + \mu a\sin\theta\cos\psi. \qquad \text{.........(19)}$$

These, together with

$$\left.\begin{aligned} \dot{x} = a\cos\psi\,\dot{\theta} + a\sin\theta\sin\psi\,\dot{\phi}, \\ \dot{y} = a\sin\psi\,\dot{\theta} - a\sin\theta\cos\psi\,\dot{\phi}, \end{aligned}\right\} \qquad \text{.....................(20)}$$

are the equations of the problem.

The equation (18) expresses the constancy of angular momentum about the vertical diameter (cf. Art. 39). The quantities λ, μ are recognized as the horizontal components of the reaction at the point of contact.

If we were to substitute the values of $\dot{x}$ and $\dot{y}$ from (20) in the expression for the kinetic energy we should obtain

$$2T = Ma^2(\dot{\theta}^2 + \sin^2\theta\,\dot{\phi}^2) + A(\dot{\theta}^2 + \sin^2\theta\,\dot{\psi}^2) + C(\dot{\phi} + \cos\theta\,\dot{\psi})^2. \quad \text{...(21)}$$

This formula is correct, but it would be improper to use it as a basis for Lagrange's equations, since the coordinates θ, ψ, ϕ which are now alone in evidence do not completely define the position of the body.

A general discussion of the above equations would be difficult, and hardly interesting, but we may at least deduce the conditions for steady motion, with the axis of symmetry at a constant inclination to the vertical.

Differentiating (20) on this hypothesis that θ is constant, we have from (16)

$$\left.\begin{aligned} \lambda &= Ma\sin\theta\,(\sin\psi\ddot{\phi} + \cos\psi\dot{\psi}\dot{\phi}), \\ \mu &= Ma\sin\theta\,(-\cos\psi\ddot{\phi} + \sin\psi\dot{\psi}\dot{\phi}). \end{aligned}\right\} \quad \text{..................(22)}$$

Substituting in (17), and dividing by $\sin\theta\dot{\psi}$, we find

$$\dot{\phi} = \frac{A-C}{C+Ma^2}\cos\theta\dot{\psi} = m\dot{\psi}, \text{ say.} \quad \text{..................(23)}$$

The equations (20) now give

$$\left.\begin{aligned} x &= -ma\sin\theta\cos\psi + \alpha, \\ y &= -ma\sin\theta\sin\psi + \beta. \end{aligned}\right\} \quad \text{........................(24)}$$

The centre of the solid therefore describes a circle of radius

$$ma\sin\theta = \frac{(A-C)\,a\sin\theta\cos\theta}{C+Ma^2}. \quad \text{......................(25)}$$

104. Principle of Least Action.

The preceding methods give us statements applicable to a dynamical system at any given instant of its motion. We now come to a series of theorems which predicate certain characteristics of its career as a whole, between any two assigned configurations through which it passes. We compare the actual motion with other imaginable (though not necessarily natural) modes of transit, differing infinitely little from it, between the same two configurations.

As in Art. 101 the symbol δ prefixed to any coordinate, or any velocity, will be used to indicate the difference between the values of that coordinate or that velocity, *at the same instant t of time*, in the varied and in the natural motion of the system, respectively. We have seen that on this supposition the operations δ and d/dt are commutative.

The symbol Δ, on the other hand, will be used to denote the variation in the value of any property of the career *as a whole.*

The best known, though not the simplest, theorem of the class now under consideration is that of 'Least Action,' originated by Maupertuis*, but first put in a definite form by Lagrange and Jacobi. The 'action' of a single particle in passing from one position to another on a given path may be defined as the space-integral of the momentum. The action of a dynamical system, i.e. the sum of the actions of its individual particles, is accordingly

$$A = \Sigma \int mv\, ds, \quad \text{......................(1)}$$

where v is the velocity of any particle m, and ds an element of its path. We may also write

$$A = \Sigma \int m(\dot{x}dx + \dot{y}dy + \dot{z}dz) = \sum_r \int p_r dq_r, \quad \text{......(2)}$$

by Art. 102 (7).

Since $ds = v\,dt$ an equivalent definition of the action is that it is the time-integral of the *vis viva* (i.e. twice the kinetic energy). Thus

$$A = \Sigma \int_0^\tau mv^2 dt = 2\int_0^\tau T dt, \quad \text{.................(3)}$$

where τ is the time of transit.

The theorem to be proved asserts that the free motion of a conservative system between any two assigned configurations is characterized by the property that

$$\Delta A = 0, \quad \text{.............................(4)}$$

provided the total energy has the same constant value in the varied motion as in the actual motion. In the case of a single particle, for instance, moving in a given field of force, we may imagine it to be started with the same velocity as in the natural motion, and to be guided by a smooth tube which begins and ends at the same two terminal positions, but deviates slightly from the free path.

* P. L. M. de Maupertuis (1698—1759), the 'first French Newtonian.'

Since the time of transit will in general be altered in the varied motion we have

$$\Delta A = 2\int_0^{\tau+\Delta\tau}(T+\delta T)\,dt - 2\int_0^{\tau} T\,dt$$

$$= 2T'\Delta\tau + 2\int_0^{\tau}\delta T\,dt, \qquad \text{.......................(5)}$$

where in the first term T' denotes the final value of the kinetic energy. A term of the second order is here neglected. Now by partial integration

$$\int_0^{\tau}\delta T\,dt = \Sigma m\int_0^{\tau}(\dot{x}\,\delta\dot{x} + \dot{y}\,\delta\dot{y} + \dot{z}\,\delta\dot{z})\,dt$$

$$= \Big[\Sigma m\,(\dot{x}\,\delta x + \dot{y}\,\delta y + \dot{z}\,\delta z)\Big]_0^{\tau}$$

$$- \Sigma m\int_0^{\tau}(\ddot{x}\,\delta x + \ddot{y}\,\delta y + \ddot{z}\,\delta z)\,dt$$

$$= \Big[\Sigma m(\dot{x}\,\delta x + \dot{y}\,\delta y + \dot{z}\,\delta z)\Big]_0^{\tau} + \int_0^{\tau}\delta V\,dt, \text{...(6)}$$

by Art. 101 (3). Hence (5) may be written

$$\Delta A = 2T'\Delta\tau + \int_0^{\tau}\delta(T+V)\,dt + \Big[\Sigma m(\dot{x}\,\delta x + \dot{y}\,\delta y + \dot{z}\,\delta z)\Big]_0^{\tau}. \qquad \text{.........(7)}$$

We have $\delta(T+V)=0$ by hypothesis. Again, the varied motion and the actual motion are both assumed to start together from the same configuration, so that δx, δy, δz vanish for $t=0$. At time $t=\tau$ the system, in the varied motion, is not in the final configuration, but will reach it in an additional (positive or negative) time $\Delta\tau$. Hence at this limit we must put

$$\delta x = -\dot{x}\,\Delta\tau, \quad \delta y = -\dot{y}\,\Delta\tau, \quad \delta z = -\dot{z}\,\Delta\tau.$$

The last term in (7) therefore cancels the first. Hence $\Delta A = 0$, as was to be proved.

It is to be noticed that the equation (4) merely expresses that the variation of A vanishes to the *first order* of small quantities. The phrase 'stationary action' has therefore been proposed as indicating more accurately what is established. The action in the free path between two given configurations is in fact not invariably a minimum, and even when a minimum it need not be the *least possible* subject to the given condition.

Good illustrations are furnished by the case of a single particle. For instance, a particle moving on a smooth surface, and subject only to the reaction of the surface, will have its velocity constant. The theorem therefore resolves itself into the statement that

$$\Delta \int ds = 0, \qquad \text{.........................(8)}$$

i.e. the path is a geodesic, or 'straightest' line. A geodesic is however not necessarily the shortest path between any two given points on it. For example, on the sphere a great-circle arc ceases to be the shortest path between its extremities when it exceeds 180°.

More generally, taking any surface, let a point P, starting from O, move along a geodesic. This geodesic will be a minimum path from O to P until P passes through a point O' (if such exist) which is the intersection with a consecutive geodesic from O. After this point the minimum property ceases. On an anticlastic surface (where the principal curvatures have opposite signs) two geodesics cannot intersect more than once, and each geodesic is therefore a minimum path between any two of its points.

The general criterion, applicable to all dynamical systems, is as follows. Let O and P denote any two configurations on a natural path* of the system. If this be the sole free path from O to P with the prescribed total energy, the action from O to P is a minimum. But if there be several distinct paths, let P vary from coincidence with O along the first-named path; the action will cease to be a minimum when a configuration O' is reached such that two of the possible paths from O to O' coincide. For instance, if O and P be positions on the parabolic path of a projectile under gravity, there will be a second path (with the same energy and therefore the same velocity of projection from O), these two paths coinciding when P is at the other extremity (O', say) of the focal chord through O. The action from O to P will therefore be a minimum for all positions of P short of O'†. Two configurations which are related as O and O' in the above statement have been called conjugate 'kinetic foci.'

In any case of motion of a particle in a conservative field of force with prescribed energy the principle takes the form

$$\Delta \int v\, ds = 0. \qquad \text{.........................(9)}$$

For instance, on the corpuscular theory of light the velocity of a corpuscle in any medium depends only on its position, being

* By the 'path' is here meant the continuous succession of configurations through which the system passes.

† The above illustrations were given by Jacobi in his *Vorlesungen über Dynamik* (published 1866). The lectures date from 1842.

proportional to the corresponding value of the refractive index (μ). Hence

$$\Delta \int \mu ds = 0. \quad \text{.......................(10)}$$

In the terminology of Geometrical Optics the integral is the 'optical distance' between two points of a ray.

105. Hamilton's Principle.

In the preceding theorem the energy of the hypothetical motion is prescribed, whilst the time of transit from the initial to the final configuration is variable. In another, and generally more convenient, theorem the time of transit is prescribed to be the same as in the actual motion, whilst the energy in the varied path will in general be different and need not indeed be constant. On this understanding we shall have

$$\Delta \int_0^\tau (T - V)\,dt = 0, \quad \text{or} \quad \Delta \int_0^\tau L\,dt = 0, \quad \text{........(1)}$$

where L is the 'Lagrangian Function' of Art. 77.

For

$$\begin{aligned}\int_0^\tau (\delta T - \delta V)\,dt &= \int_0^\tau \{\Sigma m\,(\dot{x}\,\delta\dot{x} + \dot{y}\,\delta\dot{y} + \dot{z}\,\delta\dot{z}) - \delta V\}\,dt \\ &= \Big[\Sigma m\,(\dot{x}\,\delta x + \dot{y}\,\delta y + \dot{z}\,\delta z)\Big]_0^\tau \\ &\quad - \int_0^\tau \{\Sigma m\,(\ddot{x}\,\delta x + \ddot{y}\,\delta y + \ddot{z}\,\delta z) + \delta V\}\,dt. \quad \text{...(2)}\end{aligned}$$

The integrated terms vanish at both limits, since by hypothesis the corresponding configurations are fixed; whilst the terms under the integral sign vanish by Lagrange's variational equation.

The fact that in (1) the variation does not affect the time of transit renders the formula easy of application in any system of coordinates.

Thus, to deduce Lagrange's equations, we have

$$\begin{aligned}\int_0^\tau (\delta L)\,dt &= \int_0^\tau \sum_r \left(\frac{\partial L}{\partial \dot{q}_r}\,\delta\dot{q}_r + \frac{\partial L}{\partial q_r}\,\delta q_r\right) dt \\ &= \left[\sum_r \frac{\delta L}{\partial \dot{q}_r}\,\delta q_r\right]_0^\tau - \int_0^\tau \sum_r \left\{\frac{d}{dt}\frac{\partial L}{\partial \dot{q}_r} - \frac{\partial L}{\partial q_r}\right\}\delta q_r. \quad \text{......(3)}\end{aligned}$$

The integrated terms vanish at both limits; and in order that the remainder of the last member may vanish it is necessary that the coefficients of the variations δq_r, which are independent and subject only to the condition of vanishing at the limits of integration, should severally vanish. We are thus led to Lagrange's typical equation of motion, Art. 77 (15).

It appears that the formula (1) is a convenient as well as a compact embodiment of the whole of ordinary Dynamics.

106. Varying Action. The 'Characteristic Function.'

We were concerned in Arts. 104, 105 with the free motion of a conservative system between two configurations through which it passes, as compared with other arbitrary modes of motion between the same configurations. Thus the 'action' was shewn to be unaltered, to the first order, when we substitute for the free motion any slightly different mode of transit between the same two configurations and with the same total energy.

We now investigate the change in the value of the 'action' when we compare the actual motion with another *natural* (i.e. unconstrained) motion in which the initial and final configurations, and the prescribed value of the energy, are all slightly varied.

Let
$$A = \int_t^{t'} T dt, \quad \text{.........................(1)}$$

where t and t' are the times of passing through the initial and final configurations, and let h be total energy, in the original motion. We find, by a similar investigation to that of Art. 104,

$$\Delta A = 2T'\Delta t' - 2T\Delta t + \int_t^{t'} \delta(T+V)\, dt + \Sigma m (\dot{x}'\delta x' + \dot{y}'\delta y' + \dot{z}'\delta z') - \Sigma m(\dot{x}\delta x + \dot{y}\delta y + \dot{z}\delta z), \quad \text{......(2)}$$

where accented letters refer to the instant t', and unaccented to the instant t.

According to our scheme of notation, the varied terminal configurations are passed through at the instants $t + \Delta t$ and $t' + \Delta t'$, and the corresponding positions of a particle m of the system are

$$(x + \Delta x,\ y + \Delta y,\ z + \Delta z) \quad \text{and} \quad (x' + \Delta x',\ y' + \Delta y',\ z' + \Delta z').$$

Now if A, A' be the initial and final positions of m in its original path, and B, B' the varied terminal positions, and if α, α' are the positions in the varied path at the instants t, t', respectively, we have the vector equations

$$AB = A\alpha + \alpha B, \quad A'B' = A'\alpha' + \alpha' B'.$$

Projecting these vectors on the coordinate axes we have

$$\Delta x = \delta x + \dot{x}\Delta t, \quad \Delta y = \delta y + \dot{y}\Delta t, \quad \Delta z = \delta z + \dot{z}\Delta t,$$
$$\Delta x' = \delta x' + \dot{x}'\Delta t', \; \Delta y' = \delta y' + \dot{y}'\Delta t', \; \Delta z' = \delta z' + \dot{z}'\Delta t'.$$

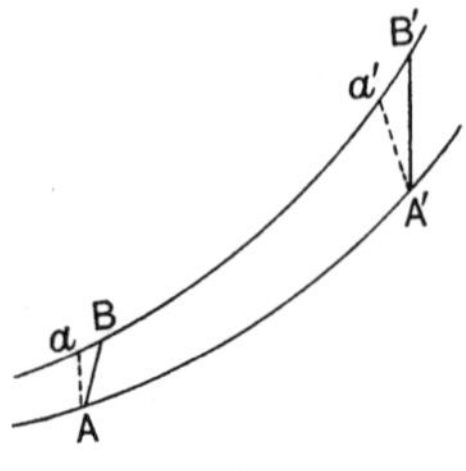

Fig. 62.

Hence, substituting for δx, $\delta x'$, ... in (2), and putting $t' - t = \tau$, we find

$$\Delta A = \tau\,\delta h + \Sigma m\,(\dot{x}'\Delta x' + \dot{y}'\Delta y' + \dot{z}'\Delta z')$$
$$- \Sigma m\,(\dot{x}\,\Delta x + \dot{y}\,\Delta y + \dot{z}\,\Delta z). \quad \text{......(3)}$$

In generalized coordinates this becomes by Art. 102

$$\Delta A = \tau\,\delta h + \sum_r p'\Delta q' - \sum_r p\Delta q. \quad \text{..............(4)}$$

Now selecting two arbitrary configurations, it is possible under certain limitations* to start the system from one of these with a prescribed value h of the total energy, so that it shall pass through the other. Regarding the 'action' A, then, as a function of the initial and final coordinates and the energy, we have

$$p_r' = \frac{\partial A}{\partial q_r'}, \quad p_r = -\frac{\partial A}{\partial q_r}, \quad \text{..................(5)}$$

and

$$\tau = \frac{\partial A}{\partial h}. \quad \text{..........................(6)}$$

The function A, as thus defined, was called by Hamilton the 'Characteristic Function.' Its form, if known, would determine all the mechanical properties of the system. Thus if the initial coordinates q_r and impulses p_r were given, the n equations

$$\frac{\partial A}{\partial q_r} = -p_r, \quad \text{..........................(7)}$$

* The kind of limitation is illustrated by the case of a projectile. The second position must be 'within range.'

together with (6), would determine h, and the values q_r' of the coordinates after any assigned interval τ. The final momenta would then be given by

$$p_r' = \frac{\partial A}{\partial q_r'}. \quad \text{..........................(8)}$$

The differential equation to determine A is given in Art. 110.

Ex. If U denotes the optical distance between the points (x, y, z) and (x', y', z') in a medium whose (variable) refractive index is μ, viz.

$$U = \int \mu \, ds, \quad \text{..............................(9)}$$

we have on the corpuscular theory $\mu = v$, the velocity of light *in vacuo* being taken as unity. Hence, from (5),

$$\mu \frac{dx'}{ds'} = \frac{\partial U}{\partial x'}, \quad \mu \frac{dy'}{ds'} = \frac{\partial U}{\partial y'}, \quad \mu \frac{dz'}{ds'} = \frac{\partial U}{\partial z'}, \quad \text{..............(10)}$$

where ds' is an element of the path of the ray. Hence

$$\left(\frac{\partial U}{\partial x'}\right)^2 + \left(\frac{\partial U}{\partial y'}\right)^2 + \left(\frac{\partial U}{\partial z'}\right)^2 = \mu^2, \quad \text{....................(11)}$$

which is a particular case of Hamilton's equation, Art. 110 (21). These equations are important in Hamilton's treatment of Geometrical Optics. The physical meaning of the function U on the wave theory is of course different; it measures time of propagation, and not 'action.' Accordingly the basis of the formulae is then Fermat's principle of 'Least Time,' instead of 'Least Action.' See Art. 111.

107. Hamilton's 'Principal Function.'

The stationary property of the integral

$$S = \int_t^{t'} L \, dt = \int_t^{t'} (T - V) \, dt,$$

when taken between fixed configurations on a natural path, and subject to a given time of transit, was proved in Art. 105. We now consider the effect of slight changes in the terminal configurations, and in the prescribed time of transit.

We have, evidently,

$$\Delta S = (T' - V') \Delta t' - (T - V) \Delta t + \int_t^{t'} \delta (T - V) \, dt. \quad \text{...(1)}$$

Now from Art. 105, we have

$$\int_t^{t'} \delta (T - V) \, dt = \Sigma m (\dot{x}' \delta x' + \dot{y}' \delta y' + \dot{z}' \delta z') - \Sigma m (\dot{x} \delta x + \dot{y} \delta y + \dot{z} \delta z). \quad \text{.........(2)}$$

Hence, substituting for δx, $\delta x'$, ... as in Art. 106, we find

$$\Delta S = -H\Delta\tau + \Sigma m(\dot{x}'\Delta x' + \dot{y}'\Delta y' + \dot{z}'\Delta z') - \Sigma m(\dot{x}\Delta x + \dot{y}\Delta y + \dot{z}\Delta z), \quad \text{.........(3)}$$

where

$$H = T + V, \quad \tau = t' - t, \quad \text{..................(4)}$$

i.e. H is the total energy, and τ the time of transit.

In generalized coordinates (3) takes the form

$$\Delta S = -H\Delta\tau + \sum_r p_r'\Delta q_r' - \sum_r p_r \Delta q_r. \quad \text{.........(5)}$$

Hence, regarding S as a function of the initial and final coordinates and the time of transit, we have

$$p_r' = \frac{\partial S}{\partial q_r'}, \quad p_r = -\frac{\partial S}{\partial q_r}, \quad \text{..................(6)}$$

and

$$H = -\frac{\partial S}{\partial \tau}. \quad \text{..........................(7)}$$

It should be noticed that in (5), and in the cognate formulæ (3) of Art. 106, the coordinates employed to specify the initial and final configurations, respectively, need not be chosen on the same plan. This appears from the fact that the expression

$$\sum_r p_r \delta q_r$$

is invariant; see Art. 102 (7).

The function S, like the function A of the preceding Art. 106, if once known, would enable us to trace completely the result of any initial conditions. If the initial coordinates q_r and impulses p_r are given, the n equations

$$\frac{\partial S}{\partial q_r} = -p_r \quad \text{..........................(8)}$$

determine the values of q_r' after a time τ. The n equations

$$\frac{\partial S}{\partial q_r'} = p_r' \quad \text{.............................(9)}$$

then determine the corresponding momenta.

For the differential equation satisfied by S see Art. 110.

108. Extension to Cyclic Systems.

The corresponding theory for cyclic systems has been given by Larmor*. We recall the definition of the function R in Art. 83, viz.

$$R = T - \kappa\dot{\chi} - \kappa'\dot{\chi}' - \kappa''\dot{\chi}'' - \ldots, \quad \text{...............(1)}$$

which is supposed expressed in terms of the 'palpable' coordinates

* *Proc. Lond. Math. Soc.* (1), t. xv (1884).

$q_1, q_2, \ldots, q_m$, and the corresponding velocities, and the constant momentum κ of the cyclic motions. It was proved that

$$p_r = \frac{\partial R}{\partial \dot{q}_r}, \quad \ldots\ldots\ldots\ldots\ldots\ldots\ldots\ldots(2)$$

and that R satisfies m equations of the type

$$\frac{d}{dt}\frac{\partial R}{\partial \dot{q}_r} - \frac{\partial R}{\partial q_r} = -\frac{\partial V}{\partial q_r}. \quad \ldots\ldots\ldots\ldots\ldots\ldots(3)$$

Hence

$$\int_t^{t'} \delta(R - V)\,dt = \sum_r \int_t^{t'} \left(\frac{\partial R}{\partial \dot{q}_r}\delta\dot{q}_r + \frac{\partial R}{\partial q_r}\delta q_r - \frac{\partial V}{\partial q_r}\delta q_r\right) dt$$

$$= \sum_r p_r'\delta q_r' - \sum_r p_r \delta q_r, \quad \ldots\ldots\ldots\ldots\ldots\ldots(4)$$

after a partial integration.

Larmor takes as the modified cyclic function

$$S_1 = \int_t^{t'} (R - V)\,dt, \quad \ldots\ldots\ldots\ldots\ldots\ldots(5)$$

supposed to be expressed in terms of the initial and final configurations (so far as these depend on the palpable coordinates q_r) and the time τ of transit. When these elements are variable, we have

$$\Delta S_1 = (R' - V')\,\Delta t' - (R - V)\,\Delta t + \sum_r p_r'(\Delta q_r' - \dot{q}_r'\Delta t')$$

$$- \sum_r p_r(\Delta q_r - \dot{q}_r \Delta t)$$

$$= (R' - V' - \sum_r p_r'\dot{q}_r')\,\Delta t' - (R - V - \sum_r p_r \dot{q}_r)\,\Delta t$$

$$+ \sum_r p_r'\Delta q_r' - \sum_r p_r \Delta q_r. \quad \ldots\ldots(6)$$

Now since

$$2T = \sum_r p_r \dot{q}_r + \kappa\dot{\chi} + \kappa'\dot{\chi}' + \kappa''\dot{\chi}'' + \ldots, \quad \ldots\ldots\ldots(7)$$

we have from (1) $$\sum_r p_r \dot{q}_r = R + T. \quad \ldots\ldots\ldots\ldots\ldots\ldots(8)$$

Hence

$$\Delta S_1 = -H'\Delta t' + H\Delta t + \sum_r p_r'\Delta q_r' - \sum_r p_r \Delta q_r$$

$$= -H\delta\tau + \sum_r p_r'\Delta q_r' - \sum_r p_r \Delta q_r, \quad \ldots\ldots\ldots\ldots(9)$$

ultimately, since terms of the second order are neglected. As in Art. 107, H stands for the total energy, which is constant along each natural path.

If $\Delta q_r = 0$, $\Delta q_r' = 0$, $\Delta\tau = 0$, we have $\Delta S_1 = 0$, which is the generalization of Hamilton's Principle (Art. 105). When the terminal configurations and τ are variable, we have

$$p_r' = \frac{\partial S_1}{\partial q_r'}, \quad p_r = -\frac{\partial S}{\partial q_r}, \quad \text{.................(10)}$$

and

$$\frac{\partial S_1}{\partial \tau} = -H. \quad \text{.......................(11)}$$

The modified form of the 'Characteristic Function' A is

$$A_1 = \int_t^{t'} (2T - \kappa\dot{\chi} - \kappa'\dot{\chi}' - \kappa''\dot{\chi}'' - \ldots)\,dt = \int_t^{t'} (R + T)\,dt. \quad \text{...(12)}$$

This is supposed expressed in a similar manner to S_1, except that the time of transit is replaced by the total energy as an independent variable. By comparison with (5)

$$A_1 = S_1 + \int_t^{t'} (T + V)\,dt = S_1 + h\tau, \quad \text{...........(13)}$$

where h is the prescribed value of H for each natural path considered. Hence

$$\begin{aligned} \Delta A_1 &= \Delta S_1 + \tau\Delta h + h\Delta\tau \\ &= \tau\Delta h + \sum_r p_r' \Delta q_r' - \sum_r p_r \Delta q_r. \quad \text{..............(14)} \end{aligned}$$

This leads to the equations

$$p_r' = \frac{\partial A_1}{\partial q_r'}, \quad p_r = -\frac{\partial A}{\partial q_r}, \quad \text{.................(15)}$$

and

$$\frac{\partial A}{\partial h} = \tau, \quad \text{.........................(16)}$$

having the same form as in Art. 106.

109. Reciprocal Theorems.

We may employ Hamilton's principal function to prove a remarkable formula connecting any *two* slight perturbations of a *natural* motion of a system. Using now the symbols δ and Δ to denote corresponding variations at time t, the theorem, given by Lagrange in the *Mécanique Analytique,* is that

$$\frac{d}{dt} \sum_r (\delta p_r \,.\, \Delta q_r - \Delta p_r \,.\, \delta q_r) = 0, \quad \text{...........(1)}$$

where the summation $\sum_r$ extends over all the degrees of freedom.

If we integrate from t to t', we have the equivalent form

$$\sum_r (\delta p_r' \,.\, \Delta q_r' - \Delta p_r' \,.\, \delta q_r') = \sum_r (\delta p_r \,.\, \Delta q_r - \Delta p_r \,.\, \delta q_r). \quad \ldots(2)$$

Let us write for shortness

$$\frac{\partial^2 S}{\partial q_r \partial q_s} = (r, s), \quad \frac{\partial^2 S}{\partial q_r \partial q_s'} = (r, s'), \quad \ldots\ldots\ldots\ldots\ldots(3)$$

where S is Hamilton's 'principal function.' Hence, by Art. 107 (6),

$$\left.\begin{aligned} \delta p_r &= -\sum_s (r, s)\, \delta q_s - \sum_s (r, s')\, \delta q_s', \\ \Delta p_r &= -\sum_s (r, s)\, \Delta q_s - \sum_s (r, s')\, \Delta q_s'. \end{aligned}\right\} \quad \ldots\ldots\ldots(4)$$

The second member of (2) thus becomes

$$= \sum_r \sum_s (r, s') \{\delta q_r \,.\, \Delta q_s' - \Delta q_r \,.\, \delta q_s'\}. \quad \ldots\ldots\ldots(5)$$

Identically the same expression is obtained in like manner for the first member. Hence the theorem.

The coordinates employed on the two sides of (2) need not belong to the same system. This follows from a remark made in Art. 107.

From the formula (2) we can derive* some remarkable reciprocal relations first obtained, in a different manner, by Helmholtz (1886).

Consider any natural motion of a conservative system between two configurations O and O' through which it passes at times t and t', respectively, and let $t' - t = \tau$. As the system is passing through O, let a small impulse δp_r be applied to it, and let the corresponding alteration in the coordinate q_s after the time τ be $\delta q_s'$. Next consider the *reversed* motion of the system, in which it would, if undisturbed, pass from O' to O in the same time τ. Let a small impulse $\delta p_s'$ be applied as the system is passing through O', and let the consequent change in the coordinate q_r after the time τ be δq_r. Helmholtz's first theorem is that

$$\delta q_r : \delta p_s' = \delta q_s' : \delta p_r. \quad \ldots\ldots\ldots\ldots\ldots\ldots\ldots(6)$$

To prove this, suppose that in (2) all the variations δq vanish, and likewise all the δp with the exception of δp_r. Further, suppose all the $\Delta q'$ to vanish, and all the $\Delta p'$ except $\Delta p_s'$. The formula (2) then gives

$$\delta p_r \,.\, \Delta q_r = -\Delta p_s' \,.\, \delta q_s'. \quad \ldots\ldots\ldots\ldots\ldots\ldots\ldots(7)$$

* *Proc. Lond. Math. Soc.*, t. xix, p. 144 (1888).

This is equivalent to (6), since we may suppose the symbol Δ to refer to the reversed motion, provided we reverse the sign of the Δp.

If the coordinates q_r, q_s are of the same geometrical character, the statement (6) may be simplified by putting $\delta p_s' = \delta p_r$, whence

$$\delta q_r = \delta q_s'. \quad \text{.........................(8)}$$

Ex. 1. In elliptic motion about the centre, if a small velocity δv in the direction of the normal be communicated to the particle as it passes through either extremity of the major axis, the tangential deviation produced after a quarter-period is easily found to be $\delta v/\sqrt{\mu}$, where μ is the 'absolute' acceleration. It is readily verified that a tangential velocity δv communicated at the extremity of the minor axis produces after a quarter-period the normal deviation $\delta v/\sqrt{\mu}$, in accordance with (8).

Ex. 2. Let O, O' be two points on the path of a particle in a conservative field of force.

First, suppose that the field is symmetrical about an axis, and that O, O' are on this axis, and let v, v' be the corresponding velocities. Let a small impulse $v\delta\theta$ be applied at O at right angles to the axis, so as to produce an angular deflection $\delta\theta$, and let the consequent lateral deviation at O' be β'. In like manner, in the reversed motion, let a lateral impulse $v'\delta\theta'$ at O' produce a lateral deviation β at O. The theorem (6) asserts that

$$\frac{\beta}{v'\delta\theta'} = \frac{\beta'}{v\delta\theta}. \quad \text{..............................(9)}$$

This has an interpretation in the corpuscular theory of Optics. An object of breadth β' at O' appears to subtend an angle $\delta\theta$ at O. If the velocity were uniform, and all paths consequently straight, we should have $\beta' = l\delta\theta$, where l is the distance OO'. In the actual case the ratio $\beta'/\delta\theta$ is called the 'apparent distance' of O' from O. The theorem (11) accordingly asserts that the apparent distance of O' from O is to the apparent distance of O from O' as v is to v', i.e. as the refractive index at O is to that at O'. If the refractive indices at O and O' are the same the apparent distances are equal*.

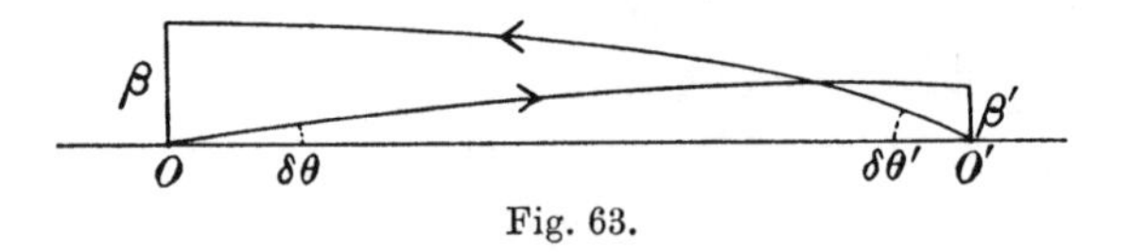

Fig. 63.

When the assumption as to symmetry about an axis is abandoned it is convenient to adopt independent systems of coordinates at O and O'. Taking these points as origins of rectangular Cartesian coordinates, the axes of z, z'

* The theorem, and the term 'apparent distance,' are due to R. Smith, *Optics*, Cambridge, 1738.

being tangential to the path, the lateral deviations at O' due to (small) impulses $\dot{x}$, $\dot{y}$ at O will be given by equations of the form

$$x' = A\dot{x} + B\dot{y}, \quad y' = C\dot{x} + D\dot{y}. \qquad \text{.....................(10)}$$

The reciprocal theorem (6) shews that the deviations x, y at O due to impulses $\dot{x}'$, $\dot{y}'$ at O' will be given by

$$x = A\dot{x}' + C\dot{y}', \quad y = B\dot{x}' + D\dot{y}'. \qquad \text{........................(11)}$$

Hence if σ' be the section at O' of a narrow stream of particles proceeding from O and included within a solid angle ω there, and if σ, ω' have similar meanings with regard to a stream from O', we shall have

$$\frac{\sigma'}{v^2\omega} = \frac{\partial(x', y')}{\partial(\dot{x}, \dot{y})} = AD - BC^*. \qquad \text{.....................(12)}$$

The same value is obtained for $\sigma/v'^2\omega'$, so that

$$\frac{\sigma'}{\omega} = \frac{\sigma}{\omega'} = v^2 : v'^2 = \mu^2 : \mu'^2, \qquad \text{........................(13)}$$

where μ, μ' are the refractive indices in the optical analogy. This gives the theorem of 'apparent distance' in a generalized form.

Ex. 3. In the most general motion of a top suppose that a small impulsive couple about the vertical produces after a time τ a change $\delta\theta$ in the inclination of the axis. The theorem asserts that in the reversed motion† an equal impulsive couple in the plane of θ will produce after the time τ a change $\delta\psi$ in the azimuth of the axis which is equal to $\delta\theta$. It is to be understood of course that the couples have no components (in the generalized sense) except of the types indicated; for instance, they may consist in each case of a force applied to the top at a point of its axis, and of the corresponding reaction at the pivot.

Most of the reciprocal relations of dynamics appear to range themselves under the theorem (6).

Thus, if the system be originally at rest, and if the time τ be infinitesimal, we may put

$$\delta q_s' = \dot{q}_s' \,.\, \tau,$$

and, on account of the linear character of the relations between the momenta and the velocities, the restriction to infinitely small impulses δp_r may be dropped. Hence the velocity of type s due to an impulse of type r is equal to the velocity of type r due to an equal impulse of type s (Art. 75).

* If with any point as centre we describe a sphere of radius v, the area included on this sphere by a cone of solid angle ω will be $v^2\omega$. Regarding $\dot{x}$, $\dot{y}$ as the coordinates of a point on the contour of this area, the ratio of corresponding areas in the plane $x'y'$ and on the sphere is

$$\frac{\partial(x', y')}{\partial(\dot{x}, \dot{y})}.$$

† This must include reversal of the spin.

Again, applying the theorem to the case of small periodic disturbances from a configuration of equilibrium, we are led to the reciprocal relation of Art. 96.

In a second reciprocal theorem, also due to Helmholtz, the configuration O is slightly varied by a change δq_r in one of the coordinates, the momenta being all unaltered, and $\delta p_s'$ is the consequent change in one of the momenta after time τ. Similarly in the reversed motion a change $\delta q_s'$ produces after a time τ a change of momentum δp_r. The theorem in question is that

$$\delta p_s' : \delta q_r = \delta p_r : \delta q_s'. \quad \text{..................(14)}$$

This follows from (2) if we assume all the δp to vanish, and likewise all the δq except δq_r, and if (further) we assume all the $\Delta p'$ to vanish, and all the $\Delta q'$ save $\Delta q_s'$.

Ex. 4. Let O, O' be the unit points, and F, F' the principal foci of an optical system symmetrical about an axis. A corpuscle entering the system parallel to the axis and at a short distance from it will emerge through F'. Considering the slight disturbance from the straight path OO', we may write

$$\delta q_r = \beta, \quad \delta p_s' = -v'\delta\theta'.$$

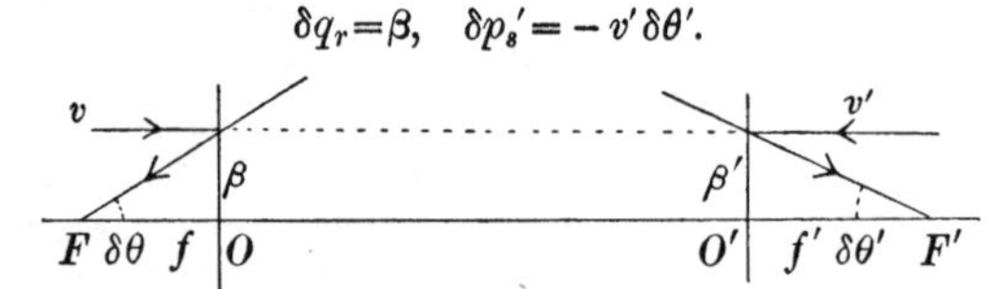

Fig. 64.

For a corpuscle entering in the opposite direction we have

$$\delta q_s' = \beta', \quad \delta p_r = -v\delta\theta.$$

Since $\beta = \beta'$, we have from (14)

$$v\delta\theta = v'\delta\theta'. \quad \text{..................................(15)}$$

If f, f' be the principal focal lengths, we have

$$\beta = f\delta\theta = f'\delta\theta',$$

whence

$$\frac{f}{f'} = \frac{v}{v'} = \frac{\mu}{\mu'}, \quad \text{...............................(16)}$$

an important result due to Gauss.

Ex. 5. Let O, O' be any pair of conjugate foci on the axis of a symmetrical optical instrument. Suppose that the δq and $\delta q'$ all vanish, and likewise all the δp and $\delta p'$ save those with the suffix r. Lagrange's formula (2) then becomes

$$\delta p_r \,.\, \Delta q_r = \delta p_r' \,.\, \Delta q_r'. \quad \text{..........................(17)}$$

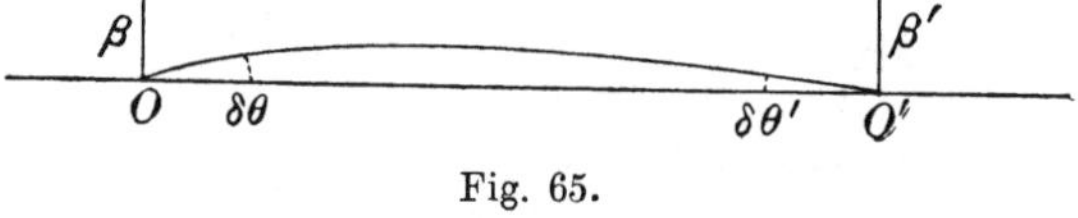

Fig. 65.

Writing $\delta p_r = v\,\delta\theta, \quad \delta p_r' = v'\,\delta\theta', \quad \Delta q_r = \beta, \quad \Delta q_r' = \beta',$

where $\delta\theta$ and $\delta\theta'$ are the divergences at O and O' of a ray between these points, and β, β' are the breadths of conjugate images, we have

$$v\,\delta\theta\,.\,\beta = v'\,\delta\theta'\,.\,\beta', \quad \text{.............................(18)}$$

or

$$\mu\beta\,\delta\theta = \mu'\beta'\,\delta\theta'. \quad \text{.................................(19)}$$

This important law connecting the relative dimensions of conjugate images with the relative divergences of corresponding rays is due to Lagrange.

In the reciprocal theorems here developed it is assumed that the reversal of the system is complete, extending to every velocity in the system. For instance, in a cyclic system the cyclic motions must be supposed to be reversed with the rest. A conspicuous instance where the theorems do not apply owing to incomplete reversal is supplied by the propagation of sound in a wind.

It may be pointed out, however, that there is no such limitation to the original formula of Lagrange. In the application to cyclic systems we may suppose that the coordinates $q_1, q_2, \ldots, q_m$ are the so-called 'palpable' coordinates, and that the cyclic momenta are invariable. Special inferences can then be drawn as before, but the interpretation is less simple, owing to the non-reversibility of the motion.

110. Hamilton's Differential Equations.

The functions A and S discussed in Arts. 106, 107 satisfy certain differential equations of the first order, first given by Hamilton in a memoir already cited.

We consider specially the case of S, defined now by

$$S = \int_{t_0}^{t} (T - V)\, dt, \quad \text{....................(1)}$$

and regarded as a function of t and of the coordinates q_r and momenta p_r at the instant t. We have, then, from Art. 107,

$$p_r = \frac{\partial S}{\partial q_r}, \quad \text{..........................(2)}$$

and

$$H = -\frac{\partial S}{\partial t}, \quad \text{..........................(3)}$$

where H is the total energy. In terms of the coordinates and the momenta we have

$$H = \tfrac{1}{2}\left(A_{11} p_1^2 + A_{22} p_2^2 + \ldots + 2A_{12} p_1 p_2 + \ldots\right) + V, \quad \text{...(4)}$$

from Art. 81 (4). Hence, substituting from (2) and (3), we have

$$\frac{\partial S}{\partial t} + \tfrac{1}{2}\left\{A_{11}\left(\frac{\partial S}{\partial q_1}\right)^2 + A_{22}\left(\frac{\partial S}{\partial q_2}\right)^2 + \ldots + 2A_{12}\frac{\partial S}{\partial q_1}\frac{\partial S}{\partial q_2} + \ldots\right\} + V = 0, \quad \ldots\ldots\ldots(5)$$

which is the first of the equations referred to.

This involves the $n+1$ independent variables $q_1, q_2, \ldots, q_n, t$, and a 'complete' integral, in the language of Differential Equations will accordingly contain $n+1$ independent arbitrary constants, one of which, however, is merely additive to S, and dynamically irrelevant. If

$$S = f(t, q_1, q_2, \ldots, q_n, \alpha_1, \alpha_2, \ldots, \alpha_n) + C, \quad \ldots\ldots\ldots\ldots(6)$$

where $\alpha_1, \alpha_2, \ldots, \alpha_n, C$ are arbitrary, be *any* such integral, the equations

$$\frac{\partial f}{\partial \alpha_1} = \varpi_1, \quad \frac{\partial f}{\partial \alpha_2} = \varpi_2, \quad \ldots, \quad \frac{\partial f}{\partial \alpha_n} = \varpi_n, \quad \ldots\ldots\ldots\ldots(7)$$

where $\varpi_1, \varpi_2, \ldots, \varpi_n$ are arbitrary, will determine $q_1, q_2, \ldots, q_n$ as functions of t and the $2n$ essential arbitrary constants. It has been shewn by Jacobi (*l.c.*) that this represents a natural motion of the system, and accordingly determines the configuration at time t consequent on any arbitrary initial conditions, viz. the n initial coordinates and the n initial momenta, since we have $2n$ arbitrary constants at our disposal.

The theorem is established by shewing that the Hamiltonian equations of motion (Art. 82) are satisfied. By hypothesis we have

$$\frac{\partial f}{\partial t} + \tfrac{1}{2}\left\{A_{11}\left(\frac{\partial f}{\partial q_1}\right)^2 + A_{22}\left(\frac{\partial f}{\partial q_2}\right)^2 + \ldots + 2A_{12}\frac{\partial f}{\partial q_1}\frac{\partial f}{\partial q_2} + \ldots\right\} + V = 0, \quad \ldots\ldots\ldots(8)$$

with the condition that the determinant formed from the constituents

$$\frac{\partial^2 f}{\partial \alpha_r \partial q_s}, \quad \ldots\ldots\ldots\ldots\ldots\ldots\ldots\ldots\ldots\ldots(9)$$

where the suffixes distinguish the rows and columns, does not vanish*.

Since (8) is assumed to hold for all values of the arbitrary

* The equations (7) would not otherwise be independent.

constants, we may differentiate with respect to them; thus differentiating with respect to α_r

$$\frac{\partial^2 f}{\partial \alpha_r \partial t} + \sum_s \left(A_{1s} \frac{\partial f}{\partial q_1} + A_{2s} \frac{\partial f}{\partial q_2} + \ldots + A_{ns} \frac{\partial f}{\partial q_n} \right) \frac{\partial^2 f}{\partial \alpha_r \partial q_s} = 0. \quad \text{.........(10)}$$

Again, the relation

$$\frac{\partial f}{\partial \alpha_r} = \varpi_r \quad \text{.........................(11)}$$

is assumed to hold for all values of t. Hence differentiating totally

$$\frac{\partial^2 f}{\partial \alpha_r \partial t} + \sum_s \frac{\partial^2 f}{\partial \alpha_r \partial q_s} \dot{q}_s = 0. \quad \text{.................(12)}$$

Comparing (10) and (12) we infer

$$\dot{q}_s = A_{1s} \frac{\partial f}{\partial q_1} + A_{2s} \frac{\partial f}{\partial q_2} + \ldots + A_{ns} \frac{\partial f}{\partial q_n}. \quad \text{.........(13)}$$

This follows from the condition imposed on the differential coefficients (9). For the n equations of the type (12) determine the velocities $\dot{q}_s$ uniquely as linear functions of the derivatives $\partial^2 f/\partial \alpha_r \partial t$, and the equations (10) in like manner determine the expressions on the right-hand side of (13) as the same functions of these derivatives.

We infer from (13) on reference to Art. 81 that the momenta in the assumed motion are given by

$$p_r = \frac{\partial f}{\partial q_r}, \quad \text{......................(14)}$$

and that the Hamiltonian equation

$$\dot{q}_s = \frac{\partial H}{\partial p_s} \quad \text{.........................(15)}$$

is accordingly satisfied. Also from (8) that the energy is

$$H = -\frac{\partial f}{\partial t}. \quad \text{......................(16)}$$

Now suppose H expressed, as in (4), in terms of the coordinates and the momenta. If in (16) we replace the momenta by their values from (14) we get of course an identity, which may be differentiated with respect to q_r. Thus

$$\frac{\partial^2 f}{\partial q_r \partial t} = -\frac{\partial H}{\partial q_r} - \sum_s \frac{\partial H}{\partial p_s}\frac{\partial p_s}{\partial q_r}$$

$$= -\frac{\partial H}{\partial q_r} - \sum_s \frac{\partial H}{\partial p_s}\frac{\partial^2 f}{\partial q_r \partial q_s}, \quad \text{.........(17)}$$

and therefore, from (15),

$$\frac{\partial^2 f}{\partial q_r \partial t} + \sum_s \frac{\partial^2 f}{\partial q_r \partial q_s}\dot{q}_s = -\frac{\partial H}{\partial q_r}. \quad \text{...............(18)}$$

The left-hand member is the total differential coefficient of $\partial f/\partial q_r$ with respect to t. Hence, from (14),

$$\dot{p}_r = -\frac{\partial H}{\partial q_r}. \quad \text{........................(19)}$$

The Hamiltonian equations (15) and (19) are thus seen to hold in the assumed motion.

The equation satisfied by the characteristic function A is obtained by substitution in the equation

$$T + V = h \quad \text{...........................(20)}$$

from (5) of Art. 106. Thus

$$\tfrac{1}{2}\left\{A_{11}\left(\frac{\partial A}{\partial q_1}\right)^2 + A_{22}\left(\frac{\partial A}{\partial q_2}\right)^2 + \ldots + 2A_{12}\frac{\partial A}{\partial q_1}\frac{\partial A}{\partial q_2} + \ldots\right\} + V = h. \quad \text{.........(21)}$$

A complete integral of this will involve n arbitrary constants, one of which is additive and may be omitted. Denoting *any* such integral by

$$A = f(h, q_1, q_2, \ldots, q_n, \alpha_1, \alpha_2, \ldots, \alpha_{n-1}), \quad \text{.........(22)}$$

Jacobi has shewn that the general solution of the equations of motion is given by

$$\frac{\partial f}{\partial \alpha_r} = \varpi_r \quad [r = 1, 2, \ldots, n-1] \quad \text{...........(23)}$$

and

$$\frac{\partial f}{\partial h} = t + \epsilon, \quad \text{...................................(24)}$$

where $\varpi_1, \varpi_2, \ldots, \varpi_{n-1}, \epsilon, h$ are to be regarded as $n+1$ additional arbitrary constants, making $2n$ in all.

The proof follows much the same lines as in the case of the equation (5).

Ex. In the case of a projectile under gravity,

$$\tfrac{1}{2}\left(\frac{\partial A}{\partial x}\right)^2 + \tfrac{1}{2}\left(\frac{\partial A}{\partial y}\right)^2 + gy = h, \quad \text{.........................(25)}$$

the mass being unity. A complete integral is obtained if we assume

$$A = \alpha x + \phi, \quad \text{.....................................(26)}$$

and determine ϕ as a function of y only from

$$\left(\frac{\partial \phi}{\partial y}\right)^2 = 2h - \alpha^2 - 2gy. \quad \text{..........................(27)}$$

Thus
$$A = \alpha x - \frac{1}{3g}(2h - \alpha^2 - 2gy)^{\frac{3}{2}}. \quad \text{.....................(28)}$$

The equations (23), (24) give

$$\left.\begin{aligned} \frac{\partial A}{\partial \alpha} &= x + \frac{\alpha}{g}(2h - \alpha^2 - 2gy)^{\frac{1}{2}} = \beta, \\ \frac{\partial A}{\partial h} &= -\frac{1}{g}(2h - \alpha^2 - 2gy)^{\frac{1}{2}} = t + \epsilon. \end{aligned}\right\} \quad \text{..................(29)}$$

Hence
$$\left.\begin{aligned} x - \beta &= \alpha\,(t + \epsilon), \\ y &= \frac{h}{g} - \frac{\alpha^2}{2g} - \tfrac{1}{2}g\,(t + \epsilon)^2. \end{aligned}\right\} \quad \text{.......................(30)}$$

111. Least Action, and Least Time.

The proof of the principle of Least Action in Art. 104 was based on the formula

$$A = \Sigma \int mv^2 dt, \quad \text{.......................(1)}$$

t being the independent variable. An alternative procedure starts from the form

$$A = \Sigma \int mv\,ds, \quad \text{..........................(2)}$$

where v is regarded as a function of the arc s measured along the path of the particle m.

This procedure may be sufficiently exemplified by the case of a single particle; but, treating the matter in the first instance as one of mere geometry, we consider the integral

$$I = \int_A^B \mu\,ds, \quad \text{.........................(3)}$$

where μ is a given function of x, y, z. The integral is supposed to be taken along a given path from one fixed point A to another B.

If the path, between the same terminal points, is slightly varied, we must assume some correspondence to be established between a point P on the original path, and a point P' on the varied path, e.g. we may regard the arc AP' as some function of the arc $AP(=s)$, the correspondence being of course such that P' coincides with B when P does. Let the symbol δ denote the effect of transition from one path to the corresponding point of the other. Then

$$\delta I = \int_A^B (\delta\mu\, ds + \mu\delta\, ds). \qquad \text{.................(4)}$$

We have
$$\delta\mu = \frac{\partial\mu}{\partial x}\delta x + \frac{\partial\mu}{\partial y}\delta y + \frac{\partial\mu}{\partial z}\delta z, \qquad \text{.................(5)}$$

and, since
$$ds^2 = dx^2 + dy^2 + dz^2,$$

$$\delta\, ds = x'\delta\, dx + y'\delta\, dy + z'\delta\, dz, \qquad \text{.................(6)}$$

where the accents denote differentiations with respect to s.

To see that the operations d and δ are commutative, let $PQ\,(=ds)$ be an element of the original path, and $P'Q'\;(=ds+\delta\, ds)$ the corresponding element of the varied path. We have, as an obvious vector equation,

$$P'Q' - PQ = QQ' - PP',$$

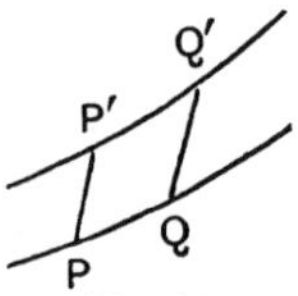

and if we project these vectors on the coordinate axes we have

$$\delta\, dx = d\,\delta x, \quad \delta\, dy = d\,\delta y, \quad \delta\, dz = d\,\delta z,$$

and consequently from (6)

P′ Q′ P Q

Fig. 66.

$$\delta\, ds = x' d\,\delta x + y' d\,\delta y + z' d\,\delta z. \qquad \text{.................(7)}$$

Hence (4) becomes, after a partial integration with respect to s,

$$\delta I = \Big[\mu\,(x'\delta x + y'\delta y + z'\delta z)\Big]_A^B + \int_A^B \left[\left\{\frac{\partial\mu}{\partial x} - \frac{\partial}{\partial s}(\mu x')\right\}\delta x \right.$$
$$\left. + \left\{\frac{\partial\mu}{\partial y} - \frac{\partial}{\partial s}(\mu y')\right\}\delta y + \left\{\frac{\partial\mu}{\partial z} - \frac{\partial}{\partial s}(\mu z')\right\}\delta z\right] ds. \qquad \text{......(8)}$$

The integrated terms vanish at both ends. Since δx, δy, δz are in other respects independent, we have the following conditions that δI should vanish for all slight variations of the path, viz.

$$\frac{\partial\mu}{\partial x} = \frac{\partial}{\partial s}(\mu x'), \quad \frac{\partial\mu}{\partial y} = \frac{\partial}{\partial s}(\mu y'), \quad \frac{\partial\mu}{\partial z} = \frac{\partial}{\partial s}(\mu z'). \qquad \text{......(9)}$$

1°. To deduce the principle of Least Action we put $\mu = mv$. Since the energy is to be unchanged

$$\tfrac{1}{2} mv^2 + V = \text{const.}, \qquad \text{....................(10)}$$

where V is the potential energy of the particle at any point of the field. Hence if (X, Y, Z) be the force due to the field

$$X = -\frac{\partial V}{\partial x} = mv\frac{\partial v}{\partial x}, \quad \text{etc., etc.} \quad \text{...........(11)}$$

Hence the conditions (9) give, on multiplication by v,

$$X = mv\frac{\partial v}{\partial x} = \frac{mv^2}{\rho} . \rho x'' + mv\frac{\partial v}{\partial s} . x', \quad \text{etc., etc.,} \quad \text{...(12)}$$

where ρ is the principal radius of curvature. Since the direction cosines of the tangent to the path are (x', y', z'), and those of the principal normal are $(\rho x'', \rho y'', \rho z'')$, the equations (12), and therefore the conditions (9), are satisfied in free motion; cf. Art. 35.

It also appears from (8) that if the terminus B (x, y, z) be displaced, we have

$$\frac{\partial A}{\partial x} = mvx', \quad \frac{\partial A}{\partial y} = mvy', \quad \frac{\partial A}{\partial z} = mvz'; \quad \text{.........(13)}$$

cf. Art. 106 (10).

2°. Next let $\mu = 1/v$. The integral

$$I = \int_A^B \frac{ds}{v} \quad \text{.......................(14)}$$

measures the *time* along the path, and the question now is: by what path must a particle, started with a given energy, be guided from A to B in a given conservative field of force, so that the transit may be accomplished in the least possible time? It is understood that the constraining forces are not to affect the energy, and must therefore be normal to the path, as in the case of a particle sliding in a smooth tube.

Now the conditions (9) require, putting $\mu = 1/v$, and referring to (11),

$$X = mv\frac{\partial v}{\partial s} . x' - \frac{mv^2}{\rho} . \rho x'', \quad \text{etc., etc.} \quad \text{.........(15)}$$

These make

$$Xx' + Yy' + Zz' = mv\frac{\partial v}{\partial s}, \quad \text{..............(16)}$$

which is by hypothesis satisfied. But for the normal force in the osculating plane, we have

$$\frac{mv^2}{\rho} = -(X . \rho x'' + Y . \rho y'' + Z . \rho z''), \quad \text{........(17)}$$

i.e. the constraining force must not only neutralize the normal component of the force of the field, but have an additional component of the same amount in the opposite direction*.

It should be added that the equations (9) are the differential equations of the path of a ray in a heterogeneous medium whose refractive index is μ, on Fermat's principle of 'Least Time.'

Ex. To find the curve of quickest descent from one point to another under ordinary gravity.

Let ψ be the angle which the tangent to the path makes with the downward vertical. The tangential resolution gives

$$\frac{\partial (v^2)}{\partial s} = 2g \cos \psi. \qquad \text{.....(18)}$$

Since the resultant normal force must be equal and opposite to that due to gravity

$$v^2 = g \sin \psi \frac{\partial s}{\partial \psi}. \qquad \text{.....(19)}$$

Hence, eliminating v^2,

$$\frac{d}{d\psi}\left(\sin \psi \frac{ds}{d\psi}\right) = 2 \cos \psi \frac{ds}{d\psi}, \qquad \text{.....(20)}$$

or

$$\sin \psi \frac{d^2 s}{d\psi^2} = \cos \psi \frac{ds}{d\psi}. \qquad \text{.....(21)}$$

Hence

$$\frac{ds}{d\psi} = C \sin \psi, \qquad \text{.....(22)}$$

a property characteristic of the *cycloid* with horizontal base†. Since $v^2 = gC \sin^2 \psi$, by (19), the velocity at any point must be that 'due to' a fall from the level of the base. Thus if the particle start from rest, the starting-point must be a cusp.

EXAMPLES. XV.

1. Deduce the equations (1) and (2) of Art. 37 from Lagrange's variational equation.

2. Deduce the equations of a slightly disturbed revolving chain (Art. 99) from Lagrange's variational equation.

3. Deduce the law of refraction of light, viz.

$$\sin \theta = \mu \sin \phi,$$

in the usual notation, from the principle of Least Action, on the corpuscular theory. (Maupertuis.)

4. Prove that in the path of a projectile under gravity the action varies as the area swept over by the radius vector from the focus.

* This general property of the 'brachistochrone,' as it is called, is due to Euler (1744).

† The determination of the brachistochrone in the case of gravity is associated with the names of J. Bernoulli, Leibnitz, and Newton (1697).

5. Prove that in a parabolic orbit about the focus the action varies as the distance described at right angles to the axis.

6. Prove that in an elliptic orbit about a centre of force at a focus the action varies as the area described about the empty focus.

Find within what limits the action, from a given position, is a minimum.

7. A number of particles without mutual influence are projected, in a conservative field of force, normally to a given surface, each with the same total energy. Prove that their paths will be orthogonal to a system of surfaces of equal action, and that the velocities with which the particles cross any such surface will be inversely proportional to the distance to a consecutive surface.

8. A particle of unit mass moves in a plane under a central force whose acceleration is $n^2 \times$(dist.). If τ be the time of transit from the position (x_1, y_1) to the position (x, y), prove that Hamilton's principal function is

$$S = \frac{n}{2 \sin n\tau} \{(x^2 + y^2 + x_1{}^2 + y_1{}^2) \cos n\tau - 2(xx_1 + yy_1)\}.$$

Verify the formulæ

$$\frac{\partial S}{\partial x} = \dot{x}, \quad \frac{\partial S}{\partial y} = \dot{y}, \quad \frac{\partial S}{\partial \tau} = -H.$$

9. With a similar notation prove that in the case of a particle under gravity

$$S = \tfrac{1}{2} \frac{(x - x_1)^2 + (y - y_1)^2}{\tau} - \tfrac{1}{2} g(y + y_1)\tau - \tfrac{1}{24} g^2 \tau^3,$$

the axis of y being drawn vertically upwards.

Verify the relations at the end of the preceding example.

10. Prove that in the case of a particle under a central attraction μ/r^2

$$S = -\alpha t + \beta\theta + \int \left(\frac{2\mu}{r} + 2\alpha - \frac{\beta^2}{r^2}\right)^{\frac{1}{2}} dr.$$

What is the meaning of the arbitrary constants?

11. If the quantity

$$Q = \int_t^{t'} (T + V + \sum_r \dot{p}_r q_r)\, dt$$

be regarded as a function of the initial and final *momenta* and the time $\tau (= t' - t)$ of transit, prove that

$$q_r' = \frac{\partial Q}{\partial p_r'}, \quad q_r = -\frac{\partial Q}{\partial p_r}, \quad H = \frac{\partial Q}{\partial \tau},$$

where H is the energy.

12. If O, O' be any two points on the axis of a symmetrical optical system, prove that the angular divergence at O' of a pencil of rays from O is to its divergence at O as the breadth at O of a parallel beam from O' is to its breadth at O', provided the refractive indices at O and O' are equal.

INDEX

[*The numbers refer to the pages*]

Printed in the United States
By Bookmasters